Operations Management - Emerging Trend in the Digital Era

Edited by Antonella Petrillo, Fabio De Felice, Germano Lambert-Torres and Erik Bonaldi

Published in London, United Kingdom

IntechOpen

Supporting open minds since 2005

Operations Management - Emerging Trend in the Digital Era
http://dx.doi.org/10.5772/intechopen.87847
Edited by Antonella Petrillo, Fabio De Felice, Germano Lambert-Torres and Erik Bonaldi

Contributors
Călin Corciovă, Doru Andriţoi, Cătălina Luca, Muralidhar Avvari, Dejene Addis Nigussie, Yang Tang, Xin Yang, Guorong Wang, Sina Joneidy, Charmele Ayadurai, Montri Wiboonrat, Bruno C.H. Silva, Islame F.C. Fernandes, Marco C. C. Goldbarg, Elizabeth F.G. Goldbarg, Jorge Laureano Moya Rodriguez, Orlem Pinheiro Pinheiro De Lima, Sandro Breval, Carlos Manuel Rodríguez Taboada, Maida Bárbara Reyes Rodríguez, Marcia Ribeiro Maduro, Paulo César Diniz De Araujo, José Carlos da Silva Lima, Nilson José de Oliveira Junior, André Luiz Nunes Zogahib, Vittal S Anantatmula, Ann Svensson, Eva Hedman, Antonella Petrillo, Fabio De Felice, Laura Petrillo, Sahand Ashtab, Yang Xing, Cheng Zheng, Eric Easton, Mario Beruvides, Andrea Jackman, Edoghogho Ogbeifun, Patrick Pasipatorwa, J. H. C. Pretorius, Jozef Stašák, Jaroslav Mazůrek, Jaroslav Kultan, Peter Schmidt, Mukhammedov Abu Urinbasarovich, Bouafia Khawla

First published in London, United Kingdom, 2021 by IntechOpen
IntechOpen is the global imprint of INTECHOPEN LIMITED, registered in England and Wales, registration number: 11086078, 5 Princes Gate Court, London, SW7 2QJ, United Kingdom
Printed in Croatia

British Library Cataloguing-in-Publication Data
A catalogue record for this book is available from the British Library

Additional hard and PDF copies can be obtained from orders@intechopen.com

Operations Management - Emerging Trend in the Digital Era
Edited by Antonella Petrillo, Fabio De Felice, Germano Lambert-Torres and Erik Bonaldi
p. cm.
Print ISBN 978-1-83881-187-7
Online ISBN 978-1-83881-202-7
eBook (PDF) ISBN 978-1-83881-203-4

For EU product safety concerns:
IN TECH d.o.o., Prolaz Marije Krucifikse Kozulić 3, 51000 Rijeka, Croatia, info@intechopen.com or visit our website at intechopen.com.

Meet the editors

Antonella Petrillo is a Professor at the Department of Engineering of the University of Naples "Parthenope", Italy. She received her Ph.D. in Mechanical Engineering from the University of Cassino. Her research interests include multi-criteria decision analysis, industrial plant, logistics, manufacturing and safety. She serves as an Associate Editor for the International Journal of the Analytic Hierarchy Process. She is a member of AHP Academy and a member of several editorial boards.

Fabio De Felice is a Professor at the University of Cassino and Southern Lazio, Italy, where he received his Ph.D. in Mechanical Engineering. His current research focuses on multi-criteria decision-making analysis (with an emphasis on AHP and ANP) and industrial, project and supply chain management. Currently, he serves as a member of the Scientific Advisory Committee of the International Symposium on the Analytic Hierarchy Process (ISAHP). He is the founder of AHP Academy that promotes the diffusion of the culture and methodologies of Decision Making, with particular reference to those based on the Analytic Hierarchy Process. He is a member of the editorial boards of several international organizations and journals and has authored/co-authored numerous articles in the areas of decision science and business management.

Germano Lambert-Torres is a Professor at the Instituto Gnarus. He received his Ph.D. degree in Electrical Engineering from the Ecole Polytechnique de Montreal, Canada, in 1990. From 1983 to 2012, he was with the Electrical Engineering Department, Itajuba Federal University (UNIFEI), where he was also the Dean of the Research and Graduate Studies, from 2000 to 2004. Since 2010, he has been the Director of R&D, PS Solucoes, Itajuba. He also serves as a consultant for many utility companies in Brazil and South America, and has taught numerous IEEE tutorials in the USA, Europe, and Asia. He is the author/editor or coauthor of nine books, more than 30 book chapters, and 50 transactions articles on intelligent systems and nonclassical logic.

Erik Leandro Bonaldi received his B.S., M.Sc., and Ph.D. degrees in Electrical Engineering from Itajuba Federal University, Itajuba, Brazil, in 1999, 2002, and 2006, respectively. He is currently the CEO at PS Solutions, in Brazil, directly commanding the largest national factory of power capacitors, wrappers, and the fabrication of sensors, smart devices, and communication systems for predictive maintenance of induction motors, transformers, and generators. His research focuses on industrial electronic automation, predictive maintenance, and artificial intelligence methodologies.

Contents

Preface

Global competition has caused fundamental changes in the competitive environment of the manufacturing and service industries. Firms should develop strategic objectives that, upon achievement, result in a competitive advantage in the market place. The forces of globalization on one hand and rapidly growing marketing opportunities overseas, especially in emerging economies on the other, have led to the expansion of operations on a global scale. The subject matter is fascinating and timely. This book aims to cover the main topics characterizing operations management including both strategic issues and practical applications. A global environmental business including both manufacturing and services is analyzed. Among the topics are safety, product and service design, decision making, maintenance, and supply chain management. The book contains original research and application chapters from many different perspectives. It is enriched through the analyses of case studies.

Chapter 1, "The Analytic Functional Resonance Analysis to Improve Safety Management", proposes the Functional Resonance Analysis Method (FRAM) to analyze the complexity of safety in industrial plants. This research integrates FRAM with the Analytic Hierarchy Process (AHP), a multicriteria technique, to overcome the limits of the FRAM. The result is a proposal of an alternative approach to risk assessment based on principles of resilience engineering. A real case study in a petrochemical company is analyzed.

In the awareness that accelerating digitization of critical infrastructures is increasing interconnection and interdependence among high-reliability subsystems, in Chapter 2, "The Modulus of Resilience for Critical Subsystems", the authors provide a scenario agnostic method to quantify resiliency by applying concepts from materials science in a generalized form. This new formulation resulted from a mapping of constructs used in tensile testing to characteristics of protracted subsystem disruptions.

Chapter 3, "Condition-Based Maintenance for Data Center Operations Management", proposes preventive and predictive maintenance (PPM), which determines the CBM as a systematic strategy of data center operations and maintenance. Use case examples of power distribution systems (PDS) of a data center have been examined to ensure their proper functionality and to reduce their deterioration rate.

It is clear that, in the digital space, the computer-based operating systems (CMMS's) enable quick and effective communication between stakeholders, facilitate improved planning, easy access to historical data, reporting and performance improvements of the maintenance function. However, success in the use of CMMS's depends on the human capacity of the users of the system, as it is explained in Chapter 4 "Harnessing the Multiple Benefits of a Computerised Maintenance Management System".

A novel adaptation of the *MAX-MIN* Ant System algorithm for the Quota Traveling Salesman Problem with Passengers, Incomplete Ride, and Collection Time is proposed in Chapter 5 "Multi-Strategy *MAX-MIN* Ant System for Solving Quota Traveling Salesman Problem with Passengers, Incomplete Ride and Collection Time". The algorithmic components proposed to consider vehicle capacity, travel time, passenger limitations, and a penalty for delivering a passenger deliverance out of the required destination.

The ant-based algorithm incorporates different sources of heuristic information for the ants and memory-based principles. Computational results are reported, showing the effectiveness of this ant-based algorithm.

Chapter 6 is entitled "System of Data Transfer from and to Social and Economic Processes via Creative Economy Networks Created Based on Cultural Heritage Administration Processes and Vice Versa" and it presents an interesting point of view on this topic. The aim of Chapter 7, "Conceptualization, Definition and Assessment of Internal Logistics through Different Approaches Using Artificial Intelligence", is to develop a new concept of internal logistics, its components parts, and how to evaluate it. To quantify the level of performance of the internal logistics of a company is an important issue to gain competitiveness.

It is also important to note that the business activities provided within any firm or company should be checked and controlled continuously, while two principal approaches should be applied: (a) qualitative monitoring, (b) quantitative evaluations, while KPI indicators play a role of the utmost importance within business quantitative evaluation in order to make adequate decisions. However, adequate applications from KPI creation and further processing seem to be very significant and important. Thus, Chapter 8, "Business Process Linguistic Modeling: Theory and Practice Part I: BPLM Strategy Creator", and Chapter 9, "Business Process Linguistic Modeling: Theory and Practice Part II: BPLM Business Process Designer", presents a conceptual model of application denoted as BPLM Strategy Creator in the form of an expert system (ES). The contribution contains such application descriptions from the qualitative, quantitative, and design point of view.

A different issue is analyses in Chapter 10, "Project Management Concepts". The chapter covers the fundamentals of project management. It introduces project management concepts and provides a systemic view of project management plans and the processes with which they are implemented. The knowledge areas include scope, time, cost, quality, and risk. In addition, the authors emphasize the interrelated nature of these knowledge areas.

Chapter 11, "A Modern Approach for Maintenance Prioritization of Medical Equipment", presents document-based methods to evaluate every aspect of the medical equipment maintenance process and to provide a correct, objective, and standardized approach that supports clinical engineering activities. Following the analysis, the results show that the combination of the use of the two methods provides an overview, in a periodic manner, of maintenance performance that indicates the use of the most appropriate procedures.

A critical issue in our world is that the aging population of the western world poses a medical challenge for the society of today and of the future. The pressure on healthcare and its organization is increasing as the demand for healthcare is growing at the same time as the costs are continuously rising. Thus, Chapter 12, "Operations Knowledge Management in Health Care", aims to analyze knowledge management mechanisms in the infrastructure of a healthcare organization. The analysis shows how mechanisms have significant impacts on the knowledge management practice in operations management. A learning and knowledge creation culture, together with an organizational architecture for adaptive and exaptive capacity, and a business model for knowledge capitalization, could support the production of smooth and effective healthcare of high quality in society.

A point of view on services is analyzed in Chapter 13, "Artificial Intelligence and Bank Soundness: A Done Deal? - Part 1", and Chapter 14, "Artificial Intelligence and Bank Soundness: Between the Devil and the Deep Blue Sea - Part 2". In detail, a conceptual analysis with the aim to provide a deeper understanding of the opportunities parted by AI from a service provider and customer perspective in the bank sector is presented.

Nowadays, productivity improvement is the concern of all industries in spite of the type of product and number of recourses or production systems that are followed by the industries. So far, productivity has been achieved by implementing various methodologies exclusively computerized maintenance and management systems. In this scenario, Chapter 15, "Implementation of Computerized Maintenance and Management System in Wine Factory in Ethiopia: A Case Study", emphasizes the effective maintenance strategy and management system as the solitary way of improving the productivity of the cause company. The foremost intention of the study is to enhance productivity through a smart maintenance management system (strategy and management tool).

Maintenance issues are well defined and analyzed in Chapter 16, "Maintenance Decision Method Based on Risk Level", which presents a framework for maintenance decision method based on risk level for mechanical equipment in the petrochemical industry.

Finally, the problem of the supply chain is explored in the last chapter, "Exploring Constituents of Short Food Supply Chains", which provides a holistic exploration of Alternative Food Networks (AFNs), which contribute to the further mobilization of locally produced products.

This book is intended to be a useful resource for anyone who deals with innovation and smart manufacturing problems. Furthermore, we hope that this book will provide useful resources ideas, techniques, and methods for further research on these issues.

As editors of this book, we very much thank the authors who contributed with their invaluable research as well as the referees who reviewed these papers for their effort, time, and invaluable suggestions. Our special thanks to Ms. Sandra Maljavac, Author Service Manager, for her precious support and her team for this opportunity to serve as guest editors.

Antonella Petrillo
Professor,
Department of Engineering,
University Parthenope of Naples,
Italy

Fabio De Felice
Professor,
Department of Civil and Mechanical Engineering,
University of Cassino and Southern Lazio,
Italy

Germano Lambert-Torres
Professor,
Departamento de Pesquisa e Desenvolvimento,
Instituto Gnarus,
Itajuba, Brazil

Dr. Erik Bonaldi
PS Soluções,
Itajubá, Minas Gerais, Brazil

Chapter 1

The Analytic Functional Resonance Analysis to Improve Safety Management

Antonella Petrillo, Fabio De Felice and Laura Petrillo

Abstract

Complex industrial plants are characterized by digitalization and innovation. In this context it is strategic to ensure the systematic design, implementation, and continuous improvement of all processes (operations management). One of the most obvious ways to improve operations performance is to reduce the risk of accidents and human errors. In this pilot study the Functional Resonance Analysis Method (FRAM) is proposed to analyze the complexity of safety in industrial plants. This research integrates FRAM with Analytic Hierarchy Process (AHP), a multi criteria technique, to overcome the limits of the FRAM. The result is a proposal of an alternative approach to risk assessment based on principles of resilience engineering. A real case study in a petrochemical company is analyzed.

Keywords: AHP, FRAM, resilience engineering, performance variability, human error, operations management

1. Introduction

Although over the years, industrial plants have improved their safety management processes, it is evident that safety systems need to be further improved [1]. This need is underlined by the many accidents that have occurred in industrial plants over recent years, arising from human causes, technical causes, or natural causes. Traditional safety management models are designed to identify negative factors and develop systems to mitigate their impact. These models allow to analyze different critical situations, but they seem ineffective for today's business needs [2]. Particularly, in modern industrial plants, only a few functions are independent of each other. Thus, analyzing them individually may not be the best model. In general, due to the complexity of the systems it is necessary to analyze all functions and tasks. In this perspective, Resilience Engineering (RE) is a useful approach to manage complex systems. This approach is a new way to think about safety and risk management [3]. Unlike the classic risk management approaches that are based on the analysis of a posteriori causes by adopting a linear cause-and-effect approach, the RE adopts a perspective that refers to the theory of complexity. RE aims to revise the analysis models to create processes that are flexible and robust. Functional Resonance Analysis Method (FRAM) proposed by Hollnagel defines complex systems through their functions and studies the interactions between these functions [4]. The main *strength* of FRAM method is based on the principle that a variation in the conditions in which an action takes place can lead to *improvements* or *worsening* that ultimately lead to its success or failure. However, this approach leads to a *qualitative result* aimed at highlighting how multiple variables combined can change the outcome of an action in a dynamic environment. The points in favor of this method and of resilience engineering are evident, but they still pose obstacles, sometimes even technical ones to overcome.

Thus, in the present research the FRAM method is used in conjunction with Analytic Hierarchy Process (AHP) to overcome the limits of the FRAM. AHP is a well-known multi-criteria decision support technique developed in the 1970s by the Prof. Thomas L. Saaty [5]. The proposed model overcomes the qualitative limits of the resilience engineering models proposed in the literature. The AHP helps to assess the subjective probability of an event or trigger cause. Furthermore, through the integration of the AHP it allows to evaluate the strength of relationship between the variability of human performance and influence of the external environment. The preset study is a pilot research. The proposed process will be tested in other situations and industrial settings. In fact, the model is extremely flexible and can be applied in different scenario.

The rest of the paper is organized as follows. Section 2 presents a general overview on resilience engineering approach and a brief state of art. Section 3 describes the proposed model based on FRAM and AHP. Section 4 describes a real case study in a petrochemical industry and its results. Finally, in Section 5 conclusion of the proposed "model" and the future research are summarized.

2. General overview on resilience engineering approach

Resilience Engineering (RE) is a multidisciplinary field of study dealing with safety in complex systems that have several interdependent elements from an economic, human, and social point of view [6]. RE is the intrinsic ability of a system to modify its functioning before, during and following a change or disruption, in order to be able to continue the necessary operations both under foreseen and unforeseen conditions [7]. In general, safety is defined as a condition that minimizes the number of negative outcomes. Thus, it is possible to understand the functioning of a system by analyzing its parts. Therefore, in this view the aim is to reduce the number of accidents by reducing their causes. This is the so-called *Safety I*. In opposition to this vision was developed *Safety II*. This approach not only focuses on adverse events, but also analyzes everyday work situations in which things are going well [8]. In this perspective, safety is defined positively as an effective daily operating situation, rather than negatively as the absence of accidents. Unlike the classic risk analysis and risk management approaches that are based on the analysis of a posteriori causes by adopting a linear cause-and-effect approach, the RE adopts a perspective that refers to the *Theory of complexity* [9]. It aims to revise the analysis models to create processes that are flexible and robust Therefore, for the RE, risk management is not aimed at reducing sources of risk, but at strengthening the ability to reduce the variability of performance both in expected and unexpected conditions [10]. In this context, Functional Resonance is a characteristic of a complex system that explains how serious consequences can arise from small variations in the performance of its parts or the environmental conditions in which it operates [11]. The Functional Resonance Analysis Method (FRAM) is a recent method developed to explore how functional variability affects the overall system [12]. An investigation on the SCOPUS, one of the most accredited databases in the scientific community (Scopus is updated periodically and offers around 25,000 articles from more than 5000 international publishers), pointed out that 47 documents have been published from 2010 to 2020. The search query used on Scopus was (TITLE-ABS-KEY (resilience AND engineering) AND TITLE-ABS-KEY (functional AND resonance AND analysis AND method)). the survey result includes only articles in which the string was found in 1) title, or in 2) abstract or in 3) key words. As can be seen from **Figure 1** there is a growing interest in this topic. It is interesting to note that there are no publications on Scopus before 2010. This means that it is a new and promising topic.

A country where there is greater scientific interest is Italy (27,6%), followed by Brazil (17%), as it is shown in **Figure 2**. The publications are mostly articles published in international journals (63,8%) and conference paper (31,9%). Furthermore, the survey pointed out that most of the research is in the engineering area (36,3%) as shown in **Figure 3**.

A recent study developed by Patriarca et al. [13] highlighted that aviation is by far the most investigated domain with the FRAM with a percentage equal to 24,87%. This is not a surprising result since FRAM was developed in the aeronautical field. Other emerging sectors are healthcare (13,99%) and industrial operations (12,44%) as demonstrated by several publications [14–16]. Furthermore, some authors pointed out that FRAM does not assess the human behavior and the human performance to analyze the human error [17, 18].

Other publications demonstrated how FRAM is a qualitative approach for accident, risk and system analysis and it does not support quantification [19–21]. Definitely, FRAM is a qualitative method. Furthermore, it does not support quantification. To overcome this issue, FRAM is used together with the Analytic Hierarchy Process (AHP). Thus, it is possible to measure the subjectivity in establishing the potential variability of functions as suggested by Rosa et al. [22]. The integration of

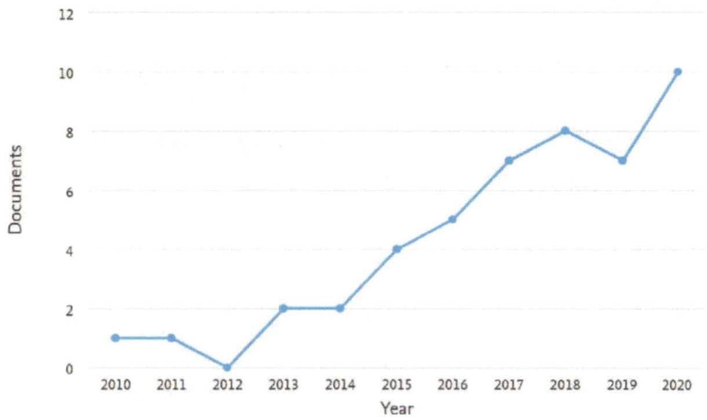

Figure 1.
Documents by year.

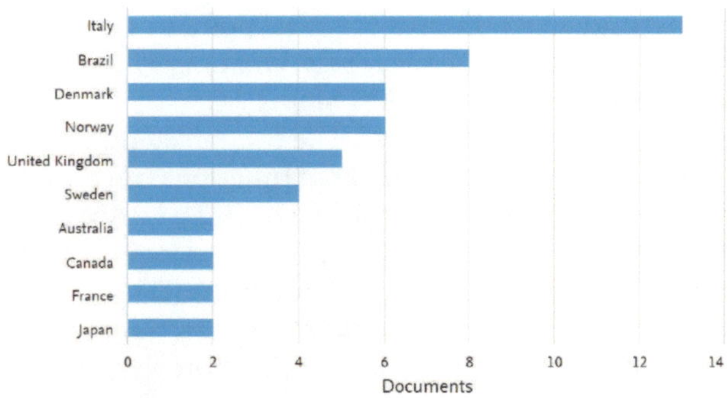

Figure 2.
Documents by country.

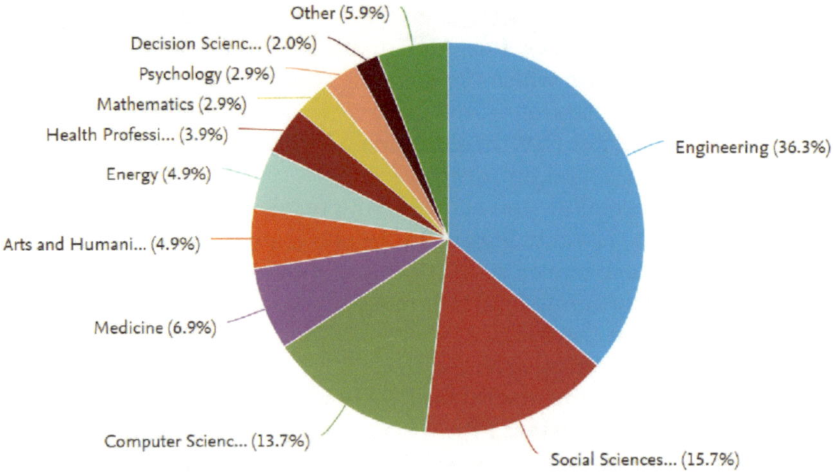

Figure 3.
Documents by subject area.

FRAM-AHP is proposed also in other two works. Both applied the hybrid to evaluate construction sustainability [23, 24].

3. Materials and methods

3.1 Functional resonance analysis method (FRAM)

FRAM methodology aims to analyze how the variability of one or more functions can be combined between them and how to prevent their resonance, which could lead to unwanted results [25]. For this purpose, FRAM method studies the system first under normal conditions, after FRAM analyzes the variability that cause to the event unwanted. The aim is obviously to be able to issue recommendations that prevent the repetition of the event. FRAM consists of four steps: 1) Identify system functions; 2) Characterize the potential variability of the functions; 4) Determine the dependencies among functions and 4) Monitor the variability. Some more details about each step are provided below [26].

Step#1 "Identification of the essential functions". The present step aims to identify the functions or the specific action that are needed to carry out a specific task [27]. Each function is described using the six aspects (as shown in **Figure 4**): INPUT (I); OUTPUT (O); TIME (T); CONTROL (C); PRECONDITIONS (P) and RESOURCES (R). Functions can have links to each other. They can typically have multiple links and dependencies. From a practical point of view, to represent the variability it is possible to use the *FRAM Model Visualiser* (FMV). FMV allow to build a graphical representation of a FRAM model.

Step#2 "Identification of variability". The present step identifies the variability of functions in order to understand how functions can become coupled and how this can lead to unexpected outcomes [28]. The FRAM assume that there are characteristic differences in the variability of technological functions (T), of human functions (M), and of organizational functions (O).

Step#3 "Aggregation of variability and define functional resonance". This step aims to analyze the variability of functions and how they interacted with each other [29]. The variability of a function depends on couplings among functions. It is not enough to evaluate the variability for the single function. It is necessary to understand how variability can be combined. This is achieved using the upstream-downstream functional coupling. The variability of the function can be the result of couplings of upstream functions that influence downstream functions. Each upstream variable can be connected to its downstream variable using the 5 available inputs (showed in **Figure 4**). Depending on the type of connection, different variability occurs (see **Table 1** as example).

Step#4 "Monitor and manage the variability". The step aims to propose ways to manage the possible occurrences of uncontrolled performance variability – or possible conditions of functional resonance – that have been found by the preceding steps [30]. The purpose is to find critical combinations and reinforce the barriers. The problems of complex systems cannot be eliminated, eliminating the variability of the performances, because this is essential to ensure the reliability of the systems. A sensitivity analysis is performed to evaluate different solutions.

3.2 Analytic hierarchy model

The main feature of Analytic Hierarchy Process (AHP) is to break down a decision-making problem in a hierarchy [31]. AHP uses a mathematical approach

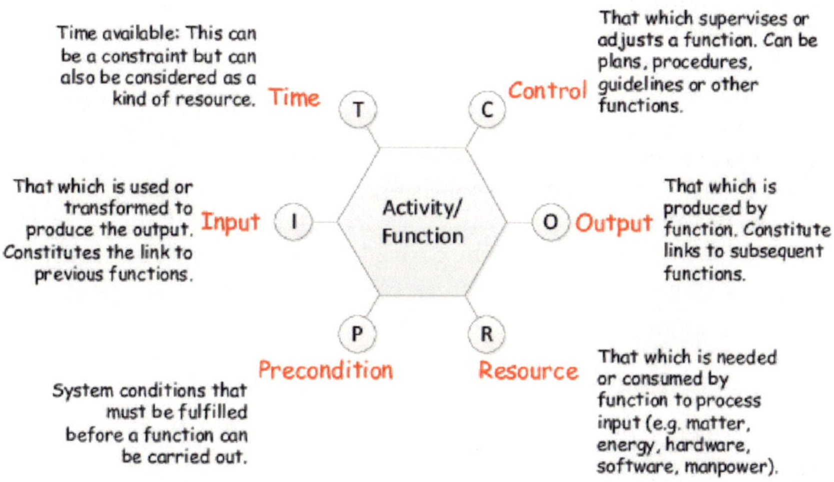

Figure 4.
FRAM hexagon: The six aspects used to characterize functions.

Output variability of upstream function		Possible effects on downstream function
Time	Too early	False start (V+)
		Possible damping (V-)
	In time	Possible damping (V-)
	Too late	Delayed activities (V+)
	Omission	Start imprecision (V+)
Accuracy	Inaccurate	Waste of time (V+)
	Acceptable	No change (V=)
	Accurate	Possible damping (V-)

Table 1.
Example of aggregation of functions (output – input).

based on matrix algebra to "measure" decisions [32]. AHP is characterized by three main phases as described below.

Phase #1 "Define hierarchy". The aim of the first step is to define the goal and the hierarchy of the decision problem. The decision maker or the experts team identifies a set of criteria for evaluating the *n* decision alternatives and assigns a percentage weight to each criterion; then assigns a score that is the impact of the criterion on the decision. The score of each decision alternative is the weighted average of the scores of each criterion on the decision by the weight assigned to each criterion. The top of hierarchy represents the goal of the decision problem. Lower levels represent criteria and sub-criteria in which the decision-making model is broken down. The bottom level represents all alternatives to evaluate in terms of the criteria [33].

Phase #2 "Perform pairwise comparison and relative weight estimation". After defining the hierarchy, the criteria are compared in pairs, the sub criteria and alternatives are compared in pairs by assigning a score of relative importance to the other. The sum of the weights must be 100%. Saaty suggested an increasing scale of values form 1 (*equal importance*) to 9 (*extreme importance*) when comparing two components [34]. The result of the comparison is the so-called *dominance coefficient* a_{ij} that represents the relative importance of the component on row (i) over the component on column (j), i.e., $a_{ij} = w_i/w_j$. The pairwise comparisons can be represented in the form of a square matrix (n x n), symmetric and diagonal. The number of pairwise comparisons grows quadratically with the number of criteria and alternatives. The score of 1 represents equal importance of two components and 9 represents extreme importance of the component i over the component j. [35].

Phase#3 "Perform consistency index". Saaty (1990) proposed utilizing consistency index (CI) to verify the consistency of the comparison matrix [36]. The CI could then be calculated by: CI = $(\lambda_{max} - n)/n - 1$. In general, if CI is less than 0.10, satisfaction of judgments may be derived. **Figure 5** shows a summary of the main steps and phases of the study.

4. Scenario modeling: a case study on a petrochemical plant

The model was applied in a real case study concerning the management of an emergency in a petrochemical company (see **Figure 6**). The plant consists of

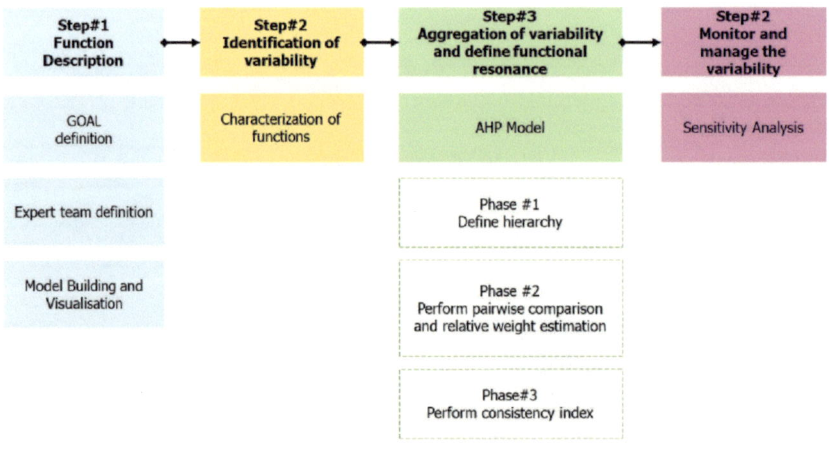

Figure 5.
Summary of the main steps and phases of the study.

Figure 6.
Petrochemical plant.

process and service plants. Plant processes include: Predistillation unit; Propane unit; Distillation unit; Catalytic hydrogenation unit and Diesel oil purification. While service facilities include: Diathermic oil system; Steam and hot water production unit; Refinery torch; Hydrogen production unit; Cooling water system and Refinery storage area. The plant preserves extremely dangerous substances in quantities equal to or greater than the limits. Thus, it is a plant with a high risk activity, where it is necessary to analyze all the deviations from the operating standards (emergency conditions) such as: gas leakage, hydrocarbon release, fire, earthquake, flood, sabotage, pollution, etc.

STEP#1 "Identification of the Essential Functions". The case study analyzes the emergency generated by the *loss of propane* gas during the transfer from tanker to tank. The **goal** of the model was to evaluate the variability of performance between upstream activities and downstream activities. An **expert team** was formed. The expert team consisted of 1 safety manager, 1 AHP expert, 1 chemical engineer, 1 mechanical engineering and 1 risk manager. The expert team analyzed the scenario and summarized the main activities are carried out during emptying the propane from the vehicle and placing it in the treatment plant. In fact, propane is a very dangerous hydrocarbon as the compound appears as a colorless and odorless gas, which can however be easily liquefied by compression and therefore highly flammable. **Table 2** describes the activities carried out during the emergency and the responsibilities.

Figure 7 shows the FRAM of the emergency management activity. FRAM Model Visualiser (FMV) was used to create a graphical representation of a FRAM model ((https://functionalresonance.com/FMV/index.html).

STEP#2 "Identification of variability". In the second step the variability of the functions was characterized and highlighted in red in **Figure 7**. In the scenario analysis, the *human functions* revealed more criticality, which could present different variability. In particular, the analysis focused on two main activities and related emergency scenarios:

- **Scenario A** "activate ESD control";
- **Scenario B** "activate hydro-foam cannon".

#	Operations	Responsible
1	Activate Electrostatic discharge (ESD) control from the control room to stop the transfer and close the cut-off valves	Desk operator
2	Activate the shower cooling system on the truck through the 10HC1V system	Desk operator
3	If possible, intercept the ATB side valve	Truck driver
4	Disconnect the unloading arms	Internal operator
5	Turn away truck	Internal operator
6	Alternatively, continue the unloading operations until the tanker is emptied, and the gas is depressurized	Internal operator
7	Activate hydro-foam cannon	Internal operator
8	Turn off furnace and cool surrounding equipment	Internal operator

Table 2.
Functions of the system.

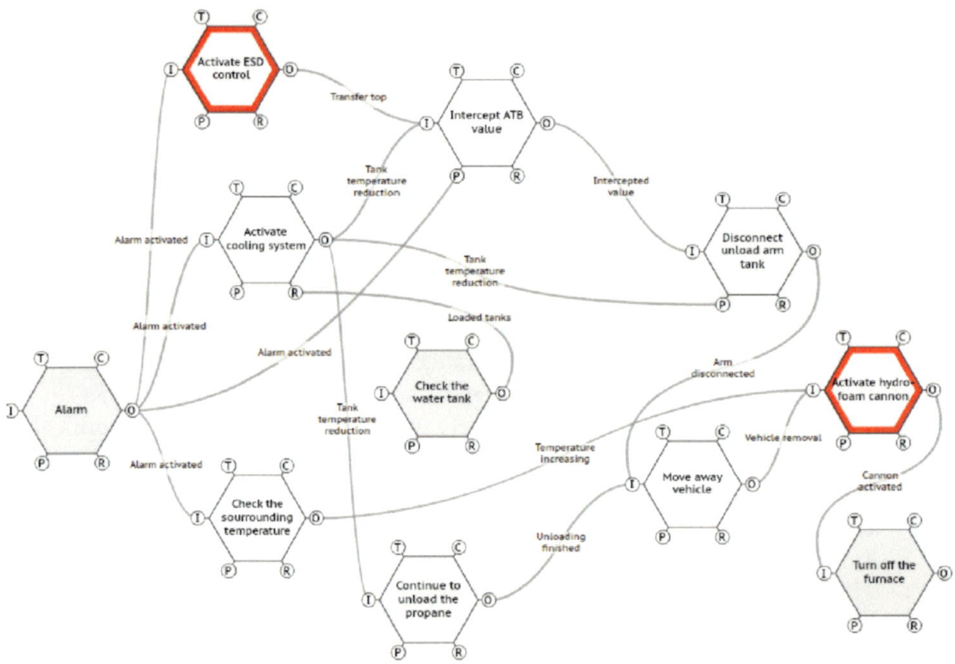

Figure 7.
FRAM representation of the system.

According to the analysis, the *expert team* characterized the environmental conditions in which the operators work. Historically, human performance is investigated through specific *performance shaping factors* (PSFs), as described below:

- PSF#1 Training and Experience. It refers to the experience and training of the operator(s) involved in the task.
- PSF#2 Procedures and Administrative Controls. It refers to the existence and use of formal operating procedures.

- PSF#3 Ergonomics and Human Machine Interaction. It refers to the adequacy or inadequacy of machine (i.e. computer).

- PSF#4 Time Available. It refers to the adequacy or inadequacy of the time available to complete a task.

- PSF#5 Complexity. It refers to the difficult of the task to perform (simple, easy, difficult, very difficult, difficult beyond standards).

- PSF#6 Workload, Stress and Stressors. It refers to mental stress or excessive workload.

- PSF#7 Work Processes. It refers to the adequacy or inadequacy of safety culture, management policies/support, etc.

STEP#3 "Aggregation of variability and definition of functional resonance". The AHP hierarchical structure created for characterizing the variability and define functional resonance is shown in **Figure 8**. AHP model was created using Super Decision Software (http://www.superdecisions.com/). When two items of the "Performance Shape Factor" level are compared with respect to the main goal, the expert answers the question "*Which PSF is more important*?". The AHP helps to assess the subjective probability of an event or trigger cause.

Table 3 summarized the weights of variability in which operators are involved according to PSFs. The weights of the factors are defined through AHP. More specifically, it emerges that PSF#1, PSF#4 and PSF#6 present a higher probability of occurrence.

A graphic representation of **Table 3** is shown in **Figure 9**.

Furthermore, the AHP model allows to define the probability of occurrence of the most critical scenario or, as the results show in **Figure 10**, the most critical scenario is scenario B: Scenario A (47%%) and Scenario B (53%).

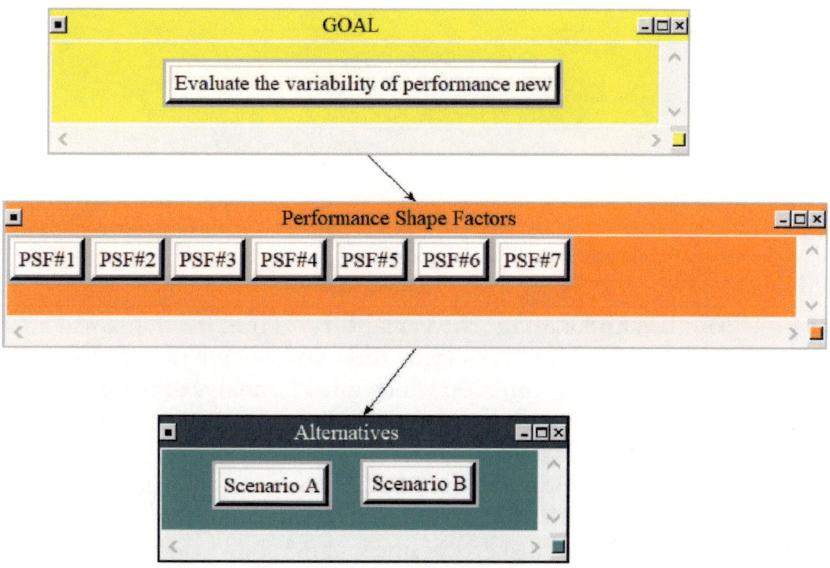

Figure 8.
AHP model.

PSFs	Weighting of variability
PSF#1	0,29,448
PSF#2	0,14,892
PSF#3	0,04571
PSF#4	0,15,991
PSF#5	0,10,006
PSF#6	0,18,167
PSF#7	0,06925

Table 3.
Weighting of output variability.

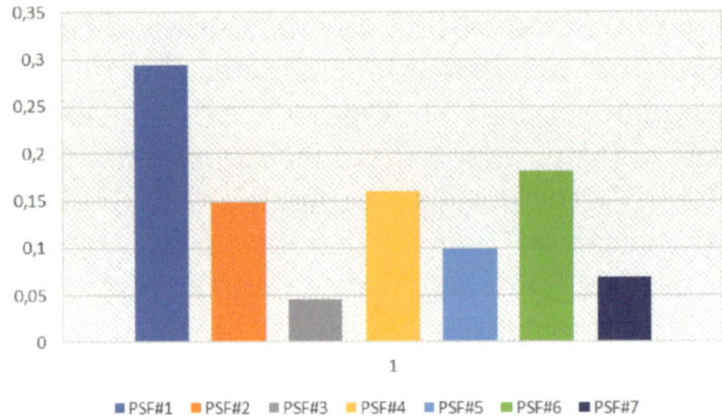

Figure 9.
Probability of occurrence of the most critical PSFs.

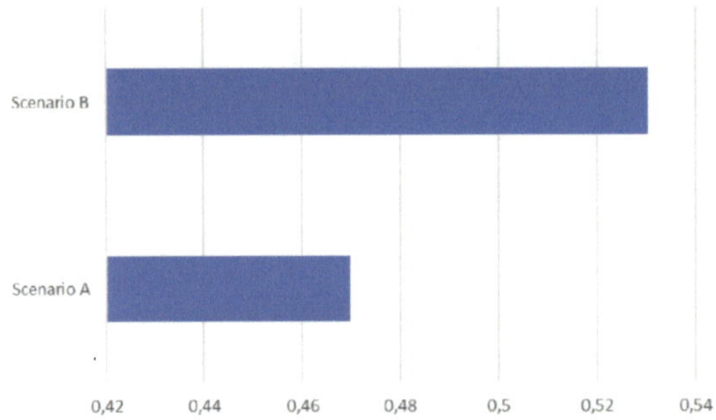

Figure 10.
Probability of occurrence of the most critical scenario.

STEP#4 "Monitor and manage the variability". From the numerical analysis FRAM emerges a critical value (considering the values shown in table n) for the "activate hydro-foam cannon" function which must be analyzed to limit its variability, which affects the downstream variable. While the "Activate ESD control" function presents a lower variability.

Considering the most critical PSF or PSF#1 a sensitivity analysis was performed to evaluate the variability of this factor and the robustness of the model. As shown in **Figure 11** it emerges that if the vertical line is at 0.5 shows the scenario A is more likely. For any PSFs greater than that, the Scenario B is the more likely.

The general result or PSF#1 highlights that *Training and Experience* is a critical point. It is an unsurprising result. In fact, it is clear that training is essential to taught

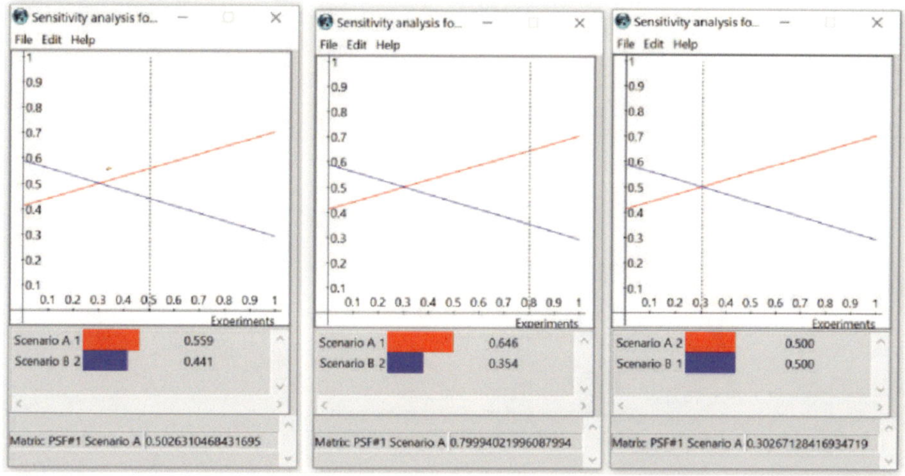

Figure 11.
Sensitivity analysis.

workers to manage complex scenarios at achieving those skills that allow them to work both by reducing risks and protecting personal and community safety. Safety training is the only "measure" that can be validly opposed to situations of residual risk.

5. Conclusions

This is a pilot study that is based on the awareness that the increasingly complexity of industrial plants and the need to analyze safety systems lead researchers to develop new methodological approach. In the present research the main gap of the qualitative approach of FRAM method was overcome with the integration of a multi-criteria decision-making method. The research proposes the integration of the traditional FRAM method with AHP. The integration of AHP with FRAM allows to investigate a new perspective in the field of risk management. The model was applied in a real case study to evaluate the performance of emergency operations in a petrochemical company. Considering the variability of each system function, the research numerically shows the level of variability generated by an upstream function on a downstream function. The results obtained are aimed at identifying function couplings that could generate high variability. Future development of research is the integration technological and organizational aspects, beyond human ones. Moreover, the model can be applied in different socio-technical systems where a high level of complexity requires the use of innovative tools. Thus, the proposed model will be tested in other situations or industrial settings.

Conflict of interest

The authors declare no conflict of interest.

Author details

Antonella Petrillo[1*], Fabio De Felice[2] and Laura Petrillo[3]

1 Department of Engineering, University of Naples "Parthenope", Naples, Italy

2 Department of Civil and Mechanical Engineering, University of Cassino and Southern Lazio, Cassino, Italy

3 S.C. Prevenzione e Sicurezza Ambienti di Lavoro - S.S. PSAL Porto Ponte Andrea Doria, Genova, Italy

*Address all correspondence to: antonella.petrillo@uniparthenope.it

References

[1] Petrillo, A., Falcone, D., De Felice, F., Zomparelli, F. Development of a risk analysis model to evaluate human error in industrial plants and in critical infrastructures. International Journal of Disaster Risk Reduction, 2017; 23, 15-24.

[2] Mechhoud, E. A., Rouainia, M., Rodriguez, M. A New tool for risk analysis and assessment in petrochemical plants. Alexandria Engineering Journal, 2016; 55(3), 2919-2931.

[3] Giardina, M., Morale, M. Safety study of an LNG regasification plant using an FMECA and HAZOP integrated methodology. Journal of Loss Prevention in the Process Industries, 2015; 35, 35-45.

[4] Hollnagel, E. FRAM: The Functional Resonance Analysis Method: Modelling Complex Socio-Technical Systems. Ashgate, 2012.

[5] Saaty, T.L. Modeling unstructured decision problems - the theory of analytical hierarchies. Mathematics and Computers in Simulation, 1978; 20(3), pp. 147-158.

[6] Akgün, İ., Gümüşbuğa, F., Tansel, B. Risk based facility location by using fault tree analysis in disaster management. Omega, 2015; 52, 168-179.

[7] Righi, A. W., Saurin, T. A., & Wachs, P. A systematic literature review of resilience engineering: Research areas and a research agenda proposal. Reliability Engineering & System Safety, 2015; 141, 142-152.

[8] Rankin, A., Lundberg, J., Woltjer, R., Rollenhagen, C., & Hollnagel, E. Resilience in everyday operations: a framework for analyzing adaptations in high-risk work. Journal of Cognitive Engineering and Decision Making, 2014; 8(1), 78-97.

[9] Hickford, A.J., Blainey, S.P., Ortega Hortelano, A., Pant, R. Resilience engineering: theory and practice in interdependent infrastructure systems. Environment Systems and Decisions, 2018; 38(3), pp. 278-291.

[10] Bergström, J., van Winsen, R., Henriqson, E. On the rationale of resilience in the domain of safety: A literature review. Reliability Engineering & System Safety, 2015; 141, 131-141.

[11] Sawaragi, T., Horiguchi, Y., Hina, A. Safety analysis of systemic accidents triggered by performance deviation. In SICE-ICASE, 2006. International Joint Conference (pp. 1778-1781). IEEE.

[12] Furniss, D., Curzon, P., Blandford, A. Using FRAM beyond safety: a case study to explore how sociotechnical systems can flourish or stall. Theoretical Issues in Ergonomics Science; 2016; 17(5-6), pp. 507-532.

[13] Patriarca, R., Di Gravio, G., Woltjer, R., (...), Ferreira, P., Hollnagel, E. Framing the FRAM: A literature review on the functional resonance analysis method. Safety Science, 2020; 129,104827.

[14] Kaya, G.K., Ovali, H.F., Ozturk, F. Using the functional resonance analysis method on the drug administration process to assess performance variability Safety Science 118, pp. 835-840.

[15] Schutijser, B.C.F.M., Jongerden, I.P., Klopotowska, J.E., (...), de Bruijne, M.C., Wagner, C. Double checking injectable medication administration: Does the protocol fit clinical practice? Safety Science, 2019; 118, pp. 853-860.

[16] Bridges, K.E., Corballis, P.M., Hollnagel, E. "Failure-to-Identify" Hunting Incidents: A Resilience Engineering Approach. Human Factors, 2018; 60(2), pp. 141-159.

[17] Toroody, A.B., Abaei, M.M., Gholamnia, R. Conceptual compression discussion on a multi-linear (FTA) and systematic (FRAM) method in an offshore operation's accident modeling International. Journal of Occupational Safety and Ergonomics, 2016; 22(4), pp. 532-540.

[18] Alvarenga, M.A.B., Frutuoso E Melo, P.F., Fonseca, R.A. A critical review of methods and models for evaluating organizational factors in Human Reliability Analysis Progress in Nuclear Energy 75, pp. 25-41

[19] Pardo-Ferreira, M.D.C., Rubio-Romero, J.C., Gibb, A., Calero-Castro, S. Using functional resonance analysis method to understand construction activities for concrete structures. Safety Science, 2020; 128,104771

[20] Kaya, G.K., Ovali, H.F., Ozturk, F. Using the functional resonance analysis method on the drug administration process to assess performance variability. Safety Science, 2019; 118, pp. 835-840.

[21] Pardo-Ferreira, M.C., Martínez-Rojas, M., Salguero-Caparrós, F., Rubio-Romero, J.C. Evolution of the Functional Resonance Analysis Method (FRAM) through the combination with other methods. Direccion y Organizacion, 2019; 68, pp. 41-50.

[22] Rosa, L.V., França, J.E.M., Haddad, A.N., Carvalho, P.V.R. A resilience engineering approach for sustainable safety in green construction. Journal of Sustainable Development of Energy, Water and Environment Systems, 2017; 5(4), pp. 480-495.

[23] Haddad, A.N., Rosa, L.V. Construction sustainability evaluation using AHP and FRAM methods IIE Annual Conference and Expo, 2015 pp. 556-565.

[24] Rosa, L.V., Haddad, A.N., de Carvalho, P.V.R. Assessing risk in sustainable construction using the Functional Resonance Analysis Method (FRAM). Cognition, Technology and Work, 2015; 17(4), pp. 559-573.

[25] De Carvalho, P.V.R The use of Functional Resonance Analysis Method (FRAM) in a mid-air collision to understand some characteristics of the air traffic management system resilience. Reliability Engineering and System Safety, 2011; 96 (11), pp. 1482-1498.

[26] Hollnagel, E., Fujita, Y. The Fukushima disaster-systemic failures as the lack of resilience. Nuclear Engineering and Technology, 2013; 45 (1), pp. 13-20.

[27] Tveiten, C.K. Resilient planning of modification projects in high risk systems: The implications of using the functional resonance analysis method for risk assessments. Oil and Gas, Technology and Humans: Assessing the Human Factors of Technological Change, 2013; pp. 147-168.

[28] Haavik, T.K. On the ontology of safety. Safety Science, 2014;67, pp. 37-43.

[29] Praetorius, G., Hollnagel, E., Dahlman, J. Modelling Vessel Traffic Service to understand resilience in everyday operations. Reliability Engineering and System Safety, 215; 141, pp. 10-21.

[30] Melanson, A., Nadeau, S. Managing OHS in complex and unpredictable manufacturing systems: Can FRAM bring agility? Advances in Intelligent Systems and Computing, 2016; 490, pp. 341-348.

[31] Saaty, T.L. Applications of analytical hierarchies. Mathematics and Computers in Simulation, 1979; 21(1), pp. 1-20.

[32] Saaty, T.L. A scaling method for priorities in hierarchical structures.

Journal of Mathematical Psychology, 1977; 15(3), pp. 234-281.

[33] Saaty, T.L. How to structure and make choices in complex problems Human Systems Management, 1982; 3(4), pp. 256-261.

[34] Saaty, T.L. Introduction to a modeling of social decision processes. Mathematics and Computers in Simulation, 1983; 25(2), pp. 105-107.

[35] De Felice, F., Deldoost, M.H., Faizollahi, M., Petrillo, A. Performance measurement model for the supplier selection based on AHP. International Journal of Engineering Business Management, 2015; 7, pp. 1-13.

[36] Saaty, T.L Eigenvector and logarithmic least squares European Journal of Operational Research, 1990; 48(1), pp. 156-160.

Chapter 2

The Modulus of Resilience for Critical Subsystems

Eric Easton, Mario Beruvides and Andrea Jackman

Abstract

Accelerating digitization of critical infrastructures is increasing interconnection and interdependence among high-reliability subsystems. The resulting dependencies create new challenges in preventing underinvestment in high impact, low probability (HILP) events which can have disastrous consequences for society's critical subsystems. These more impactful events highlight the differences between reliability and resiliency, with the latter applicable to black swans. A number of approaches for quantifying resiliency have been proposed; however, a review of literature identified conceptual gaps when applied to empirical event data. This chapter provides a scenario agnostic method to quantify resiliency by applying concepts from materials science in a generalized form. This new formulation resulted from a mapping of constructs used in tensile testing to characteristics of protracted subsystem disruptions. Based on the mapping and gap analysis, a resiliency index calculation was developed and applied using examples based on empirical data from high impact events.

Keywords: resiliency, critical infrastructures, high impact, low probability (HILP), reliability, digital systems

1. Introduction

Digitization is occurring in many industries in many different forms; however, regardless of the application, a common set of enablers are employed. As the proliferation of digital transformation continues, decision makers will need to distinguish between reliability and resiliency in the planning, design, and operation of these subsystems. Tightly coupled common hardware and software platforms potentially increase the breadth of accidental failures as well as the impact of intentional sabotage. Beyond end use applications is an overall reliance on electricity which these digital subsystems require to function. Hardware, software, and electricity form the foundation upon which digitalization rest. The increased interdependence and interconnection can lead to common failure modes of previously isolated subsystems, resulting in increased probability of high impact events. Interconnection results in the establishment of a singular system with all other structures existing as subsystems. Evaluation of subsystems will need to include internally and externally initiated disruptive events. Highly impactful events, sometimes termed black swans, cannot only disrupt subsystems but fundamentally change their structure. Impactful as they are, rarity can make these events prone to underinvestment due to heuristics and biases, most prominently the availability heuristic. A quantifiable metric can aid in our ability to appropriately allocate resources to study, adapt, and mitigate these high impact, low probability events before they unexpectedly fracture the established subsystems we rely on. The avoidance of fracture is central to the application of the modulus of resilience in critical subsystems. The chapter will review the differences between the reliability and resiliency as well as the importance of distinguishing between the concepts. Additionally, ideals related to resilience are identified and expressed in a concise operational definition. The research utilized the progression shown in **Figure 1** for the investigation.

Borrowing concepts from materials science allows for an isomorphic application where analogous structures are leveraged to represent HILP event scenarios. In this chapter, the isomorphic application is presented to provide a method of quantifying resiliency or its absence based on the intended aim of the subsystem. This concept is consistent with select portions of previous literature, but divergent in others. Following a review of previous research, a gap analysis was completed to identify opportunities for new considerations in quantifying resiliency. Lastly, an example in applying the modulus of resilience for critical subsystems is provided to demonstrate the computational process.

2. The increasing case for resiliency

Reliability and resiliency are sometimes discussed in a similar context with respect to subsystem performance; however, they differ conceptually in both the events they measure and the characteristics they quantify. The measures which define reliability provide insights as to the context of the metrics use. Many of the most common reliability metrics utilize mean-based calculations from reoccurring failures over time. These metrics include mean time between failure (MTBF), mean time to failure (MTTF), and mean time to repair (MTTR). These metrics require successive failures in order to quantify subsystem performance. Mean time between failure (MTBF) is used in reliability to provide the number of failures per million hours for a subsystem. Mean time to repair (MTTR) is the time needed to repair a failed subsystem. Mean time to failure (MTTF) measures reliability for a subsystem which cannot be repaired. It is the mean time expected until the first failure of a subsystem. MTTF is a statistical value and represents the mean over a long period of time and a large number of operations. The reliability metrics can effectively represent common cause events which produce reoccurring failures; however, these calculations are less applicable to low probability special cause events. A special cause is something special, not part of the system of common causes. It is detected by a point that falls outside the control limits [1]. Often, subsystems have an allowable level of tolerance to minor disruption preventing sustained impairment in accomplishing the aim of the subsystem. Plotting the number of events by type versus percent of subsystem output disrupted graphically displays the relationship between common cause and special cause events. The allocation of events is closely represented by a pareto distribution **Figure 2**.

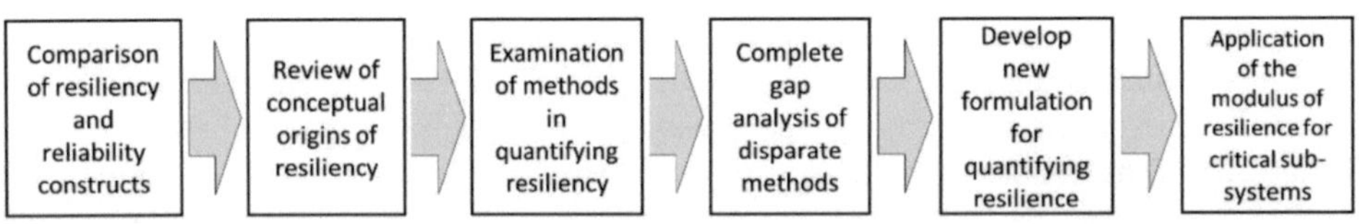

Figure 1.
Research phases.

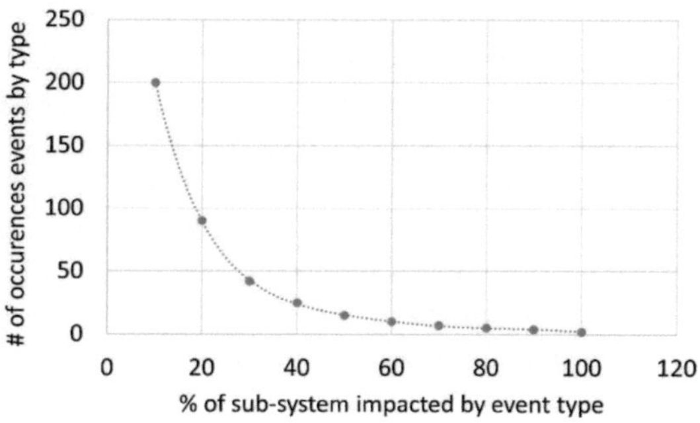

Figure 2.
Representative plot of event type distribution.

Resiliency events reside at the tail of the distribution as rare events resulting from extraordinary scenarios. Such events have been produced by multiple failures within a single subsystem as discussed in the book Normal Accidents by Charles Perrow. His work examined failures in highly complex operating environments. The increasing interdependence results in an interconnected ecosystem where a failure in a single subsystem can create failures in multiple subsystems. When interactive complexity is joined with tight coupling, the risk of a system accident is considerably increased. Interconnectedness and complexity among contemporary subsystems is increasing at a rapid pace as technologies develop faster than assessments can be made regarding their risks. As we move away from individual events and account for the larger system, we find the "eco-system accident," an interaction of systems that were thought to be independent but are not because of the larger ecology [2]. As systems grow in size and in the number of diverse functions they serve, and are built to function in ever more hostile environments, increasing their ties to other subsystems, they experience more and more incomprehensible or unexpected interactions [2]. Common mode failures, first included in analytical models in 1967, can contribute to unexpected actions from complex systems. In addition to common mode failures, proximity and indirect information sources are two additional indications of interconnectedness. Ultimately, the probability of a subsystem being subjected to significant disruption is dependent on the cumulative probability of both internal and external risks. Inevitably, the probability of significant disruption will increase as interdependence increases. While increases in events causing significant disruption are expected, their count is not expected to be significant enough for the application of mean-based reliability metrics. Therefore, resiliency-based metrics are needed which match the periodicity and scale of high impact, low probability events.

3. Quantifying high impact, low probability events

HILP events require a subsystem to bounce back to normalcy following major disruption. The goal is to regain pre-disruption levels of output as quickly as possible; however, recovery time is not the only metric of importance. The shape of the recovery curve is also of significance. Resiliency aids in defining a disaster response paradigm which differs from previous approaches such as resistance and sustainability by emphasizing return to normal. Nonetheless, the literature frequently uses the concept of resilience to imply the ability to recover or bounce back to normalcy after a disaster occurs [3]. Review of scholarly work related to the resiliency concept identified three main ideals: no assumption that disaster prevention is always possible, recognition of the need to include social variables, and the necessity to include disciplines outside the physical sciences and engineering. The term resiliency has increased in usage over the past decades. A multitude of definitions have been proposed whose interpretations can align with either resistance or sustainability. Although the resilience construct offered advantages in many areas relative to competing paradigms, the ambiguity associated with its meaning and scope hindered consensus. The multiplicity of definitions is a reflection of the philosophical and methodological diversities that have emerged from disaster scholarship and research [4].

Resilience first came to prominence in the English language in the early 19th century when Tredgold used the term to describe a property of timber [5]. In his essay "On the transverse strength and resilience of timber," Tredgold tested the properties of timber to be used in ship making. Tredgold cites resilience as the power of resisting a body in motion [5]. The statement is foundational in establishing the concept of resilience as more than recovery but instead as an ability to first withstand an applied force. Furthermore, Tredgold varied the weight and height of objects dropped on the test samples and recorded the effects to different forces on various wood pieces. These effects ranged from no effect, broke to curved. A second reference to the consideration of force can be found in the 1858 work, "On the Physical Conditions Involved in the Construction of Artillery, and on Some Hitherto Unexplained Causes of the Destruction of Cannon in Service," by Robert Mallet. He states the modulus of resilience of other writers, referred to hereafter, depends, is much greater for gunmetal, and hence a given force produces a greater proportional distortion of form [6]. The modulus of resilience was further formalized by materials science using stress/strain testing.

4. Methods in quantifying resilience

The range of methods for defining resilience include qualitative, quantitative and probabilistic. A quantitative method can be used to compare outcomes using data from different actual events. A number of researchers have explored quantifying resilience to move beyond qualitative representations. Henry and Ramirez-Marquez [7] proposed a quantitative approach for system resilience as a function of time. The formulation was a ratio of the recovery and losses using a figure-of-merit function. A disruptive event (e^j) at time, t_e, impacts the system until time, t_d.

$$Я_{\varphi}\left(t|e^j\right) = \frac{\varphi\left(t|e^j\right) - \varphi\left(t_d|e^j\right)}{\varphi(t_0) - \varphi\left(t_d|e^j\right)} \tag{1}$$

As shown, the numerator relates to the recovery until time t and the denominator represents the total loss due to disruption. Hosseini et al. [8] reviewed definitions and measures of system resilience. Their literature review was based on multiple domains including organizational, social, economic, and engineering using papers published between 2000 and April 2015. The major categories of assessment approaches are qualitative and quantitative with quantitative measures further defined as either probabilistic or deterministic.

The intent to analyze protracted subsystem disruptions leads to a focus on quantitative deterministic methods of calculating resiliency. The literature review by Hosseini et al. [8] included 11 deterministic methods of quantification. Bruneau et al. [9] utilized a method of integration based on the degradation in quality of infrastructure during recovery period of Eq. (5). Larger RL values indicate lower resilience while smaller RL imply higher resilience. Hosseini et al. [8] RL is calculated based on the formulation in Eq. (2).

$$RL = \int_{t_0}^{t_1} [100 - Q(t)]dt \tag{2}$$

Zobel [10] proposed a method based on the total possible loss over some suitably long-time interval (T^*), percentage of functionality lost after disruption (X), and time required for full recovery (T). An effort was made to analyze different combinations of X and T which result in the same level of resilience as shown in Eq. (3).

$$R(X,T) = \frac{T^* - XT/2}{T^*} \tag{3}$$

This metric is based on a linear recovery making it unrealistic for some scenarios.

Alternative methods were proposed by Cox et al. [11] based on economic resilience using the difference in disruption ($\%\Delta DY^{max}$) between the expected disruption ($\%\Delta Y$) and maximum potential disruption ($\%\Delta Y^{max}$). Therefore, an estimate of performance degradation is required. Such an estimation may be a challenge to precisely develop; however, the formulation is shown in Eq. (4).

$$R = \frac{\%\Delta Y^{max} - \%\Delta Y}{\%\Delta DY^{max}} \tag{4}$$

Alternatively, Rose [12] considered time effects using a concept of dynamic resilience. The quantification of dynamic resilience is the difference in system recovery with hastened system recovery (SO_{HR}) and without hastened system recovery (SO_{WR}). This calculation is utilized over the total number of time steps (N) considered. The dynamic resilience calculation is shown in Eq. (5).

$$DR = \sum_{i=1}^{N} SO_{HR}(t_i) - SO_{WR}(t_i) \tag{5}$$

Wang et al. [13] explored resilience in information systems based on the number of operations in the enterprise information system (m). The ratio of the demand time (d_i) and completion time of operation (c_i) are weighted by the importance of operation (z_i).

$$R = \max \sum_{i=1}^{m} z_i \frac{d_i}{c_i} \tag{6}$$

The larger the value of the metric the more resilient the system is determined to be. The calculation requires the assignment of a weight and assumes the number of operations is known. When attempting to quantify unknown events the number of operations can be difficult to estimate.

Chen and Miller-Hooks [14] quantifies the "post-disruption expected fraction of demand that, for a given network, can be satisfied within pre-determined recovery budgets" (Hosseini et al.). The measure was based on transportation networks and compares the maximum demand that can be satisfied before disruption (D_w) and after disruption (d_w) for pair (w).

$$Resilience = E\left(\sum_{w \in W} d_w / \sum_{w \in W} D_w\right) \tag{7}$$

Orwin and Wardle [15] considered the instantaneous and maximum disturbance in the quantification of resilience. The maximum absorbable force without upsetting system function (E_{max}) and effect of the disturbance on safety (E_j) at a given time (T_j) are used to define resilience.

$$Resilience = \left(\frac{2 \times |E_{max}|}{|E_{max}| + |E_j|} \right) - 1 \tag{8}$$

Frameworks for local and global resilience were introduced by Enjalbert et al. [16] for modeling system safety in public transportation systems. A safety indication function (S(t)) is used to calculate resilience either instantaneously or over time, representing local and global, respectively. Global resilience is calculated from the time of disturbance (t_b) to the end of the disturbance (t_e). The calculations are as follows:

$$Local\ resilience = \frac{dS(t)}{dt} \tag{9}$$

$$Global\ resilience = \int_{t_b}^{t_e} \frac{dS(t)}{dt} \tag{10}$$

Francis and Bekera [17] introduced a metric for dynamic resilience. The calculation uses the speed of recovery (S_p), original performance level (F_o), performance level at new stable level (F_r) and performance level immediately after disruption (F_d). The speed of recovery variable assumes exponential growth for a maximum acceptable recovery time (t_δ), total recovery time (t_r) to a new equilibrium state, time to complete initial recovery (t_r^*), and a decay in resilience (a). The resilience metric is calculated using Eq. (11).

$$\rho_i = S_p \frac{F_r}{F_o} \frac{F_d}{F_o} \tag{11}$$

$$S_p = \left(t_\delta / t_r^*\right) exp\left[-a\left(t_r - t_r^*\right)\right] for\ t_r \geq t_r^* \tag{12}$$

Otherwise,

$$S_p = \left(t_\delta / t_r^*\right) \tag{13}$$

Cimellaro et al. [18] utilized quality of service to represent resilience. The method uses before disruption quality of service ($Q_1(t)$), post disruption quality of service ($Q_2(t)$), a control time (T_{LC}) and a weighting factor (α) in developing a healthcare resilience metric.

$$R = \alpha \int_{T_{LC}}^{\cdot} \frac{Q_1(t)}{T_{LC}} dt + (1 - \alpha) \int_{T_{LC}}^{\cdot} \frac{Q_2(t)}{T_{LC}} dt \tag{14}$$

Aside from the works investigated by Hosseini et al. [8], Dessavre et al., [19] introduced a new model and visual tools adding a stress dimension representing the force and stress of disruptive events. Defining the stress of the events is not a trivial task and completely domain dependent [19].

A review of the concepts found in literature was completed for elements consistent with the modulus of resilience. Methods were limited to quantitative approaches which could be utilized with empirical data sets. Although the use of scaling factors was identified in literature [13, 18], such methods are not desired in the development of subsystem-based methods due to the subjectivity associated with them. A ratio-based approach has merit in its ability to normalize event effects and resulting recovery. Area-based calculations using integration are preferred to point calculations based on their ability to compensate for nonlinear restoration curves; however, complexity beyond the resilience triangle [9] would be necessary to capture differences in event magnitude and restoration response in disparate events.

The concept of a yield point was not identified in existing literature. A return to normal operation was typically used to identify the end of the restoration time period; however, this approach does not set the time based on the aim of the subsystem. Evaluations of subsystems beyond a critical point with respect to use of the subsystem output could lead to poor decision-making. One of the main weaknesses of the current resilience metric is that they do not relate the effects of a disruptive event to any of the event characteristics, unlike materials science [19]. Materials science utilizes a change in length for evaluation of stress and strain; however, the difference in recovery response to a common cause and special cause event was not found in the literature review. These distinctions serve to highlight the differences between reliability for normally occurring events and resiliency to low frequency events. Additionally, the need for utilizing subjective variables [10, 11, 12, 14, 15] does not lend well to empirical study.

The ability to normalize responses to different events is beneficial for evaluating the resiliency of different subsystems or different events on the same subsystem. The literature reviewed began analysis of the event from the start of restoration [7] or by treating the entire curve from time of event to the completed restoration as a single integral [16]. This approach can confuse the quantities of force, stress and strain. An equal force can result in different stress and strain based on the subsystem being reviewed. As a result, the descending slope and associated area prior to the start of recovery may prove informative of stress. Strain is more associated with the total area under the curve. The review of literature did not identify a bifurcation of the curve to delineate stress (prior to start of recovery) and strain (total area). Therefore, the assumption of instantaneous loss and exponential recovery [17] are not representative of many empirical cases.

In reviewing the concepts of resilience, a force is applied to a subsystem, the subsystem absorbs a portion of the force, experiences stress, and adapts to recover to a pre-disruption state. These references highlight an importance of considering the stress on the subsystem in determining the resiliency of a subsystem. Three primary points of measure for use in quantifying resiliency were identified including: stress, total area of event and change in length. Stress is a foundational variable of resiliency, as the term resiliency implies a response to a significant disruption. Therefore, only events of significance from a subsystem level are commonly referred to in terms of resilience. Additionally, the ability to compare resiliency events needs some level of normalization based on the associated stress for each event. Force continues to be applied until the subsystem decay ceases, allowing for subsystem assessment and initiation of recovery. The rate of subsystem decay influences the stress applied to the subsystem and the subsystem ability to bounce back. This connection exists due to the role of adaptation in the resiliency process. A slow evolving scenario (i.e., slow subsystem decay) presents the subsystem opportunity to adapt, resist, and recover in ways an acute decay will not. Therefore, when considering the normalization process of resiliency both the decay (i.e., stress proxy) and recovery portion of the resiliency curve must be independently considered. The delayed decay provides an opportunity for improved response from the subsystem.

Total area of recovery best quantifies recovery and resiliency by compensating for the nonlinearity in the response function. As the subsystem attempts to recover, disruptions in the recovery process may cause discontinuities not captured by linear slope calculations. Similarly, time to recovery (i.e., 3 days to recovery) calculations may fail to represent intermediate progress in recovery.

Consideration of a failure point based on the aim of the subsystem aids in representing real-world scenarios. Recovery which occurs after a critical point of the subsystem would indicate a lack of resiliency. As an example, if a water subsystem requires 10 days to restore operation post contingency but the consumers of the water can only survive 4 days without water; the subsystem lacks resiliency. Attempts to quantify the subsystem's resilience should stop at 4 days. Calculations beyond the 4-day time period no longer support the aim of the subsystem or the practical operation of the subsystem.

Lastly, change in length was included in the materials science calculation of the modulus of resilience. The change in length from the original length to the length under stress could be translated to a subsystem resilience construct to allow consideration of how subsystem recovery under lower stress common cause events and high stress special cause events are related. The consideration of a change in length may aid in joining concepts associated with reliability in the quantification of resilience.

Comparing these constructs with the reviewed literature results in the identification of conceptual gaps. The resulting resiliency values should reflect the subsystem performance for practical cases. Units are required based on subsystem parameters. The x-axis utilizes units of time, while the y-axis measures the units associated with the aim of the subsystem.

The methods of quantification reviewed begin the process of quantification at the point of recovery or assume no time delta between the initiating event and start of recovery. To support the incorporation of stress in the quantification of resilience, a bifurcation of the event curve is used as shown in **Figure 3**.

The use of ratio methods may provide consistency in scenarios of similar characteristics. When disparate characteristics are present, computed values may prove inconsistent with event outcomes. Depending on the event characteristics, either ratio methods or area-based methods may identify a less resilient subsystem response as more resilient. **Figure 4** depicts the concept of less recovery time for less disruption. The scenario of **Figure 4** is representative of a minor difference in subsystem response and would provide consistent rankings for resilience outcomes in many cases, where less area is representative of increased resilience.

Conversely, cases may exist where a longer recovery results from a less impactful initial event. The delayed recovery to a less impactful event could result from many factors including a lack of preparedness, inability to adapt, etc. In such cases, observation would assume that the subsystem which took longer to recover from a less impactful event is less resilient. However, present formulations may suggest the opposite. **Figure 5** illustrates this scenario, where the smaller area is not representative of the more resilient outcome.

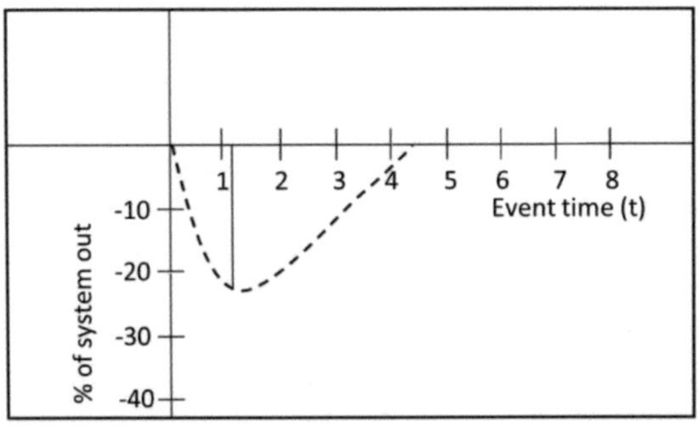

Figure 3.
Bifurcation of event curve.

The fracture point should be set based on the aim of the subsystem. For example, if a drinking water subsystem failure requires a 7-day restoration period but 4 days is the survival period without water; the calculation of subsystem resiliency should be limited to a 4-day period. In some cases, the acknowledgement of a fracture point will result in the calculation of resiliency stopping prior to the subsystem returning to pre-disruption output levels. **Figure 6** represents a case where the subsystem recovery takes longer than the subsystem failure point.

Calculations to quantify resiliency which consider values beyond the failure point are theoretical as opposed to practical in nature. The failure point should be given priority in quantifying resiliency.

An operational definition is derived from the combination of literature review and isomorphic adaptation of the modulus of resilience. Hence, resiliency is defined as the ability to limit proportional stain from abnormal stress to less than the subsystem yield point, through the achievement of recovery in less than the subsystem critical timeframes. This definition allows the use of quantitative

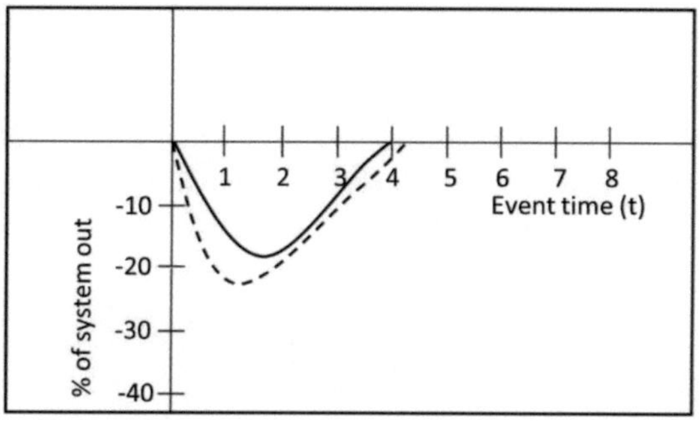

Figure 4.
Recovery curves with similar characteristics.

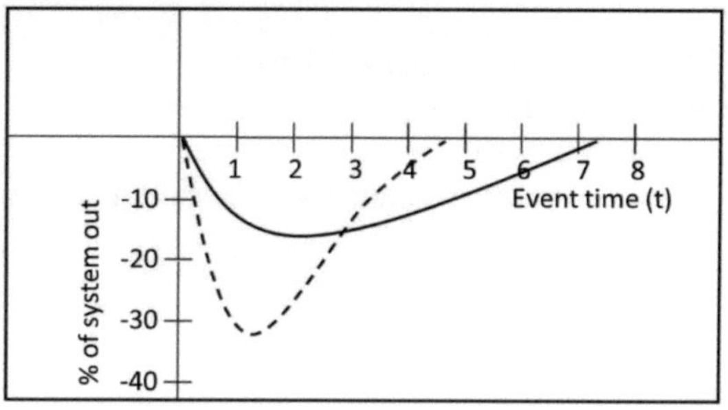

Figure 5.
Recovery curves with dissimilar characteristics.

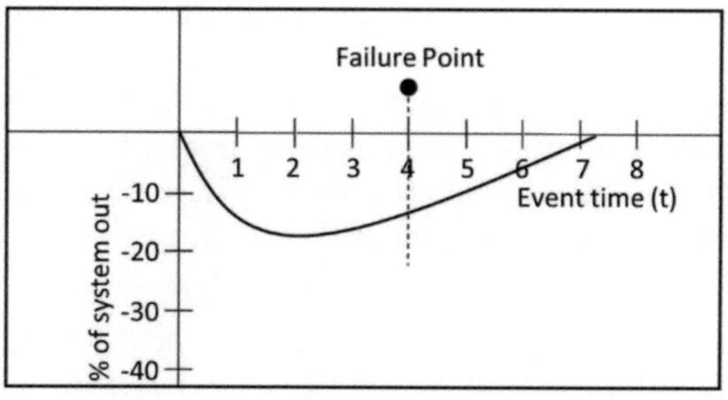

Figure 6.
Representation of failure point.

measures in the calculation of resilience in a deterministic and normalized approach based on concepts from materials science.

An evaluation between two groups can result in an isomorphic application of findings from one structure to another. This mapping between groups can yield opportunities to apply known methodologies in an inter-disciplinary manner. The process of verifying an isomorphism requires the identification of elements in each structure and evaluating their equivalence. If equivalence is identified an opportunity for applying the computational framework may exist. The quantification of subsystem resilience was compared to resiliency as used in materials science. Materials science's definition of resiliency includes the concepts of per unit volume, maximum energy, and integration from zero to the elastic limit. The modulus of resilience (Ur) is found from the stress-strain curve measured during the tensile test. Stress (σ) in the stress-strain curve is "the applied force per unit original undeformed cross-sectional area of the specimen" [20] as delineated in Eq. (15).

$$\sigma = \frac{F}{A_0} \tag{15}$$

where F = force; A_0 = cross sectional area.

Young's modulus (E) serves as a measure of stiffness for a solid material. "Because of the difficulty in determining the elastic limit, it is commonly replaced by the proportional limit, which is the stress at which the stress-strain curve is out of linearity" [20].

$$E = \frac{F/A}{\Delta L/L_0} \tag{16}$$

And,

$$F = \sigma \times A \tag{17}$$

where F = force; A = actual cross-sectional area; ΔL = amount of change in length; L_0 = original length of the object.

"The modulus of resilience is the strain energy per unit volume absorbed up to the elastic limit for a tensile test and equals the area under the elastic part of the stress-strain curve" [20].

$$U_r = 1/2\,(\sigma^2/E) \tag{18}$$

"This quantity indicates how much energy a material can absorb without deforming plastically" [20]. Plastic deformation occurs when a material undergoes non-reversible changes in response to applied forces. The use of the stress-strain curve from materials testing is similar to conditions faced by disrupted subsystems regardless of type. Stress is the impact to the material under test, while strain is the resulting effects of the stress.

Based on the desire of applying a consistent methodology to quantify resilience regardless of disruption magnitude or subsystem size, the percentage of subsystem disrupted is proposed to achieve a per unit value. The area under the curve will then be integrated from the beginning to end of the disruptive event. Calculus to determine area under the curve is shown in Eq. (19).

$$\text{Total Area under the Curve} = \int_{E_i}^{E_r} f(x)dx \tag{19}$$

where E_i = Event initial; E_r = Event restored.

Materials science.	Protracted subsystem disruption	Comparisons
Stress applied	Peak percent of subsystem out of service	Percent out of service is equivalent to stress
Cross-sectional area	Area under curve from the origin to peak subsystem out of service	Area from zero to peak subsystem out of service is point where curve loses linearity
Actual cross-sectional area	Area under curve for entire disruptive event	Represents total strain experienced by subsystem
Change in length	Delta between subsystem's average duration of disruptions and event disruption duration	Use of change in duration accounts for the change in length between average and protracted event
Original length	System's average duration of disruptions	Accounts for average non-protracted disruptions events

Table 1.
Parallels between materials science test and protracted subsystem disruptions.

The area under the curve will then be applied to the maximum percentage of subsystem disrupted.

$$\text{Resiliency index (RI)} = \frac{1}{2}\left(\frac{(SD)^2}{\frac{(SD\times A_t)/A_{nl}}{(D_e-D_a)/D_a}}\right) \quad (20)$$

where SD = % of subsystem disrupted; A_{nl} = Area under the curve to nonlinearity; A_t = Total area under the curve; D_a = Duration of average disruption; D_e = Duration of event disruption.

Protracted subsystem disruptions create stress and strain due to an inability to complete the subsystem aim. The similarities between tensile strength test used in materials science and the need to measure stress and strain subsystems create an isomorphic relationship. **Table 1** shows the parallels between materials science and protracted subsystem disruptions.

The application of the modulus of resilience to a specific subsystem requires the identification of an aim the subsystem exist to accomplish. "Without an aim, there is no system" [21]. The aim should be quantifiable with metrics available for analysis. The data must be accessible in order to serve as the basis for the resilience calculations and will vary based on the subsystem under study. Examples include percentage of successful operations or percentage of end users receiving service. The next section provides an empirical example in applying the modulus of resilience.

5. Application of the modulus of resiliency

The power industry was selected to provide an example for applying the modulus of resiliency using empirical data. The aim of the electric subsystem is to deliver electricity to all end use customers; therefore, data regarding the number of customers out of service can be used to quantify subsystem performance. The use of customers out of service in quantifying subsystem performance was supported by a review of regulatory reliability metrics used by Public Utility Commissions. For major electric utility disruptions, DOE situation reports provide customer outage information for and are publicly available from the DOE website. One of the most prominent events to challenge utilities is hurricane, and as a result, multiple hurricane events have data on the DOE website. Following data collection, plots can be constructed of the electric utility response in restoring customers. The inflection points were identified, and a yield point designated by reviewing disaster preparedness data from the Capital Region Study [22]. The study indicated that 73% of survey respondents had less than 10 days of food stored. Therefore, an event lasting greater than 10 days would most likely result in scarcity from food spoilage and diminished retail capabilities. With a known bifurcation and yield point, analysis can be completed.

Hurricanes Wilma and Irma presented an opportunity to compare resiliency of separate events in the same region. Following Wilma, the ability of several infrastructures to recover from severe events was reviewed in the Florida region. "[M] ore than $141.5 million has been obligated by FEMA for 119 Hazard Mitigation Grant Program projects to build stronger, safer more resilient communities in Florida" [23]. Florida was once again subjected to a hurricane when Irma came ashore 12 years later. More than six million customers lost power as a result of Irma; compared to 4 million from Wilma. Although more than a decade apart, these two storms provide an opportunity to compare the recoveries following significant investment in resiliency. The comparison of the two resiliency indices can present an opportunity to calculate a cost per unit of resiliency and explore concepts such as diminishing returns or optimization from multi-hazard investment. Multi-hazard resiliency actions would provide an ability to address multiple HILP scenarios with a single investment. A resiliency index for each of the scenarios would be computed in order to create a composite change in resiliency for a given investment. The goal of this composite approach is to provide a means for justifying highly adaptable subsystem structures based on resiliency benefits.

The example demonstrates the process of calculating the resiliency index for a power utility scenario and comparing the response before and after the investment in resiliency. The values shown in **Table 2** were extracted from United States Energy Information Administration (EIA) data. The additional data points associated with 0.5 and 1.5 days were included due to nonlinearities in customer outages associated with Hurricanes Wilma and Irma, respectively. Similarly, day 9 for Hurricane Wilma was approximated for the purpose of this analysis. The data required to calculate the change in length was available by collecting System

Day.	% Out of service (Hurricane Wilma 2005)	% Out of service (Hurricane Irma 2017)
0	0	0
0.5	34	20
1	35	40
1.5	34	64
2	31	56
3	28	40
4	21	31
5	18	20
6	12	11
7	10	7
8	9	4
9	6	1

Table 2.
Outages for Hurricanes Wilma and Irma.

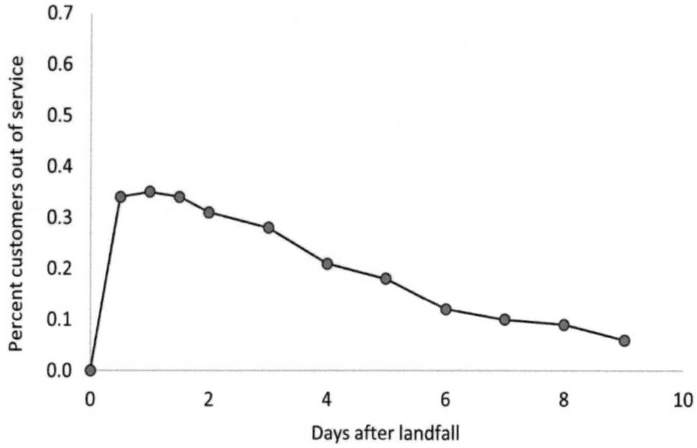

Figure 7.
Hurricane Wilma restoration plot.

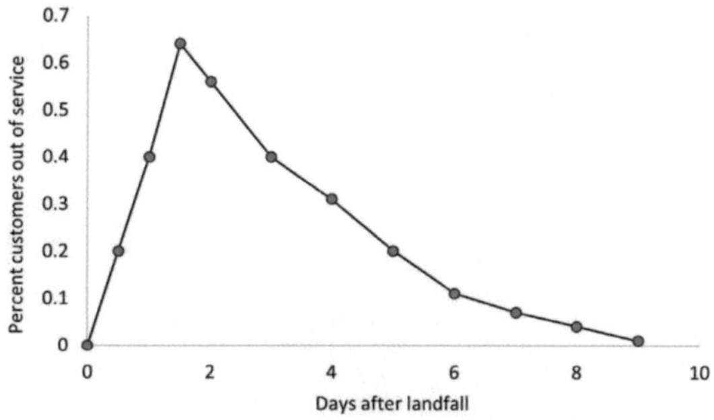

Figure 8.
Hurricane Irma restoration plot.

Average Interruption Duration Index (SAIDI) data. SAIDI data provides a basis for the average duration a customer faces and can be compared to the protracted system disruption as a change in length.

Following the collection of empirical data, the total area under the curve was calculated by dividing the outage curve into time steps and summing the areas of each time step as shown in **Figures 7** and **8**, respectively.

The study region had a SAIDI of 60 minutes and a protracted outage duration of 12,960 minutes. Therefore, the resiliency index (RI) for Hurricane Wilma is determined as shown in Eq. (7).

$$\text{Resiliency index (RI)} = \frac{1}{2}\left(\frac{(0.35)^2}{\left(\frac{(0.12\times 1.758)/0.258}{(12,960-60)/60}\right)}\right) = 16.07 \tag{21}$$

The study region had a SAIDI of 57 minutes and a protracted outage duration of 12,960 minutes. Therefore, the resiliency index (RI) for Hurricane Irma is determined as shown in Eq. (22) based on EIA data [24] (**Tables 3** and **4**).

$$\text{Resiliency index (RI)} = \frac{1}{2}\left(\frac{(0.64)^2}{\left(\frac{(0.11\times 2.175)/0.56}{(12,960-57)/57}\right)}\right) = 89.14 \tag{22}$$

Day.	% Out of service (Hurricane Wilma 2005)	Area
0	0	0.085
0.5	0.34	0.173
1	0.35	0.173
1.5	0.34	0.163
2	0.31	0.295
3	0.28	0.245
4	0.21	0.195
5	0.18	0.150
6	0.12	0.110
7	0.10	0.095
8	0.09	0.075
9	0.06	[redacted]
	Total area under curve	1.758
	Area under curve to nonlinearity	0.258
	Maximum % of customers out	0.350

Table 3.
Resiliency index calculation for Hurricane Wilma.

Day.	% Out of service (Hurricane Irma 2017)	Area
0	0	0.050
0.5	0.20	0.150
1	0.40	0.260
1.5	0.64	0.300
2	0.56	0.480
3	0.40	0.355
4	0.31	0.255
5	0.20	0.155
6	0.11	0.090
7	0.07	0.055
8	0.04	0.025
9	0.01	[redacted]
	Total area under curve	2.175
	Area under curve to nonlinearity	0.560
	Maximum % of customers out	0.640

Table 4.
Resiliency index calculation for Hurricane Irma.

Change in resiliency is found by Eq. (3).

$$\Delta RI = \frac{RI_{final} - RI_{initial}}{RI_{initial}} = \frac{89.14 - 16.07}{16.07} = 4.55 \quad (23)$$

The determination of a change in resiliency allows for a quantitative measurement related subsystem response. The use of resiliency indices can aid in quantifying the efficacy of resiliency investment.

6. Conclusions

In this chapter, a comparison to mean-based reliability was contrasted with the use of resiliency calculations for HILP events. Resiliency calculations are required, given the infrequent nature of protracted subsystem disturbances. Following a review of resiliency computations, a gap analysis was used to identify the opportunities for ensuring a resiliency calculation can capture the nonlinearities observed in empirical data. Parallels are provided between the modulus of resilience construct from materials science and an isomorphic application defined. In conclusion, an example is presented for the power utility sector demonstrating the methods of collecting the inputs and completing the computations. These inputs include defining the aim of the system and failure point, data collection, determination of bifurcation point, and the use of reliability data for calculating a change in length.

The ability to calculate resiliency regardless of the subsystem or scenario can assist in the evaluation of resiliency actions already taken or planning for new investment. The ability to compute resiliency on a common base may also offer opportunities to optimize investment based on interconnectedness to the subsystems which yield the greatest improvement. A more integrated approach may lead to increased systemic resiliency as opposed to more common heuristics-based subsystem specific approaches. The proposed method more closely adheres to the ontological and conceptual frameworks associated with initial references of resiliency. Furthermore, subjective inputs are avoided increasing the replicability and repeatability of associated research. By acknowledging a yield point specific to the aim of the subsystem, results from the resiliency index better represent the outcomes of real-world subsystems. Lastly, bifurcating the event curve allows the onset characteristics of the disruptive event to normalize the resiliency performance metric.

Further research on the distribution of events by type will be conducted to validate the anecdotal evidence regarding common cause and special cause events. This additional data will assist in the development of statistics for assessing the correlation between increasing interdependence and HILP events for critical subsystems. In order to test a wider array of empirical data sets, resiliency indexes will be calculated using both historical and future HILP event data. The results of these analyses will be used to continually evaluate the efficacy of the metrics and identify opportunities for enhancements.

Acronyms

HILP	high impact, low probability
MTBF	mean time between failure
MTTF	mean time to failure
MTTR	mean time to repair
DOE	department of energy
FEMA	Federal Emergency Management Agency
EIA	energy information administration
SAIDI	system average interruption duration index
RI	resiliency index

Author details

Eric Easton[1*], Mario Beruvides[1] and Andrea Jackman[2]

1 Texas Tech University, Lubbock, United States

2 IBM, Pittsburg, United States

*Address all correspondence to: eric.easton@ttu.edu

References

[1] Deming E. The New Economics. 2nd ed. Cambridge, MA: MIT, Center for Advanced Education; 1994

[2] Perrow C. Normal Accidents Living with High-Risk Technologies. New York: Basic Books.nal Services; 1984

[3] McEntire A, Fuller C, Johnston C, Weber R. A comparison of disaster paradigms: The search for a holistic policy guide. Public Administration Review. 2002;**62**(3):269

[4] Manyena SB. The concept of resilience revisited. Disasters. 2006; **30**(4):434-450

[5] Tredgold T. Chapter XXXXVII: On the transverse strength of timber. Philosophical Magazine: A Journal of Theoretical, Experimental and Applied Science. London: Taylor and Francis; 1818;**1**:214-216

[6] Mallet M. On the Physical Conditions Involved in the Construction of Artillery: An Investigation of the Relative and Absolute Values of the Materials Principally Employed and So Hitherto Unexplained Causes of the Destruction of the Canon in Service. London: Long, Brown, Green, Longmans and Roberts; 1856

[7] Henry E, Ramirez-Marquez J. Generic metrics and quantitative approaches for system resilience as a function of time. Reliability Engineering and System Safety. 2012;**99**:114-122

[8] Hosseini S, Barker K, Ramirez-Marquez J. A review of definitions and measures of system resilience. Reliability Engineering and System Safety. 2016;**145**:47-61. DOI: 10.1016/j.ress.2015.08.006

[9] Bruneau M, Chang SE, Eguchi RT, Lee GC, O'Rourke TD, Reinhorn AM, et al. A framework to quantitatively asses and enhance the science the seismic resilience of communities. Earthquake Spectra. 2003;**19**(4):733-752

[10] Zobel CW. Representing perceived tradeoffs in defining disaster resilience. Decision Support Systems. 2011;**50**(2): 394-403

[11] Cox A, Prager F, Rose A. Transportation security and the role of resilience: A foundation for operational metrics. Transport Policy. 2011;**18**(2): 307-317

[12] Rose A. Economic resilience to natural and man-made disasters: Multidisciplinary origins and contextual dimensions. Environmental Hazards. 2007;**7**(4):383-398

[13] Wang JW, Gao F, Ip WH. Measurement of resilience and its application to information systems. Enterprise Information Systemss. 2010; **4**(2):215-223

[14] Chan L, Miller-Hooks E. Resilience: An indicator or recovery capability in intermodal freight transport. Transportation Science. 2012;**46**(1): 109-123

[15] Orwin KH, Wardle DA. New indices for quantifying the resistance and resilience of soil biota to exogenous disturbances. Soil Biology and Biochemistry. 2004;**26**:1907-1912

[16] Enjalbert S, Vanderhaegen F, Pichon M, Abel Ouedraogo K, Millot P. Assessment of transportation system resilience. In: Cacciabue C, Hjälmdahl M, Luedtke A, Riccioli C, editors. Human Modeling in Assisted Transportation. Italy: Springer-Verlag Italia Srl; 2011. pp. 335-341

[17] Francis R, Bekera B. A metric and frameworks for resilience analysis of engineered and infrastructure systems.

Reliability Engineering and System Safety. 2014;**12**:90-103

[18] Cimellaro GP, Reinhorn AM, Bruneau M. Seismic resilience of a hospital system. Structure and Infrastructure Engineering. 2010;**6** (1–2):127-144

[19] Dessavre D, Ramirez-Marquez J, Barker K. Multidimensional approach to complex system resilience analysis. Reliability Engineering and System Safety. 2016;**149**:34-43. DOI: 10.1016/j.ress.2015.12.009

[20] Pilkey W. Formulas for Stress, Strain, and Structural Matrices. Hoboken: John Wiley & Sons, Inc; 2005

[21] Deming E. The New Economics. 2nd ed. Cambridge, MA: MIT, Center for Advanced Educational Services; 1994

[22] Williams M, Saathoff G, Guterbock T, MacIntosh A, Bebel R, et al. Community Shielding in the National Capital Region: A Survey of Citizen Response to Potential Critical Incidents [Internet]. 2005. Available from: http://www.medicine.virginia.edu/community-service/services/ciag/programs/community_shielding/publications/community_shielding_report_body.pdf [Accessed: 10 October 2015]

[23] FEMA. Hurricane Wilma: Ten Years Later. Retrieved [Internet]. Available from: https://www.fema.gov/news-release/2015/10/22/hurricane-wilma-ten-years-later [Accessed: 24 March 2017]

[24] U.S. Energy Information Administration (EIA). Hurricane Irma cut power to nearly two-thirds of Florida's electric customers. Retrieved [Internet]. Available from: https://www.eia.gov/todayinenergy/detail.php?id=32992 [Accessed: 28 March 2018]

Chapter 3

Condition-Based Maintenance for Data Center Operations Management

Montri Wiboonrat

Abstract

This chapter presents data center operations management by giving four case studies of power distribution systems (PDS) of data centers (Tier I, Tier II, Tier III, and Tier IV). The four topologies of PDS have defined by the design of single points of failure and redundant equipment and systems. The concepts of Mean Time between Failures (MTBF) and Mean Time to Repair (MTTR) apply during PDS design for reduced system downtime. Moreover, MTBF and MTTR use for estimating system availability of each Tier classification. Human factors consider as critical part of data center operations that need to quantify and qualify on knowledge and skills such as certified levels. For sustainable data center operations, the new software for data center operations called Data Center Infrastructure Management (DCIM) has deployed for monitoring and controlling entire system operations that interact among system of systems and human interfaces by deployed condition-based maintenance (CBM) as preventive and predictive conditions. Moreover, CBM performs as long-term cost saving for total cost of ownership (TCO) and energy efficiency.

Keywords: data center, condition-based maintenance, DCIM, power distribution systems, operations management

1. Introduction

Any system fault of data center is decreasing in total system performance when consider with the minimum requirements of system specification. Therefore, the fault may incur from many reasons such as design error, erroneous installation, machine malfunctions, device defectiveness, miss operations, human error, over operating conditions, or an amalgamation of all of those incidents. In case if the error is not detected within a timely manner and correct response, system failure may happen. Mostly, data center downtime had occurred from cascading failure from devices to sub-system and system. As the results, preventive and predictive mechanism probe to detect the error before it become a failure. The best practice of data center operations, corrective maintenance is not acceptable, for instance in case of New York Stock Exchange in 2015, within the 4 hours downtime, after an upgrade failed, at the stock exchange will result in the consequence of one's action at least $2.5 million per hour. The data center downtime is not only costly in financial compensated but also ruin reputation that sometime cannot be evaluated.

The research from Ponemon Institute [1] reports that the total average cost of data center downtime soared by 38 percent in 2010 from $505,502 to $740,357 per unplanned downtime in 2016. Thereby, to evade these costs of data center downtime; they require deploying more procedures of intensive training and operations, modern maintenance strategies, and experiencing data center's operators.

Downtime costs are a part of operating expenditure (OPEX) subject to lawsuit or penalty costs that result of any incident. The legal punishment can avoid by PPM approach or called insurance investment, that help reduce TCO in long-term operations. TCO consists of the sum total of operational and capital expenses involved in erecting and maintaining a data center. PPM approaches is not just only protecting downtime costs but also preventing reputation costs of the company that may not be estimated.

The traditional approach to avoid a downtime is applying the action plan through time-based maintenance (TBM). This means that the maintenance team plan for maintenance or upgrade systems by monitoring and controlling up on the schedule time of weeks, months, or annually based on the supplier's recommend. Moreover, TBM approach prevents the system downtime by following these maintenance schedules; regular inspection, easy to deployment, no condition monitoring needed; decision-maker control (maintenance age or MTBF) maintenance performed when the device reaches MTBF. On the other hand, the condition-based maintenance (CBM) strategic approach relies on an online/offline data collection and continuous measurable condition of devices or systems entirely during they are executing. By applying sensor devices and tools, gathering information that can perform to establish database system for trend analysis, gathering information prediction, and estimated remaining useful lifetime (RUL) of a device or system. The CBM takes action when reaches over the condition of the measurable point that system performance is directly degrading or most likely failure. A prognostic approach of online performance monitoring needs for the throughout degrading processes, from the outset of the system design, installation, operations, and until system failure. This difference approach from scheduled intervals recommends with preventive maintenance.

Since 21st century, the technological advancement, data-driven approach to PDS is predictable and precise. For this reason, many of these data center outages can avoid or mitigate with the properly maintenance approaches and deploying sensing technologies. Predictive maintenance is the complementary of preventive maintenance. Predictive maintenance imposes on the device working condition and tracking operating environment before system breakdown happens. With online condition monitoring system, the predictive maintenance takes action when the deterioration level M reached. (Decision variable: M/threshold deterioration level).

In this research, researcher proposes the preventive and predictive maintenance (PPM) which determines the CBM as systematic strategy of data center operations and maintenance. Use case examples of PDS of data center had examined to ensure their proper functionality and to reduce their deterioration rate. PPM approach can insure devices, sub-systems and systems operating safety, operate as their functional reliability and efficiency, reduce failure rates, and prevent unscheduled downtimes.

2. Background

2.1 Preventive and predictive maintenance

Preventive maintenance implies to regular maintenance or TBM that maintains devices and systems up and operating as normal condition, prevent any unplanned downtime, uneconomical costs from unpredicted system failure, and preserve the operation running efficiency and effectiveness.

CBM comprehends as predictive maintenance. It is a useful mechanism of strategic approach for preventive maintenance that collaborates with monitoring and controlling conditions of critical devices and equipment parameters. This process will operate in order to predict device failure, to assess the RUL, and to avoid system risks, which could be happened if minimum conditions are exceeded. This strategy demonstrates the economical savings over observation of lessons or time-based preventive maintenance, because exertion will execute only when guaranteed.

A RUL defines based on the maintenance policy for single unit deteriorating system that all conditions are continually monitoring with deploying A-B-C analysis to device criticality build up on early successful diagnostics. The A-B-C analysis will diagnose and categorized level of system maintenance into 3 groups; reactive maintenance and excessive repairs and failures; proactive maintenance; and excessive PM and no failure and no repairs [2], as presented in **Figure 1**.

2.2 Condition-based maintenance

The CBM imposes as the predictive maintenance strategy, which executes device or system maintenance based on setting up conditions, performance, parameter monitoring and the subsequent actions before device or system failures happened. The CBM is a maintenance pattern that advises for maintenance decisions refer to the data and information collecting from condition monitoring system processes. During operating condition, CBM is executing as monitoring appliance through sensing device, which can gauge parameter based on various monitoring attribute s, for example temperature, humidity, vibration, noise levels, contaminants, CO_2 and CO scale, and lubricating oil concentration. The usefulness of CBM is the application of the condition monitoring process, where the signals and data are online monitoring by applying many types of sensors inform of wire and wireless technologies. The core of CBM is executing in a real-time assessment of devices and systems conditions in order to analyze all data to perform the decision analysis for maintenance conditions and solutions, while reduces an planned or unplanned

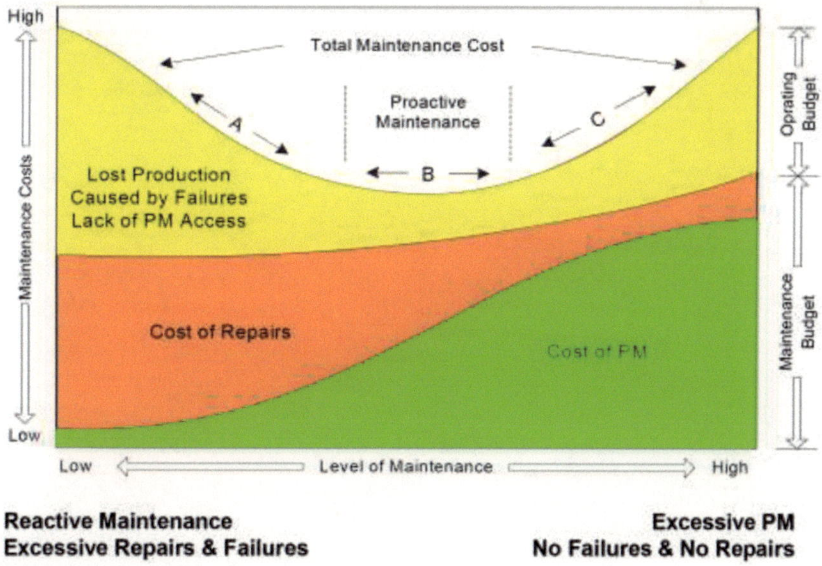

Figure 1.
Total maintenance related costs.

downtime, eliminates unnecessary maintenance, and cuts related costs. Thereby, maintenance activities require only when they need after the decision analysis for maintenance conditions such as repairs or replacements before the failure [3].

There are various techniques and technology to implement for data collecting, processing, diagnostics, and prognostics for performing CBM through the system performance operations. Lee (1998) [4] describes CBM strategic approach into three scenarios: data-driven, model-based, and knowledge-based.

First, the data-driven scenario has applied historical and statistical data to comprehend a numerical model of systematic determinants such as mean time between failure (MTBF), mean time to repair (MTTR), and maximum tolerable period of disruption (MTPD) [5]. However, this scenario has depended on the accuracy of sensing devices, operational data, data interpretation, and perceived condition of stressful situation.

Second, model-based scenario has deployed an analytical algorithm such as simulation modeling to demonstrate the system reliability, system degradation, and system efficiency. Mostly, this m0del-based need high-level application software for simulated models such as MATLAB or reliability block diagram (RBD).

Last, knowledge-based scenario has depended on human experience by applying from the past real case based analysis or deriving data from the past project information related to data collecting, gathering, analyzing, decision, and execution. Moreover, they are systematic approach of engineering knowledge and maintenance attention to system facilities to guarantee their proper functions and to reduce their deterioration rate. Sometime knowledge-based can be perform through machine learning or AI in the future.

CBM approaches provisioning load or trend profile the earliest probable prediction of device or system failure, with optimal advantage by reduced maintenance time, labor and inventory costs, eliminated downtime, increased device or system life, and cut capital expenditures. The P-F Curve in **Figure 2** depicts the performance condition of device or system, which declines overtime series, this condition leads to functional failure or potential failure. The CBM system is an on-line monitoring, controlling, and inspecting that prepare the greatest P-F Intervals, which are scarcely interrupting than traditional TBM. This helps inspector for a planning

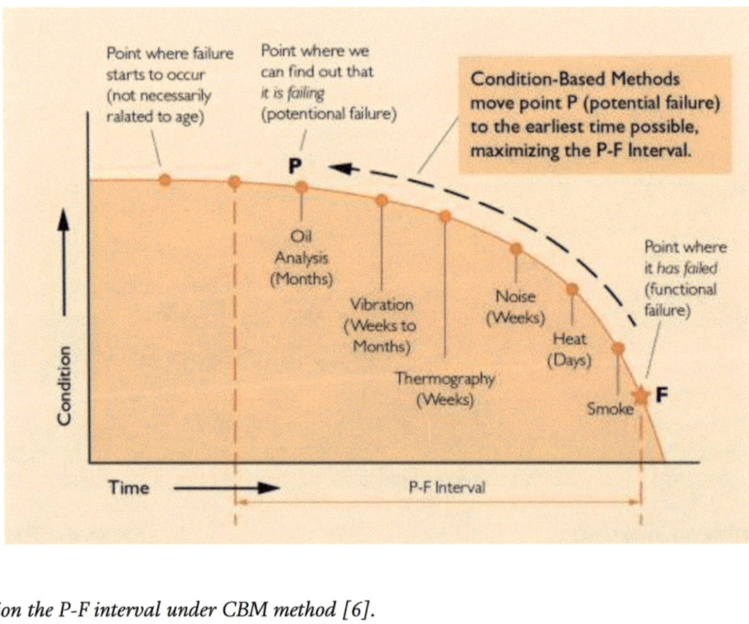

Figure 2.
Optimization the P-F interval under CBM method [6].

for downtime inspections. The process and routine of inspection defines as difference in the length of time manner, therefore it creates the utility of the P-F Interval. The evasion of off-line inspections, which frequently cause of data center downtime and ruin reputation, can apply CBM methods for economically feasibility. The most usually applied techniques of CBM monitoring are:

- Lubricant Sampling and Analysis
- Corrosion Monitoring
- Motor Current Analysis
- Acoustic Emissions Detection (e.g., ultrasound)
- Vibration Measurement and Analysis
- IR Thermography
- Process Parameter Trending (e.g., flows, rates, pressures, temperatures, etc.)
- Process Control Instrumentation (measurement and trending)
- Visual Inspection (look, listen and feel).

2.3 Data center reliability

Data center reliability is reinforced by creating redundant topology to each system such as utility supplies, backup power supplies (generators and UPSs), fiber optic communication connections, networking connectivity, environmental controls, and security devices. The report from Emerson [7], as presented in **Figure 3**, is described some critical devices that related to system failure. The racking top 3 incidents are UPS battery, over capacity of UPS, and human error.

The prognostics method, the condition monitoring process can be performed either continuously or periodically. Sensing devices and data collection systems may be required for continuous monitoring through DCIM [8, 9]. Graphically, how the prognostics method performs is demonstrated in **Figure 4**. The deterioration trend of the device condition is represented via the horizontal and vertical axes, which present the operating times, trend monitoring, condition levels, and forecast point

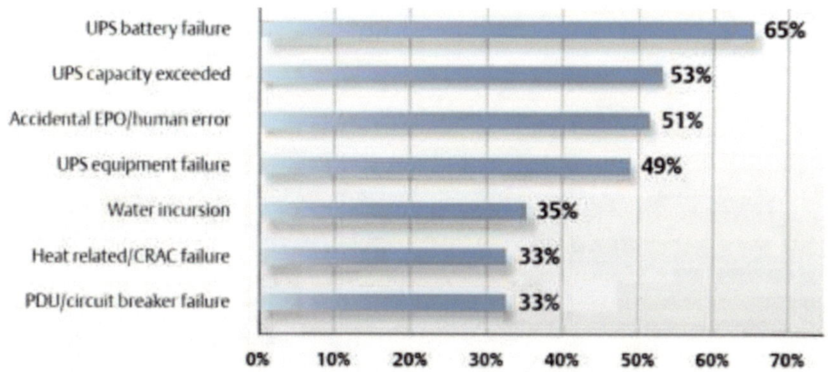

Figure 3.
Root causes and failure analysis inside data center operations.

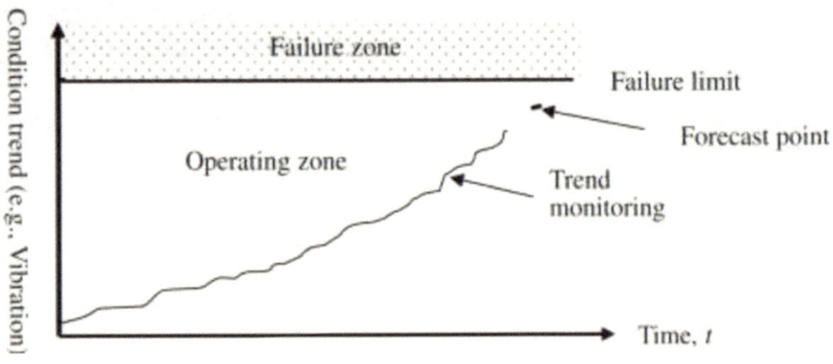

Figure 4.
The principle of the prognostics method.

respectively. The failure limit line determines the borderline between the operating and failure zones. If the forecasted trend line reaches or exceeds the failure limit, appropriate maintenance may be planned and scheduled ahead of time before the forecast point [10]. The ability to predict the future deterioration trend is the core of the prognostics method in the preventive maintenance strategy.

PPM can be defined as a strategic approach to improve the availability and reliability performance of a particular data center device or system. CBM is one type of PPM that extrapolates and predicts device or system condition over time, utilizing probability equations to assess and predict the downtime risks.

How to prevent those courses of data center failures? First, redundant system design is the first solution to prevent primary failure while selected devices and systems with highest MTBF rate is other best option. Uptime Tier Classification [11] and BICSI-002 [12] are classified the solution to prevent against the causes of failure. **Figure 5** presents the level of prevention of Uptime that Tier 4 is the highest level and Tier 1 is the lowest level of system protection while **Table 1** represents the level of prevention of BICSI 002 that Class F0 is the lowest level and Class F4 is the highest level of system protection respectively. The annual allowable planned for maintenance is the crucial factor to prevent data center downtime.

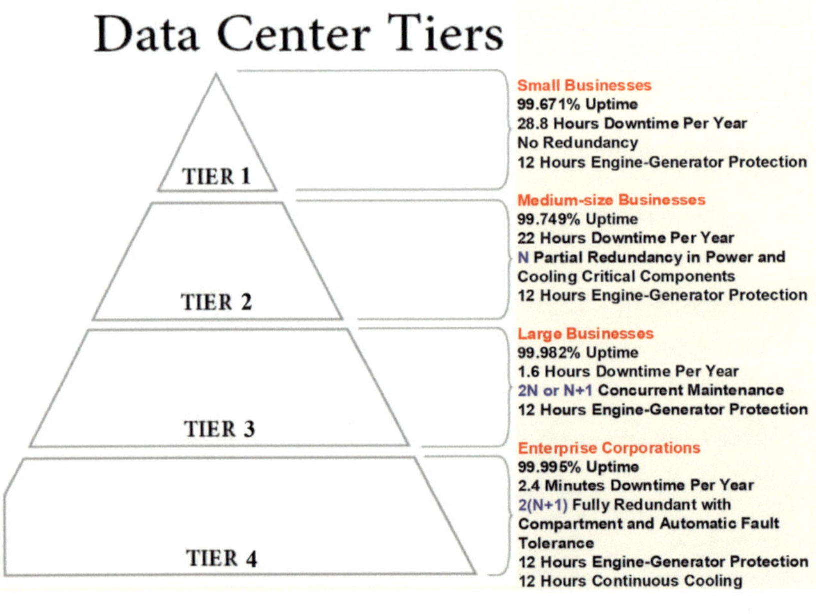

Figure 5.
Uptime data center tier classification.

System/class	Class F0	Class F1	Class F2	Class F3	Class F4
Description	Single path without any one of the following: alternative power source; UPS; proper IT Grounding	Single path	Redundant component/ single path	Concurrently maintainable and operable	Fault tolerant
Utility	Single feed	Single feed	Single feed	1 source with 2 inputs of 1 source with single input electrically devise from backup generator input	Dual feed from different utility substations
Topology	N or <N	N	N + 1	N + 1	2 N, 2 (N + 1)
Redundancy	No requirement	N	N	N + 1	Greater than N + 1
Generator fuel run time	No requirement	8 hrs.	24 hrs.	72 hrs.	96 hrs.
Impact of downtime	Sub-local	Local	Regional	Multi-regional	Enterprise wide
Annual allowable planned maintenance (hours)	>400	100–400	50–99	0–49	0
Availability as %	>99.00	99.00–99.90	99.90–99.99	99.99–99.999	99.999–99.9999

Table 1.
BICSI 002 system reliability classification.

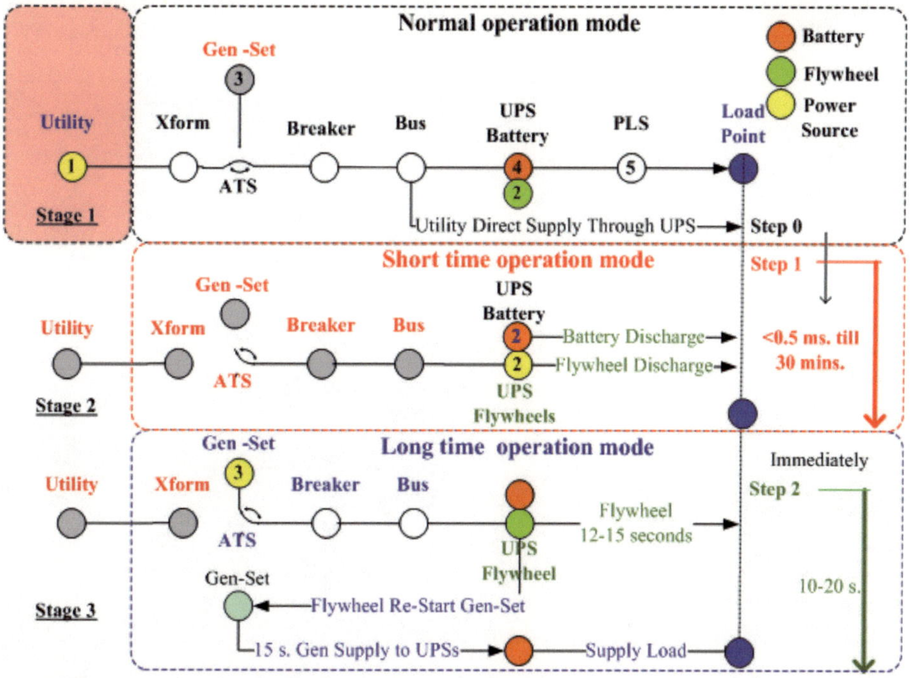

Figure 6.
Condition failure mode of power distribution systems in data center.

For reinforcement of system reliability, the Class F4 and Class F3 are designed for system reliability of PDS for 2(N + 1) and 2 N or N + 1 topology respectively that help more robust on CBM for tolerant maintaining operations with minimal down-time effect to entire system.

Second, how deep to understand consequence of device/system protection of power distribution system. The failure mitigation map illustrates, for each primary

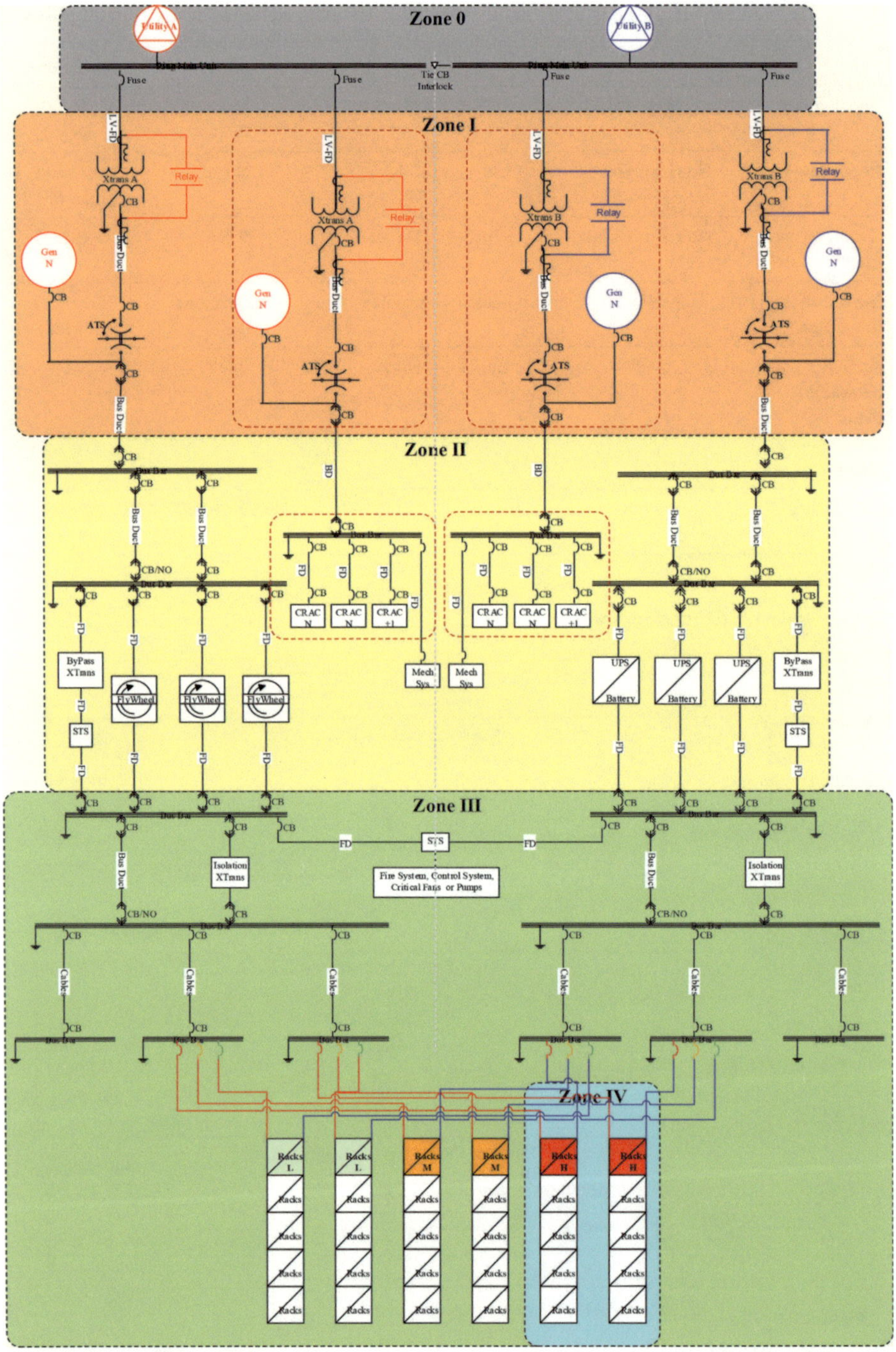

Figure 7.
Zone preventive approach for CBM.

failure, the extent to which that failure is mitigated by functional redundancy (or some other design consideration) to prevent it from acting as a single point of failure [13]. A protection design of system reliability for data center can be classified to three stages, which imply as the sources of power protection, as demonstrated in **Figure 6**.

Stage 1: On normal condition, data center is operating with power utility sources as primary power.

Stage 2: On utility outage condition, at short duration with less than <0.5 millisecond to 15 second UPS with flywheel systems can capable handle critical IT loads immediately, while the UPS with battery systems will continuous take action to protect critical IT equipment after flywheel already discharged within 30 seconds. The design capacity of batteries loads is depended on critical IT application and equipment needs, mostly designer or consultant has designed for 15 to 30 minutes. This important information must be given for IT team and data center consultant for calculation design for predicted solution for critical loads [14].

Stage 3: During operation of Stage 2, generator will start after detected utility outage within 12–15 seconds, if the power utilities still not recover on normal function, after generator control sensor detected utility outage within 15 seconds power standby system is already to takeover load from Stage 2 (UPSs).

Last, power distribution system of data center designs for isolating and dividing CBM into 4 groups or zones: Zone 0, Zone I, Zone II, Zone III, and Zone IV, as presented in **Figure 7**, by:

Zone 0: Utilities (2 N) Preventive Approach, CBM can be performed to utility service level agreement (SLA) and remote monitoring and controlling.

Zone I: Generators 2(N + 1) Preventive Approach, CBM can be performed to software DCIM and main contractor SLA or 3rd parties contract for SLA.

Zone II: UPSs 2(N + 1) Preventive Approach, CBM can be performed to software DCIM and main contractor SLA or 3rd parties contract for SLA.

Zone III: Dual Power Paths (2 N) Preventive Approach, CBM can be performed to software DCIM and main contractor SLA or 3rd parties contract for SLA.

Zone IV: Load Shedding Preventive Approach, CBM can be performed to software DCIM and in house training to handle load shedding (within 10 minutes), main contractor SLA or 3rd parties contract for SLA.

3. Research methodology

The power distribution system (PDS) of data center has exanimated as case studies for this research. They are 4 topology prototypes of Uptime (Tier I, Tier II, Tier III, and IV) and 5 topology prototypes of BICSI (Class F0, F1, F2, F3, F4) of demonstration on operations and maintenance management. Plan-Do-Check-Act (PDCA) has been applied through PPM model. This process has established more data collection from earlier cycles as the same time this process has certified data training for fault diagnostics and prognostics. The fault diagnostics perform through auto-discovery in DCIM software. StruxureWare software [15] had deployed as auto-discovery subject to ability to detect a device, model it and measure that relevant data points of that equipment. PPM approach has examined by system flow diagram (SFD), as depicted in **Figure 8**.

The SFD begins with data collection from sensing devices at condition monitoring state; data processing and data analytic; feature selection to form statistic modeling before pass through fault diagnostics and prognostics. Output of prognostic process constructs data set and transfers to estimate RUL for input data for predictive maintenance [16]. Predictive maintenance and CBM are synchronized processing with the

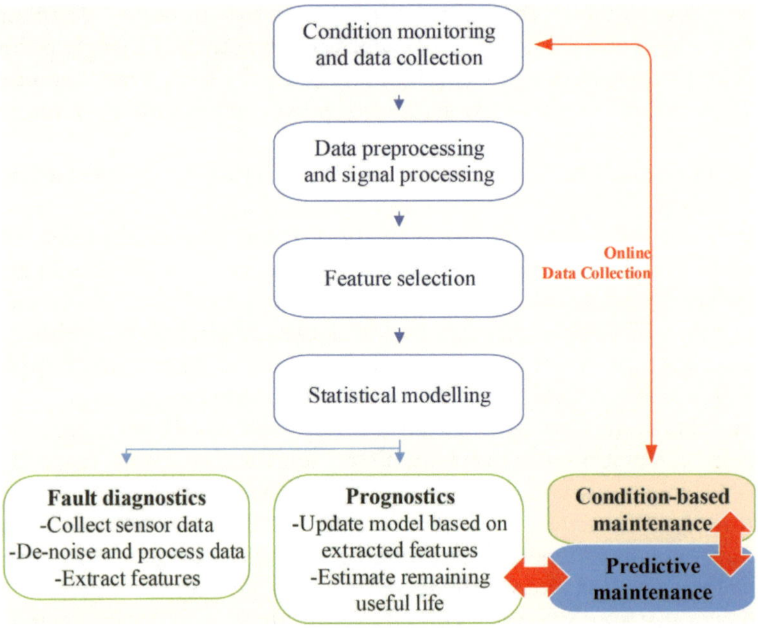

Figure 8.
PPM system flow diagram of data center operations management.

same data set from RUL and providing data set loopback to the outset of data collection and condition monitoring as plan-do-check-act (PDCA) continuous process. The PPM produces data set for CBM database at the first round and the next rounds will generate data training for fault diagnostics, prognostics and predictive maintenance. CBM can leverage as the strategic approach to guarantee the availability of the entire PDS of data center by monitoring from the device level down as transformers, generators, transfer switches, breakers and switches, UPSs, batteries, PDUs, and PSUs. CBM will manipulate as recursive function of data collection process.

The PDS of data center Tier IV had deliberated as maintenance model management (MMM) for constructing CBM of PDS, as illustrated in single line diagram of **Figure 9**. The critical devices and systems, which simulate to MMM all data derive from IEEE 493 Gold Book [17] and former research models of Wiboonrat [18, 19].

The devices and systems list, in **Table 2**, presents the quantifying characteristics of unit produced per year, number of failure, failures rate per year, MTBF, and MTTR. The following list of power devices in **Table 2** (active and supported distribution path) concentrates on the online monitoring data, which desire as input data for CBM and prognostic process for RUL [20].

The power reliability assessment of PDS needs to measure throughout the overall statuses of the PDS devices and systems of data center that comprise as the following [21]:

- Transformer
- Entrance switchgear
- Automatic transfer switch (ATS)
- Diesel generator
- Uninterruptable power supply (UPS)

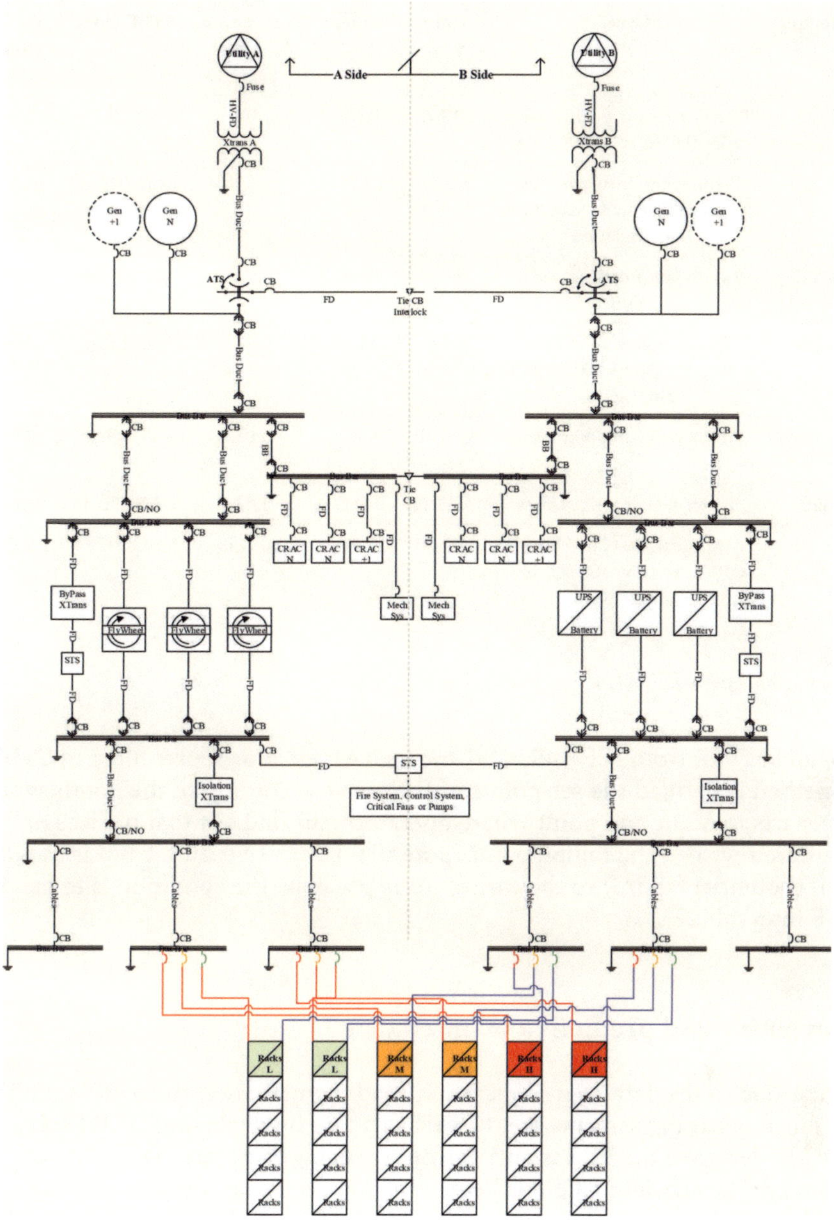

Figure 9.
Single line diagram of PDS of datacenter tier IV.

- Leaded acid batteries

- Distribution switchgear

- Power distribution unit (PDU)

- Rack-Power supply unit (PSU)

The capacity analysis of power systems has investigated to diagnose and analyze of all power devices and systems as above list. The MMM designs to perform as PPM of PDS of data center Tier IV. All critical devices have been derived data set of

Category	Class	Unit/ year	Failures	Failure rate (failures/ year)	MTBF (hrs.)	MTTR (hrs.)
E38-113	Transformer, dry, air cooled, >1500 kVA < =3000 kVA	840.20	0.00	0.00	14,432,242.40	0.00
E36-230	Switchgear, insulated bus, >5 kV, all cabinets, ckt. bkrs. not included	732.50	3.0	0.00	2,139,024.00	37.33
E34-110	Switch, automatic transfer, >600A	690.30	22.00	0.03	274,853.50	1.64
E18-121	Diesel engine generator, packaged, 250 kW to 1.5 MW, continuous	266.00	115.00	0.58	15,033.80	25.74
E39-200	UPS, small computer room floor	426.40	4.00	0.01	933,708.00	2.00
E2-120	Battery, lead acid, strings	3215.30	24.00	0.01	1,173,590.30	32.13
E36-210	Switchgear, insulated bus, < =600 V, all cabinets, ckt. bkrs. not included	322.70	0.00	0.00158	5,543,247.10	0.00

Table 2.
IEEE 493 active equipment MTBF.

MTBF and MTTR from IEEE 493 [17] for each category as represented in **Table 1**. This method is defined the set-points of P-F curve according to the points where failure starts to occur and point where operators can find out that devices or systems are revealed the failing point (potential failure) because CBM is moving point P (potential failure) to the earliest time possible, the condition is to maximize the P-F interval [22].

4. Preventive and predictive maintenance

According to the data center operations and maintenance under PPM, online condition-monitoring systems are the best scenario by deploying DCIM software. The DCIM design of the PDS is option from reducing long-term operating costs and complexity. The efficient DCIM is being evolution to the automatic processes as the critical success factor for maintaining downtime. By self-diagnosis of DCIM, PDS devices and systems can track age, operating hours, working statuses, warning alarms, MTBF, MTTR, and the last modified or upgraded by who and when.

In this deliberation, researcher has installed StruxureWare [15], a DCIM software from Schneider Electric as sensing instrument for data collection. StruxureWare performs as points of online data collection by measuring all values at set points on the devices or systems, as shown in **Figure 10**. These data are online and real-time verifying with outset-determined data from CBM database to impose the critical levels as basic criteria. Control levels (before critical level) are ordinarily imposed for apprising automatic warnings before system shutdown. The types of automatic warning are depending on the severe consequence of the cascading failure. It has a process to send warning message to each personal mobile or e-mail by configuration. The foundation of StruxureWare is relied on transducers, sensors, networking and intelligent electronic devices (IED) for collecting data throughout the PDS in data center devices [23].

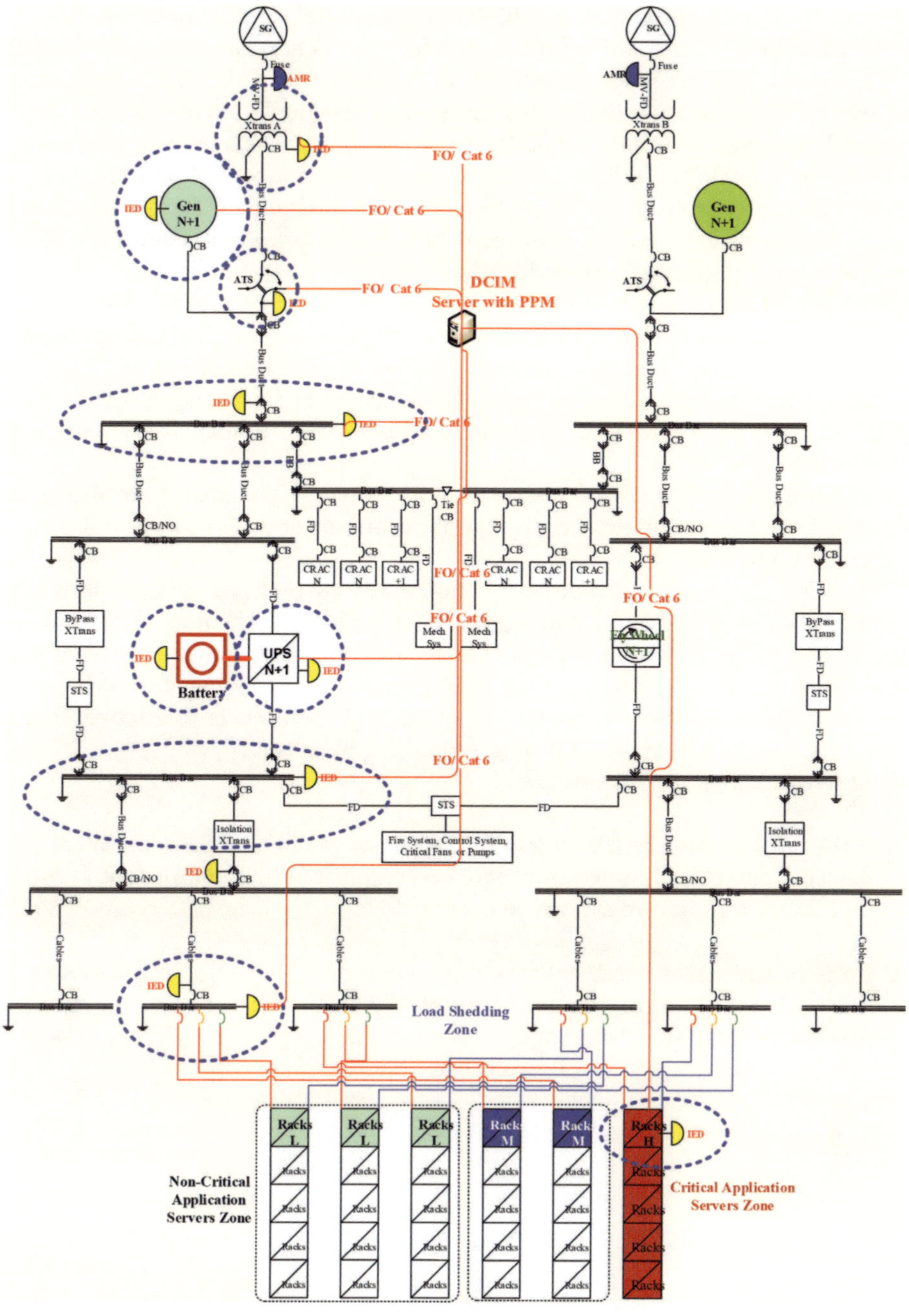

Figure 10.
Data collection from PDS of data center tier IV.

4.1 CBM model for StruxureWare (DCIM)

Tracking the increasing probability of future failure of device or system is primary function of CBM. Extrapolating and predicting system condition over time will help to analyze particular devices that could possibly to have defects requiring repairs. A CBM method also diagnoses, through statistics and data, which devices or systems most likely will remain in acceptable condition without the requirement for maintenance.

Since, Uptime Institute [11] and BICSI [12] have defined the data center Tier IV and Class F4 as the standard design for the data center site availability at 99.995 percent. The investigation of system reliability of PDS data center is an objective for this research model. Researcher has designed 12 sensor points by installed IED devices for data collection points throughout the PDS of data center [24]. The StruxureWare had installed and applied the concept of CBM to verify PDS of data center in only one single line diagram. Each devices and systems are differed functions in electrical and mechanical design proposes. Therefore, each device and system needs different location for installing and collecting data at the level of physical contact. All IED data collection must be measured in term of instantaneous and trending of all electrical status such as voltage, amperage, phase, total harmonic distortion (THD); and mechanical status; alarm, vibration, noise, temperature, leakage, oil level or other status; equipment aging, run-time, failure history, degradation percentage, abnormal events [25], as presented in **Figure 10**.

This CBM design proposes for extending P-F interval. StruxureWare shows data collecting from the last point at critical application server zone or Rack PSU, as depicted in **Figure 11**.

This helps data center administrator realizes the current power conditions when compares (Left PSU is 0.5 kW and Right PSU is 1.3 kW) to the maximum power capacity of each rack (4 kW) such as voltage, ampere, frequency, phase balance, temperature of the rack, space of rack available, and the last time audit. Moreover, this monitor from the device level up, from PSUs of each server to discover idle servers that are quietly draining power and taking up space.

The research presuppositions are:

1. If the failure status befall after device aging or MTBF and StruxureWare has detected and the administrator team can repair it before component failure, thereby system failure cannot be occurred

2. If the failure status befall before device aging or MTBF and StruxureWare detected and the administrator team can repair it before component failure, thereby system failure cannot be occurred

Figure 11.
Data collection from rack PSU of data center tier IV by StruxureWare.

3. Replacement of parts, changing lubrication or changing spare parts could be executed during operations as supplier's recommendation for critical devices without interrupting system operations (Concurrent Maintenance)

4. Extending aging for non-critical devices benefits when move point P (potential failure) to the earliest time possible maximizing the P-F interval before it has failed (functional failure).

4.2 Value and status of data collections

4.2.1 Value and status from condition monitoring systems

Field data collection is the beginning of CBM process. As the single line diagram of PDS of data center appointed 12 equipment installations for StruxureWare by set-point value as specify in **Table 1**, and status monitoring as specify in **Table 3**.

The maintenance set-point value at the beginning refers from IEEE 493, MTBF, plus condition of P-F interval. Mostly, device status condition comes from supplier data sheet's for maintenance. Both of data collection sources are sending to StruxureWare, which intends for manipulating after; condition monitoring and data collection process; and data processing and signal processing. DCIM will execute function selection as operator's requirement and create statistic modeling for fault diagnostics and prognostics for calculating RUL. All data collection will input through the predictive maintenance function for setting up the new value and status as the beginning of condition monitoring, PDCA process, as represented in **Table 3**. Almost 12 months of data collection by StruxureWare and PPM model, there are no blackout in PDS of data center Tier IV. No blackout does imply no any device or system failure but Tier IV topology designs as fully redundancy 2(N + 1), therefor, some devices or systems can be failure but the other still perform without system interruption. The StruxureWare can detect and discover before sending information to administrator team to repair it under MTTR condition. Because data center Tier III is designed as 2 N and Tier IV is designed as 2(N + 1) topology. It allows more fault tolerance to devices and systems failure. The system warning occurs a few times but data center administrator can fix the problems by warning instruction from StruxureWare monitor guides. The StruxureWare has designed for easing to understand and predict any device or system failure and resolve it before it fails, which implies CBM help decrease planned and unplanned downtime, labor hours, and spare part inventory, while increases throughput of system productivity. Moreover, CBM supports the provision and early warning system for all devices and systems failure functions, StruxureWare has capable to controls inventory level much more effectively and no need as many emergency spare parts [26].

4.2.2 Value and status from idle servers

Idle server is a physical server that is still running but has no perform any computing resources or any transaction processing, that it consumes power but is serving no useful purpose. The Uptime Institute survey reports around 30 percent of global data center servers are either underutilization or completely idle. This server can consume power an impressive 175 watts when it is idle mode. A survey of server PSUs [27] reports the range of efficiency related to load of PSUs, as illustrated in **Figure 12**.

In the red zone, power loaded of PSU is lower than 20 percent the efficiency drops off precipitously. In the yellow zone, 20–40 percent, PSU efficiency begins to drop but typically exceeds 70 percent. In the green zone, the PSU operates above 40

	Components/ systems	Infrared thermography	Precise timing and trending	Visual inspection	Insulation resistance	Motor circuit analysis	Polarization index/ dissipation factor	Cable condition monitoring	Oil and gas levels	Vibration monitoring	Lubricant analysis	Wear particle analysis	Bearing temperature analysis	Leakage detection	Performance monitoring	Ultrasonic monitoring
1	Transformer							✓	✓		✓				✓	
2	Entrance switchgear	✓	✓	✓	✓	✓		✓								
3	Automatic transfer switch (ATS)	✓	✓	✓	✓	✓		✓								
4	Diesel generator			✓			✓	✓		✓	✓	✓	✓		✓	✓
5	Uninterruptable power supply (UPS)	✓		✓				✓							✓	
6	Leaded acid batteries			✓				✓						✓	✓	
7	Distribution switchgear	✓	✓	✓	✓	✓		✓								
8	Power distribution unit (PDU)	✓	✓	✓	✓	✓		✓								
9	Rack-PDU	✓	✓	✓	✓	✓		✓								

Table 3.
Values and status of data collection from condition monitoring systems.

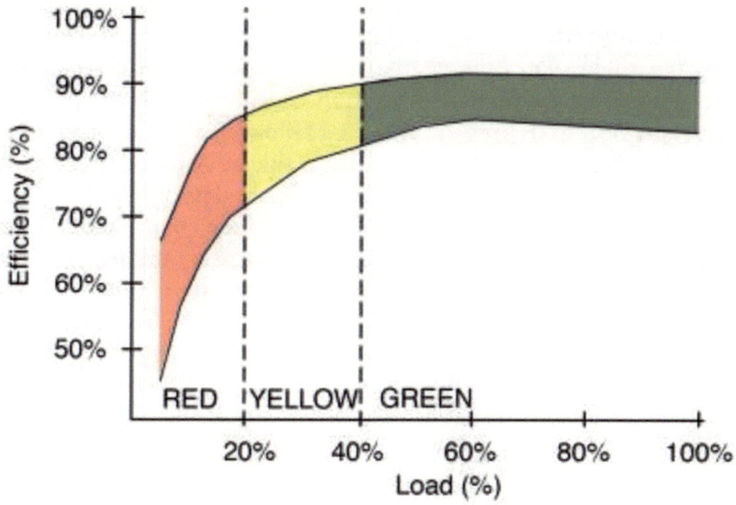

Figure 12.
Power supply efficiency.

Power supply size (Watts)	400	400	400	400	400
Idle power draw (kW)	0.6	0.6	0.6	0.6	0.6
Power waste (Watts)	240	240	240	240	240
Hours per year	8760	8760	8760	8760	8760
Cost of electricity per kW/hr ($)	0.08	0.1	0.12	0.14	0.15
Savings ($)	**168.19**	**210.24**	**252.29**	**294.34**	**315.36**

Table 4.
Idle server and electricity costs.

percent loaded, where their efficiency is at or above 80 percent. At idle mode, current servers still draw power about 60 percent of peak load electricity. In normal data center operations, average server utilization is only 20–30 percent [27]. Now data center operators deal with growing cost restraints and energy efficiency goals, it is become primal objective to identify and eliminate these severs promptly. **Table 4** shows the saving costs due to idle power draw of each server per year compare to range of cost of electricity per kW/hour.

Locating and identifying an idle server is performed function through DCIM solution. The DCIM applies database from field data collection is the beginning of CBM process at device level of PSUs and PDUs. The DCIM and intelligent PDU can give data center operator the insights which data need to gain complete control of power usage, load profile or utilization of servers, and cost-efficiency IT environment.

5. Results and discussion

After design the single line diagram of PDS, in **Figure 10**, all main devices and systems had monitoring through IT sensing devices such as transformer, entrance switchgear, automatic transfer switch (ATS), diesel generator, uninterruptable power supply (UPS), leaded acid batteries, distribution switchgear, power distribution unit (PDU), and rack-Power supply unit (PSU), for measurement of the instantaneous and trending of all electrical status; voltage, amperage, phase, total harmonic distortion (THD); and mechanical status; alarm, vibration, noise, temperature, leakage, oil level or other status. All data collection had recorded through DCIM system for define set-point or condition-based maintenance (CBM) of each critical device and system to prevent potential failure or P-F Curve. The results from installed and operations data center with StruxureWare software show system warning of DCIM reduce data center operator time in day-by-day to fine out root causes of the problems such as location of devices or systems, history condition of operations device, with device is broken first and cascading failure to which system, and more easy for operator to make decision with completely information for future provision.

6. Conclusions

Total cost of ownership (TCO) is an excellent measure of the value of data center uptime. System uptime is momentous for the success of mission crucial for data center business. More data center uptime defines lower operating costs and higher customer satisfaction and trust. Data center downtime leads to high TCO due to issues such as increased penalty costs, recovery data and systems costs, and reputation costs. The data center Tier IV proposes for high system reliability by applying fault tolerance topology or fully redundancy 2(N + 1) strategy. Consequently, during operations and maintenance they needs fully fault protection from system failure. Therefore, preventive and predictive maintenance (PPM) has considered for monitoring and detecting all possible potential devices and systems failures before data center failure happened. In this research chapter, The StruxureWare as a DCIM software has deployed for PPM model to eliminate PDS downtime and trace the idle servers. The benefits of data center system maintenance when deployed DCIM properly are reduced downtime costs, increased uptime productivity, eased for online and real time management, reduced inventory costs, reduced fix costs in long-term operations and maintenance. The condition-based maintenance (CBM) has the advantage to deal with 2 crucial determinants, detecting error or faults before devices or systems failure (MTBF) and predicting the time between maintenance processes and time to repair (MTTR) while impacts on saving penalty costs of downtime, saving labor hours, inventory costs, increasing data center uptime, and reducing overall TCO.

Author details

Montri Wiboonrat
KMITL Business School, King Mongkut's Institute of Technology Ladkrabang, Bangkok, Thailand

*Address all correspondence to: montri.wi@kmitl.ac.th

References

[1] Ponemon Institute, Cost of Data Center Outages: Data Center Performance Benchmark Series, Sponsored by Vertive, January, 2016.

[2] Vitucci F, Predictive maintenance in a mission critical environmant, Emerson: White Paper CSI 2130 Machinery Health Analyzer, 2017

[3] Keizer O, Teunter H, Veldman J, Babai Z, In: Condition-based maintenance for systems with economic dependence and load sharing Int. J. of Production Economics, vol.195, 2018. p.319-327

[4] Lee J, Teleservice engineering in manufacturing: challenges and opportunities. In: International Journal of Machine Tools & Manufacture, 38 (8), 1998. p. 901–910

[5] Sobral J, Soares G, Preventive maintenance of critical assets based on degradation mechanisms and failure forecast, IFAC-Paper Online, 49-28, 2016. p. 97-102

[6] Blann R, Maximizing the P-F Interval Through Condition-Based Maintenance, http://www.maintworld.com/Applications/Maximizing-the-P-F-Interval-Through-Condition-Based-Maintenance, 2018

[7] Emerson, Addressing the Leading Root Causes of Downtime, White Paper: SL-24656-R10-10, Liebert Corporation, 2010.

[8] Shin H, Jun B, On condition based maintenance policy, In: J. of Computational Design and Engineering, vol.2, 2015, p.119-127

[9] Compare M, Bellani L, Zio E, Reliability model of a device equipped with PHM capabilities, Reliability Engineering and System Safety, vol.168, 2017, p.4-11

[10] Jonge B, Teunter R, Tinga T, The influence of practical factors on the benefits of condition-based maintenance over time-based maintenance, Reliability Engineering and System Safety, vol.158, 2017, p.21-30

[11] Uptime Institute, Data Center Site Infrastructure Tier Standard: Topology. Uptime Institute Professional Services, LLC. Uptime Institute, 2014.

[12] BICSI-002, Data Center Design and Implementation Best Practices, BICSI 002-2014, 2014.

[13] Martorell P, Marton I, Sanchez I, Martorell S, Unavailability model for demand-caused failures of safety devices addressing degradation by demand-induced stress, maintenance effectiveness and test efficiency, Reliability Engineering and System Safety, vol.168, 2017, p.18-27

[14] Keizer O, Teunter H, Veldman J, Joint condition-based maintenance and inventory optimization for systems with multiple devices, In: European J. of Operational Research, vol.257, 2017, p.209-222

[15] StruxureWare, StruxureWare Data Center Operation 8, Schneider Electric, 2018

[16] Gregory V, Brian W, Moneer H, A review of diagnostic and prognostic capabilities and best practices for manufacturing. In: J. Intell. Manuf., 2016. p.1-17

[17] IEEE 493, IEEE Std. 493-2007 (Revision of IEEE 493-1997), Recommended Practice for Design of Reliable Industrial and Commercial Power System, Gold Book, 2007.

[18] Wiboonrat M, Transformation of system failure life cycle, In: Int. J. of Management Science and Engineering

Management, Vol. 4, No. 2, 2008, p.143-152

[19] Wiboonrat M, An empirical study on data center system failure diagnosis, In: The 3rd Int. Conf. on Internet Monitoring and jProtection, 2008, p. 103-108

[20] Shin J, Jun H, On condition based maintenance policy, 2, 2015, p. 119-127

[21] Cole D, Developing a priventive maintenance plan for your data center, White Paper, PTS Data Center Solution, Rev 2010-1.1, 2010

[22] Vogl W, Weiss A, Helu M, A review of diagnostic and prognistic capabilities and best practices for manufacturing, In: J. Intelligent Manufacturing, 2016

[23] Zavoda F, Sensors and IEDs required by smart distribution applications: Smart grid and distribution automation, In: The 1st Int. Conf. on Smart Grids, 2011, p. 120-125

[24] Hor C, Crossley A, Knowledge Extraction from Intelligent Electronic Devices, Springer-Verlag, Berlin Heidelberg, 2005

[25] Bayle T, Preventive maintenance strategy for data centers, Schneider Electric, White paper 124, Rev 1, 2011

[26] ECOS, EPR, Efficient power supplies for data center, ECOS and EPR, Tech. Rep., Feb. 2008.

[27] Barroso L, Holzle L, The case for energy-proportional computing, IEEE Computer. Jan, 2007

Chapter 4

Harnessing the Multiple Benefits of a Computerised Maintenance Management System

Edoghogho Ogbeifun, Patrick Pasipatorwa and Jan-Harm C. Pretorius

Abstract

Those involved in maintenance operations are enjoying the benefits of information and communication technology in the planning and management of maintenance activities, resource management and planned production. In the digital space, the computer-based operating systems, commonly referred to as computerised maintenance management systems (CMMSs), enable quick and effective communication between stakeholders, facilitate improved planning, easy access to historical data, reporting and performance improvements of the maintenance function. However, success in the use of CMMSs depends on the human capacity of the users of the system. In practice, many organisations use the CMMS tool for planning, operations management and reporting, without the aid of detailed analysis of operational information in the CMMS database. They fail to harness all the possible benefits. Three case studies were used to illustrate the situation. Two of them refer to academic institutions and the third is a manufacturing company. In the academic institutions, the CMMS was used for maintenance planning, management and periodic reporting. The manufacturing company included analysis of the information in the operational database, which culminated in identifying the level of the reliability of machines in the production network through benchmarking. The conclusion is that the quality of the human capacity enables organisations to harness and make maximum use of the potentials inherent in typical CMMS software.

Keywords: analysis of operational information, benchmarking, computerised maintenance management systems, effective communication, reliability of machines, reporting

1. Introduction

Maintenance operations have over time continued to evolve and, in the digital age, are maximising the opportunities offered through the application of information and communication technology. 'Information systems that support maintenance functions are referred to as Computerised Maintenance Management Systems (CMMSs)' ([1], p. 269). Some of the objectives of a 'CMMS are effective maintenance of machines, rational asset management, and consequently higher productivity of a company' ([2], p. 277). The application of CMMSs in maintenance management allows the effective documentation of maintenance operations, planning and execution of maintenance operations, communication of periodic reports to stakeholders, analysis of operational history and the development of performance improvement strategies. Typical CMMS software contains different modules, which enable the maintenance operatives to execute different activities. Some of the operational modules include asset management, work order management, preventive maintenance management and inventory control and report management [1]. Others include service management, contract management, materials management and procurement management [2]. Any or multiple modules can be activated, depending on the quality of the human capacity of each organisation. They can be used for the planning and execution of scheduled operations and the development of functional periodic reports. It can also be used for the analysis of the operational history and for the development of performance improvement strategies. Therefore, the continuous explorative use of the content of a CMMS tool in maintenance operations enables the maintenance unit to achieve the 'just-in-time' (JIT) response to a maintenance request, address the problem and restore the facility or machine to functional use [3]. Achieving JIT facilitates a reduction in the cost of maintenance, a reduced duration in repair time and consequently the length of downtime, achieves improved customer satisfaction and improvements in production. Improvements in JIT require the deliberate action of collecting and analysing the data of maintenance operational history and the subsequent development of performance improvement strategies. This cycle of operations is in tandem with the requirements of smart maintenance [4].

The focus of this chapter is to explore the vast potential of typical CMMS software and to show that this vast potential can be harnessed in each module. The use of CMMSs in three maintenance units serves as an illustration of how the potential of CMMSs is being harnessed or neglected. The chapter covers the literature review in Section 2, which highlights the use and potential of a typical CMMS, and progresses to Section 3, which explores the place of the case study research strategy in qualitative research. Section 4 provides information on the research findings and their discussion, while Section 5 provides the synthesis of the content of the chapter in the form of conclusions and recommendations.

2. Literature review

A literature review provides the platform for researchers to explore and evaluate their efforts against previous research endeavours to harness existing information, models and methodologies. The literature reviewed here provides general information on CMMSs and its use in maintenance planning, operations, periodic reports, performance improvements and benchmarking.

2.1 Computerised maintenance management systems

The concept of maintenance has transformed over time. In the past, maintenance was the act of replacing a broken component in a process, machine, or facility [5]. However, maintenance is now considered as a vital component in the complex management process and is associated with several organisational processes like production, quality, environment, risk analysis and safety [6]. Considering that maintenance is an important function of organisations, maintenance management requires a multidisciplinary approach with a business perspective and the use of digital technology [7]. Adopting computer systems to support maintenance operations is commonly referred as Computerised Maintenance Management Systems (CMMSs) [1]. CMMSs allow the effective documentation of maintenance operations, communication of scheduled activities, enhance the ability to develop and monitor the forward planning and execution of maintenance operations. Typical CMMSs have assigned a set of functions and applications, including asset management, work order management, preventive maintenance management, inventory control and report management [1]. It also includes service management, contract management, materials management and procurement management [2]. Each of these functions is represented as specific modules in a typical CMMS software package and each module is activated according to the need and human capacity of the organisation. Many of the systems available in the market do not perfectly match the specific requirements of some organisations. Therefore, they prefer to develop their own software or buy commercial packages and activate the modules of interest [1].

The introduction of CMMSs in maintenance operations facilitates the tracking of progress in the execution of requests, the effective development and dissemination of periodic reports and the analysis of the operational history in the database of each facility [8]. The outcome of the analysis allows maintenance operatives to know the state of the facility, identify the causes of failure and develop suitable maintenance and performance improvement strategies. Furthermore, the CMMS enables the automation of maintenance procedures in terms of communicating with maintenance units, planning and executing maintenance operations and communicating (real time) with all relevant stakeholders [9]. The ease of use and usefulness of CMMSs are influenced by the level of training of the user before and during the implementation. The proposed training may end up with negative impacts if the focus of the training is on the technology itself and not on how the technology enhances personal satisfaction, facilitates and supports the user's method of executing relevant tasks [10].

The maintenance unit of any organisation spends considerable time in the development, operations and maintenance of the facilities that enable the performance of the core functions of the organisation. They do not pay adequate attention to documentation, reporting or providing extended information to the customer from the operational history for each facility in its portfolio. However, in the digital age, with adequate human capacity, the CMMS tools enable the maintenance unit to communicate easily with all relevant stakeholders in a user-friendly format and to include appropriate visual displays [11, 12]. Briefly, the CMMS documentation enables senior management to know the state of the facilities in the portfolio or production network. It helps to identify possible constraints to the effective performance of the core function of the organisation, resources and financial management. The details and structure of each report should reflect the hierarchy of the recipient. The executive summary of report is useful for leaders at the strategic level, but leaders at the tactical level require more details. The effective use of the CMMS tools facilitates the production and dissemination of periodic information to relevant stakeholders and the analysis of operational details, which guide senior management in taking objective decisions [9, 11, 13, 14].

2.2 Analysis of the content of a computerised maintenance management system (CMMS) database

The analysis of operational history is an extension of periodic reports over a long period with the objective of determining the functional state of the whole facility or its component parts. Unfortunately, the operational information about a facility, in many maintenance organisations, is stored in their computer or files for many years without objective analysis being done to determine the functional state of the facility or its components [8]. The periodic analysis of a facility's history enables maintenance units to effectively educate its stakeholders, especially senior management, on the state of the facilities or machines, in the case of production lines of manufacturing industries. The analysis provides the needed guidance for the effective maintenance management and the development of performance improvement strategies [15]. The progression in the analysis of a facility's history includes the identification of distress recorded against each facility; the classification of the distress recorded according to the constituent component sections; and determining the frequency of distress, the status of execution, outstanding requests not attended to and cost incurred [8, 16]. Each periodic report and the associated analysis should include explanatory notes to guide the customer and senior management in decision-making [11, 17]. Furthermore, the adequate archives of periodic reports, the analysis and accompanying notes form useful background information for data collection when developing long-term plans for maintenance, renovation, or rehabilitation exercises. Therefore, it enables senior management to decide whether to continue to maintain or to replace parts or the whole facility.

Detailed analysis of the CMMS database can save organisations from making costly mistakes when deciding on the change of use of a facility, upgrade, renovate or respond to legislative requirements. Research [15] cites the examples of two universities where detailed operational analysis and assessment guided the decision on change of use:

> *At a small urban university, officials wanted to add two stories to a historic building. An assessment determined that the existing structure could not support new floors; the work could be done, but it would require significant capital to do so. In another institution, officials needed to know if a 1960s science building could accommodate a program expansion. An assessment of the facility's operational history concluded that the best option was to build a new structure ([17], p. 311).*

The detailed analysis of a facility's history, which is disseminated to the relevant stakeholders, is a useful tool for effective communication, facilitates timely decision-making and enhances improved relationships between the customers and the maintenance unit. It also provides an objective assessment of the performance of a maintenance unit by their customers.

In a typical manufacturing industry, the analysis of the CMMS database enables the maintenance unit to know the frequency of breakdowns of a specific machine or machines in the production network and the delay before repairs are carried out and the duration of machine downtime in the production network of the industry and provides information for benchmarking [15]. The periodic analysis provides useful information on the productivity, the profitability of the industry and the development of performance improvement strategies.

2.3 Performance improvement

The continuous performance improvements in any endeavour, be it service or manufacturing industries, can be enhanced through the practice of the discipline of periodic analysis of operational records in the CMMS database. Such analysis helps to validate the performance measurement (PM) system in place, identify factors responsible for low performance and helps to develop performance improvement strategies. Performance assessment is the comparison of performance results (assessment) against the expectations of the measuring system in operation [18]. The assessments should be timely, accurate and relevant. The exercise should be undertaken in ways easily understood by the employees using the performance measuring system being evaluated [18]. PM or its tools are not an end in themselves but road maps for more effective management. It requires the effective analysis of results and honest attempts at improving performance [19]. The effective analysis of the content of the CMMS database and the coordinated feedback from stakeholders, facilitate the identification of the magnitude and source of variance, which require improvements. The differences in measurements are harnessed in coordinated feedback; this in turn is used to develop suitable strategies for improved performance. The indicators (in PM) are designed to achieve the goals of the organisation while the feedback from periodic observations (assessments) is used to improve the production or work process.

To achieve performance improvement, the outcome of the analysis of the CMMS database must be accompanied by action steps, according to reference [20]. These action steps are summarised as follows:

1. After analysis of PM, develop broad areas of performance targeted for improvement.

2. Continuously test the performance improvement strategies to confirm if they are working, and if not, why.

3. Establish the right structure which facilitates the effective use of PM results.

4. Use the PM results to bring about change in the organisation.

The developed performance improvement strategies challenge the relevant stakeholders to either confirm or change the current policy or ways of doing things to meet the goals of the organisation and to progressively refine and improve its operations [19].

The outcome of the analysis of operational history, which leads to the development of performance improvement strategies, provides intelligent information for decision-makers at all levels to assess the achievement of predetermined goals [19]. It facilitates the tracking of past progress, helps to learn about the future and challenges maintenance operatives to practise the art of continuous data collection, analysis and the interpretation of feedback information [21]. It enhances oversight and compliance activities, supports proposals for change and requests for additional resources [21, 22]. The practice of performance improvement is reinforced through contextual benchmarking.

2.4 Benchmarking and performance improvement

The general use of the word benchmark involves identifying a point of reference (a benchmark) that serves as a standard against which relative performance may be measured. The point of reference may be internal to an organisation or external in relation to competitors or 'best practice' [23]. Benchmarking has been referred to ([24], p. 42) as the 'continuous process of measuring one's products, services and business practices against the toughest competitors or those companies recognised as industry leaders'. Benchmarking promotes superior performance by providing an organised framework through which organisations learn how the 'best in class' do things. In essence, 'benchmarking is the process of borrowing ideas and adapting them to gain competitive advantage' ([24], p. 41). In a nutshell, benchmarking is identifying 'best buy' or 'best practice' and making deliberate efforts to emulate it, devoid of unhealthy practices [25]. The implications of benchmarking could be summarised as a process of constantly comparing own performance against superior performances within a peer group of best practice [26]. To achieve positive results from benchmarking requires commitment and investment from both senior management and operations personnel of the maintenance unit. The requirements of a successful benchmarking exercise include, but are not limited to the following ([26], p. 41):

- A clear understanding of organisational goals, knowing what needs improving and by how much.
- An authentic and dynamic database for computation, analysis and comparison with a peer group.
- A constant reminder that since 'the best does not stand still', improvement should be a continuous process.
- The selection of peer group members, which is critical to the success of the exercise. The peer group must have identical features and be the best in the chosen field from anywhere in the world.
- The selection of appropriate parameters for the benchmarking exercise.

It is worth noting that benchmarking is not a 'quick-fix' solution but an exercise that requires commitment to succeed [27]. The success is influenced by the level of competence, capacity and capabilities of the operating personnel, quality of the data and commitment to their analysis [26].

Specifically, in the manufacturing industry, the effective analysis of a CMMS database and benchmarking provide information for effective maintenance operatives. The information is used to monitor the impact of the frequency of machine breakdowns, the length of time between repairs and the duration of downtime on the availability and reliability of machines in the production network. This in turn enhances the productivity and profitability of the industry [28].

Literature is awash with the vast potential inherent in a typical CMMS tool. It can be used for asset management, work order management, preventive maintenance management, inventory control, report management, service management, contract management, materials management and procurement management. The benefits gained are influenced by the number of modules activated, the human capacity and training provided to the operators of the system.

3. The research method

The qualitative research method was adopted for the case studies discussed in this chapter. The case research strategy allows the detailed, in-depth and broad-based investigation of situations or phenomena in its context [29, 30]. The approach also enables the researcher to relate to the officials directly involved in the subject matter being investigated. The population and samples used for the research were 'purposively' selected [31] from the maintenance operatives and other relevant stakeholders associated with the operation of the maintenance units. The data were collected using a semi-structured questionnaire used as an interview guide and the detailed review (document analysis) of periodic reports emanating from the maintenance units of the institutions and manufacturing industry were used for the case study [30].

3.1 Interviews

In qualitative research, the interview (one-on-one or in groups as in the case of focus group sessions) is one of the methods of obtaining information from research respondents [30]. The one-on-one interview approach was adopted in this study, using a semi-structured (open-ended) questionnaire. A semi-structured questionnaire enables the researcher to ask follow-up questions, allowing respondents to provide additional information to emphasise or add to information provided to the lead questions. The transcribed information collected from each respondent was returned to them for confirmation of accuracy and interpretation [30]. The principle of content analysis was used for analysing the qualitative data, the synthesis of the ideas generated, as answers to the survey questions, lead to the development of suitable themes discussed in the section for the discussion of findings [32].

3.2 Document analysis

Document analysis allows researchers to examine operational records, reports, archival materials, or statutory information related to the subject of the research in hard or soft copies. These records provide authentic historical information about the research [33]. The documents analysed for this research were the periodic reports on maintenance operations of the academic institutions and the operational record in the CMMS database of the manufacturing company. The analysis of the periodic reports from the academic institutions shows that the reports contain generic information on maintenance operations, which do not serve as a tool of effective communication with their customers. However, the analysis of the operational history in the CMMS database of the manufacturing industry facilitated the identification of areas for improvement and the development of performance improvement strategies through benchmarking.

The findings revealed that, while the academic institutions limited their use of the CMMS to maintenance planning, operations and periodic reporting, the manufacturing industry went a step further. They included the detailed analysis of the operational history in the CMMS database. The analysis and benchmarking led to the improvements on the reliability and availability of the machines in its production network. The section on findings and discussion explains how the research strategy was executed and presents the resulting outcomes.

4. Findings, analysis and discussion

The three case studies used for the discussion on harnessing the potential of CMMSs in maintenance operations involve two academic institutions and a manufacturing industry. In the academic institutions, the CMMS is used for maintenance planning, operations and the development of periodic reports. The maintenance operations at Institution 1 are executed by service contractors, while Institution 2 adopts the combination of using in-house personnel and service contractors. Most of the maintenance operations in these institutions are executed through the breakdown maintenance approach. The customers register their maintenance requests through the call centre that assigns a unique code to the request according to the facility's fabric, component, or services. The maintenance request is sent to the trade supervisor, who raises a work order and assigns the request to the appropriate personnel or service contractor. When the request is addressed (or closed), the completed work order is signed off by the customer, returned to the supervisors and the call centre for record purposes. Similarly, the work requests that are not addressed (or open) are also documented. The maintenance units, from these institutions, produce their periodic reports from the information in the CMMS database and do nothing further.

In contrast, the maintenance unit in the manufacturing industry adopts the combined approach of a planned and breakdown maintenance system. In this regard, all machines in the production line are scheduled for maintenance at predetermined times. Any breakdown or other maintenance requests are sent to the maintenance unit for proper attention. Maintenance operations in this industry are executed through in-house personnel and in limited cases, through specialist service contractors. Many operational modules of the CMMS software are activated in this industry. These allow the maintenance unit to work in synergy with the stores department for inventory control of spare parts and the finance department for timely purchase of necessary materials and services. In addition to the periodic reports, the unit conducts detailed analyses of the information in the CMMS database and sets benchmarks for measuring performance. The evidence from the analyses brought to the fore the current performance, identified areas requiring improvement and developed suitable performance improvement strategies; thus, harnessing the vast potentials inherent in a CMMS, improving on the JIT principle and approaching the practice of smart maintenance.

In some maintenance organisations, the same personnel manage maintenance operations and the call centres, while they are separated in other organisations. The quality of the human capacity in both units influences how the potential of a typical CMMS can be harnessed.

Hereafter, this section presents, firstly, the use of the CMMS in the maintenance units of the academic institutions, and secondly, the use of the CMMS in the manufacturing industry followed by the discussion sub-section.

4.1 Operational report from institution 1

The development of periodic reports is the third component of the CMMS module activated by the operatives of the maintenance unit of this institution. The reports include the weekly report used for management meetings of the campus managers and a monthly report produced for the director. The typical weekly/monthly report contains the information about the work requests received in the period under review, the status of execution and the cost incurred. An excerpt of the format used for the development of these reports is shown in **Table 1**. The information provided includes the code for the request, reference number for the work order, description of the work request, date the customer made the request, the service contractor assigned to execute the request, date the work was completed and the cost incurred. The monthly reports are usually produced on the first Monday of the following month. The report of the work requests for March, examined for this research, was produced on 4 April 2010, and was made available in a 13-page document [34].

In **Table 2**, the analysis of the report shows that a total of 2995 requests were lodged with the maintenance unit during March 2010, and 813 or 27.20% were addressed by 4 April 2010. The report is silent about the over 70% of the work requests not yet addressed. The quality of this report can be improved by extrapolating the result of work done until 30 April. This shows that 2013 requests (67.21%) of the total work requests for March have been attended to, leaving an outstanding balance of 982 (32.79%). Furthermore, the analysis helps to present a fair representation of the performance of the maintenance unit.

Building code	Assigned work order	Work description	Date work requested	Service contractor's code	Date work completed	Total cost
127	70,792	Remove, investigate and quote on repair of leaking pump. Replace packing with mechanical seal	2010/03/01	PUMDATA	2010/05/10	R5462.88
127	70,794	Repair noisy pump motor fan	2010/03/01	MJL	2010/03/29	R53865
127	70,795	Professional service to HVAC. Supply and install 1 × 24,000 BTU York midwall unit in room GH525	2010/03/01	PERFECTAIR	2010/04/12	R10180.20
131	70,796	Supply and instal 1 × 18,000 BTU York midwall unit in room 236	2010/03/01	PERFECTAIR	2010/04/12	R9234.00
446	70,797	Repair/ replace broken toilet soap dispenser in room 2B34. Urgent	2010/03/01	SUPERCARE	2010/03/18	R0.00

Table 1.
Typical structure of monthly report.

4.2 Periodic reports from institution 2

Like Institution 1, the periodic reports from this institution are in the form of monthly and annual reports on general maintenance operations. The reports are too technical, economical with details and are only understood by those who prepared them. The summary of the monthly report provides information on the quantity of requests lodged with the unit, the quantity resolved and the outstanding number. **Figure 1** presents the scorecard of the unit from all the campuses of the university for the year 2013, showing the number of requests received and the number addressed or closed.

Furthermore, the unit provides separate information on the number of unresolved requests; it is an indication that these outstanding issues are kept in perspective, as shown in **Figure 2**. However, this report is silent on what the unit is doing about these outstanding requests, the effect of deferred maintenance on the functional state of the facilities they represent and the financial implications of

2995			March				April		
Day ending	7	14	21	28	4	11	18	25	30
Quantity completed	9	85	286	463	813	1114	1621	1914	2013
% completed	0.30	2.84	9.55	15.46	27.15	37.20	54.12	63.91	67.21

Table 2.
Typical monthly report ([34], p. 117).

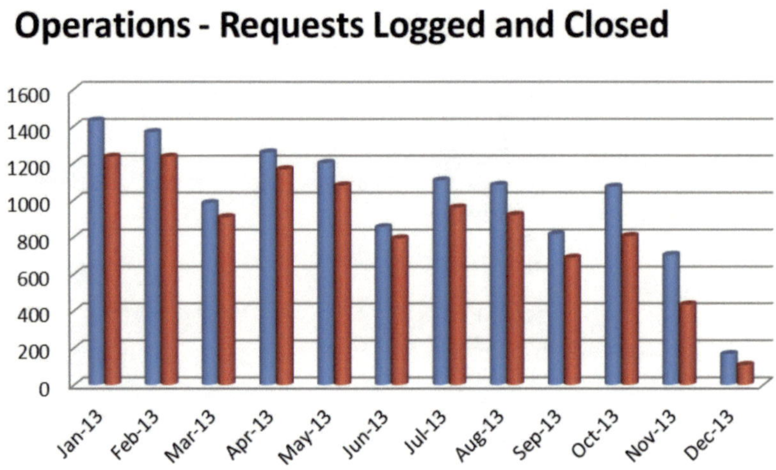

	Jan-13	Feb-13	Mar-13	Apr-13	May-13	Jun-13	Jul-13	Aug-13	Sep-13	Oct-13	Nov-13	Dec-13
No of Requests	1432	1369	984	1260	1201	854	1107	1082	815	1072	704	167
Closed	1237	1236	908	1167	1080	793	960	920	689	806	434	107

Figure 1.
Summary of performance on logged requests for the year 2013 ([35], p. 14).

addressing them. These reports provide generic information on maintenance operations without specifics or costs incurred.

The module of the CMMS tool activated by the maintenance units in these academic institutions is capable of managing maintenance planning (preventive, schedule and breakdown maintenance), work order, contract and procurement management, asset management, documentation of maintenance operations and analysis of historical records, report management and many more. However, **Table 3** provides a summary of the limited use and the latent potential of the CMMS tool, within the module being used by these institutions.

4.3 Discussion of findings

In a typical maintenance unit, there may be separate or integrated personnel for the call centre and maintenance operations. The quality and quantity of information developed from the CMMS database depends on the human capacity of the operatives in both units. This influences the ability to use the CMMS tool effectively [10]. Although maintenance operations have progressed from manual to digital systems, many maintenance units are underutilising the potentials available in a typical CMMS. The maintenance units in these academic institutions used the CMMS for work request management, maintenance operations and periodic reporting. **Table 3** provides the limited use of the CMMS tool. These components, currently being

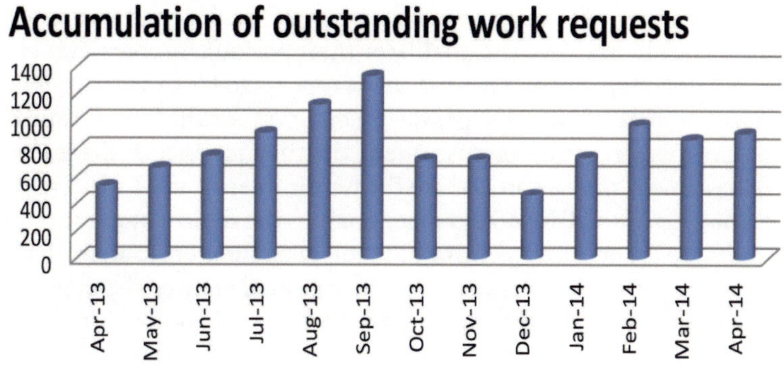

	Apr-13	May-13	Jun-13	Jul-13	Aug-13	Sep-13	Oct-13	Nov-13	Dec-13	Jan-14	Feb-14	Mar-14	Apr-14
■ Accumulative	535	667	756	922	1126	1336	730	731	470	745	979	874	918

Figure 2.
Summary of outstanding work requests.

Institution	Limited use of the CMMS tool	Remark: Latent potential not harnessed
Institution 1	*Recording of maintenance requests	*Developing planned maintenance *Effective maintenance budget and management of deferred or backlog maintenance *Producing customer friendly periodic report, accompanied by appropriate visuals *Detailed analysis of operational history, identifying the functional state of facilities, planning rehabilitation or disposal; educating customers on requests for change of use
	*Management of work requests	
	*Periodic reports on status of execution and cost incurred	
Institution 2	*Same as above, except that there are no records of cost incurred	
	*Record of work requests not attended to	

Table 3.
Summary of findings.

activated and used by these maintenance units, are insignificant fraction of the content of the maintenance management module and the inherent potentials of a typical CMMS software. The CMMS software has elastic capacity in different modules, which includes, but is not limited to asset management, work order management, preventive maintenance management, inventory control and report management [1]. Others include service management, contract management, materials management and procurement management [2]. Two major challenges were observed in the structure of the current periodic report. They are, firstly that the reports are not helpful in educating or communicating with the relevant stakeholders, because the customers cannot identify the component of the report that reflects the situation of the facilities in their portfolio or the status of execution of the work request emanating from their units. The second challenge is that the report is silent on the status of the work not completed at the time of reporting. Therefore, the units are not able to develop and monitor the forward planning, deferred maintenance, effective renovation, or rehabilitation scheme, provide suitable information for decisions on change of use or disposal of facilities. Consequently, they are not able to effectively communicate the performance of the maintenance units to their customers.

It is important to note that the maintenance and call centre operatives of these institutions should improve the quality of their periodic reports. Generally, a periodic report is the first step. Subjecting the set of reports to further analysis will enable maintenance operatives to manage the effectiveness of the loss of time before they respond to the work request of customers, the time taken to address the work request and to determine if they approximate the requirements of JIT [3]. Furthermore, detailed analysis of the CMMS database enables the maintenance unit to produce reports suitable for effective communication with their customers on the status of execution on their work requests and the functional state of the facilities in the customers' portfolio. To demonstrate this, the manager of the call centre of Institution 1, was requested to provide the comprehensive information on the requests lodged with the call centre from the School of Civil and Environmental Engineering for the period January to March 2010; arranging them in table format and providing a visual representation of the status of work and financial expenditure [34]. The eight-page report was reduced to a table as shown in **Table 4**. In a nutshell, **Table 4** provides the essential information contained in the eight-page report in a user-friendly format. **Figure 3** presents the status of the work requests for each month.

The visual presentation of the financial commitment, presented in **Figure 4**, shows that plumbing cost the most, while items under quotation cost the least.

The additional information, which is necessary, but was not available during the time of this research, includes explanatory notes on outstanding work requests, alternative suggestions on how to execute the outstanding work, the cost implications and the effect of the deferred maintenance on the functional state of the facilities of the customer. Despite these shortcomings, the Head, School of Civil and Environmental Engineering commented as follows: 'The layout is easy to determine the state of maintenance and it is easy to read. It also indicates that the maintenance unit is concerned about maintenance of the facilities in our School'. Furthermore, the Dean of the Faculty added: 'It is a good start. I would also like to see an age analyses (10 days, 30 days, 60 days, etc.) of addressing complaints or requests'.

Consequently, the vast potential inherent in a typical CMMS tool can be harnessed if the maintenance operatives could ask the right questions and the operatives in the call centre had the capacity, patience and commitment to answer the questions. This underlines the fact that the quality of the human capacity influences how much of the latent potential of a typical CMMS tool can be harnessed.

4.4 Analysis of maintenance database of a manufacturing company

The manufacturing industry used for this research is Adcock Ingrams Critical Care (AICC), a pharmaceutical manufacturing industry in South Africa. Maintenance requests are communicated directly to the unit as it doubles as call centre. In this industry, the maintenance unit uses the CMMS tool for scheduled and breakdown maintenance management. Furthermore, the unit conducts periodic detailed analyses

Problem type	Jan	Feb	Mar	Total issued	Total completed	Cost
Electrical	9	5	5	19	18	10,837.80
Plumbing	6	3	3	12	11	15,763.90
Quotation	1	1	2	4	1	136.80
Building		1	1	2	1	695.14
HVAC		1	3	4	3	2547.90

Table 4.
Summary of periodic report on work requests.

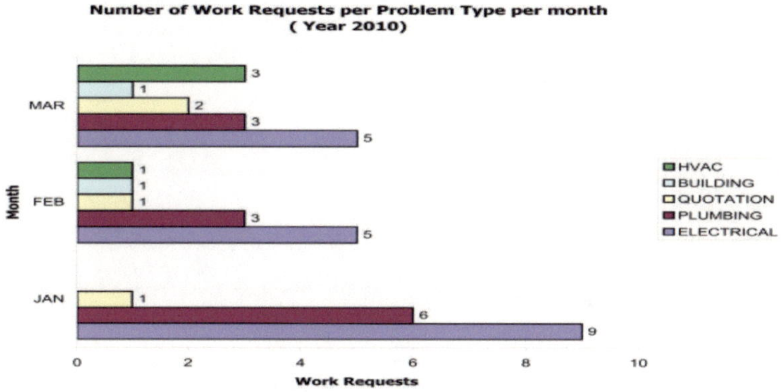

Figure 3.
Graphic presentation of the status of work requests.

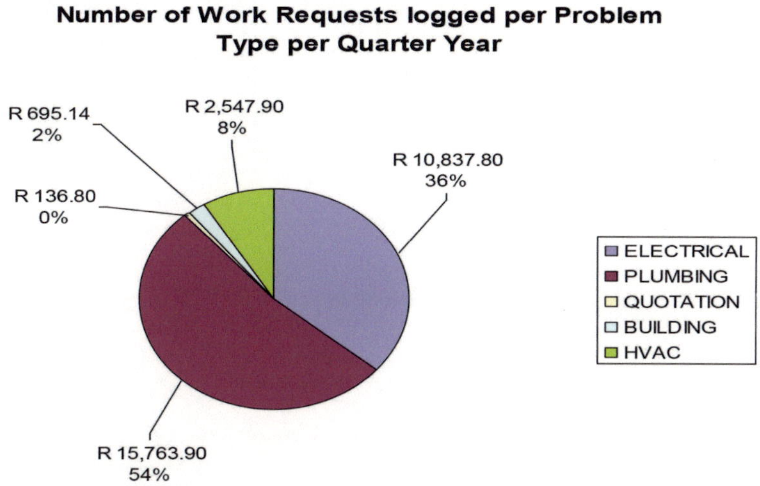

Figure 4.
Financial involvement of the work requests.

of the information in the operational database of the CMMS. The information in **Figures 5–8** present the findings of the analysis of the information in the CMMS database for all the machines in the production line of the company, for a period of 21 months between January 2017 and September 2018. The analysis identified the number of total breakdowns, mean time to repair (MTTR), total downtime and plant reliability. The reliability of the machines in the production line of the AICC was benchmarked with those in the operation of a sister company, Adcock Ingram Health Care (AIHC), to identify the level of performance improvements required.

4.4.1 Total breakdowns

The number of breakdowns per month in all the machines in the production line over the 21-month period is presented in **Figure 5**. The target number of 130 breakdowns, per month, was adopted as baseline for 2017 and 120 was set as target for 2018. The results from the analysis show that the average number of breakdowns in 2017 was 145, an increase of 15 breakdowns per month. Similarly, in September 2018, the average number of breakdowns was 128, showing an increase of 8 breakdowns above the benchmark of 120. However, in March, April, June and September 2018, the number of breakdowns was lower than the benchmark of 120, and in August, the number

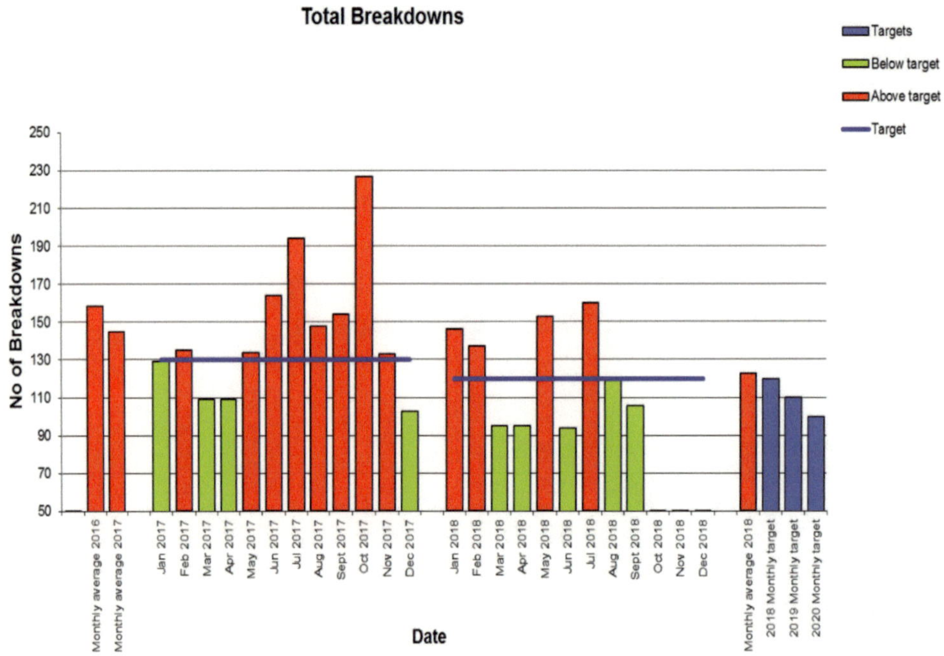

Figure 5.
Total number of breakdowns.

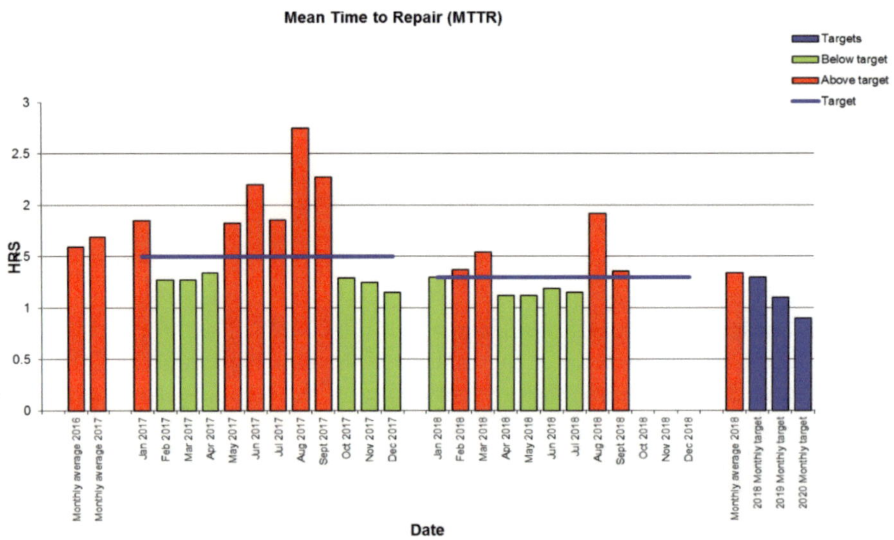

Figure 6.
Mean time to repair (MTTR).

of breakdowns was the same as the benchmark set for 2018. It was hoped that in the remaining months of 2018, the plants would record a lower number of breakdowns. Another concern was the length of time it takes to conclude repairs or known as MTTR.

4.4.2 Mean time to repair (MTTR)

The MTTR is the average time taken to complete the repair of each breakdown, restore the plant(s) to functional use and resume production. The shoter

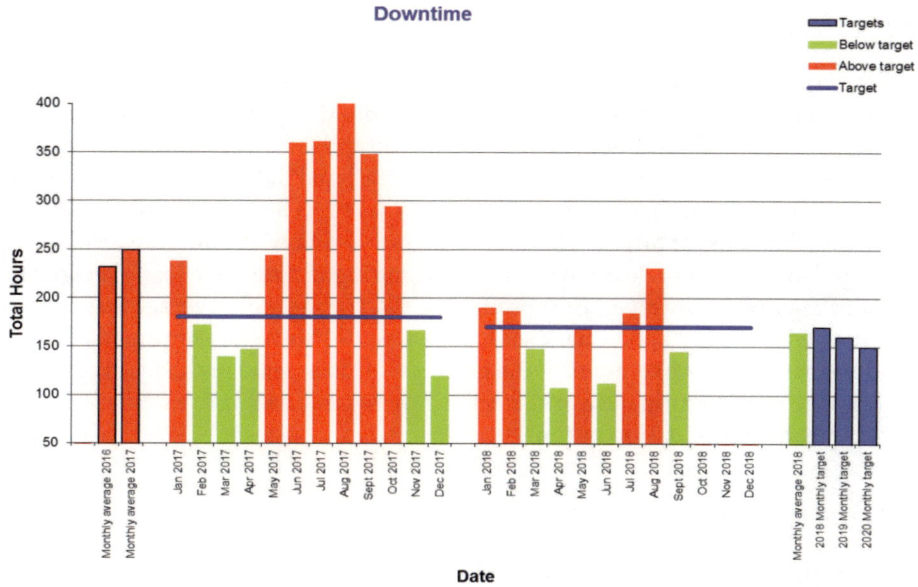

Figure 7.
Total duration of downtime of plant equipment.

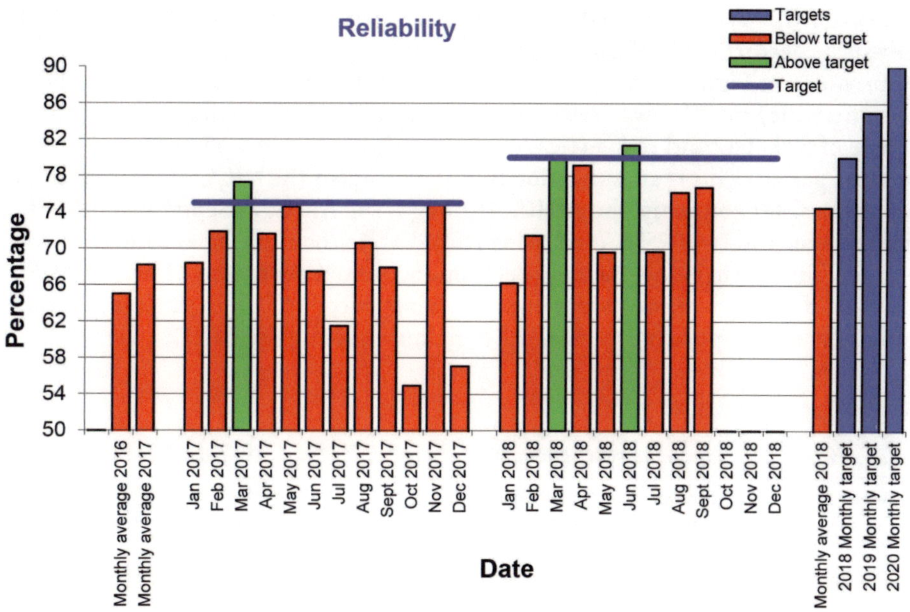

Figure 8.
Plant reliability.

the MTTR, the better. This is because the MTTR influences the duration of the downtime of machines in the production line. Similar to other factors, the monthly and annual averages for 2017 exceeded the benchmark of 1.5 hours, as is evident from **Figure 6**. Similarly, in 2018 the average in September is marginally above the benchmark of 1.35 hours. However, the monthly performance in January, April, May, June and July was impressive. It was hoped that, in the last quarter of the year, the performance would improve.

4.4.3 Total downtime

The graph in **Figure 7** summarises the duration of downtime when any or a combination of the plant or equipment in the production network breaks down. The target of 180 and 170 hours per month was adopted for 2017 and 2018, respectively. The analysis of the information in the operational database showed that the monthly average duration of downtime for 2017 was 250 hours per month, which represents 38.8% above target. This translates into 2.9 days' production lost per month and an average of 34.8 days per year, resulting in a revenue loss equivalent to 1 month per year. However, in September 2018, the monthly average was about 170 hours, which is the benchmark set for the year. If this trend continued, there would be marked improvements in the duration of downtime in the plants on the company's production network, ensuring increased machine reliability and improvements in production output.

4.4.4 Plant reliability

Plant reliability can be described as the probability that the plant(s) in the production network will be available for effective production in a manufacturing industry. Plant reliability is influenced by the number of breakdowns, the length of time before repairs are concluded and the duration of downtime. Machine reliability is usually expressed in percentages. Positive improvements in the MTTR (like JIT) translate in the reduced duration of downtime as well as a reduction in the number of breakdowns. The plant reliability performance of the AICC was benchmarked against the performance of a sister industry, the AIHC.

As shown in **Figure 8**, the plant reliability benchmark of the AIHC (the blue horizontal line) for 2017 was 75%. In comparison, the AICC achieved a monthly average of 68% (second red vertical bar from origin of the graph in **Figure 8**) for the same period. This is an indication that the plants of the AICC performed below that of the sister company, the AIHC. Although, the AICC, in September 2018, attained a reliability of 75%, (the red vertical bar, to the right, next to 2018 yearly average), this is still below the new benchmark of 80% set by the AIHC for 2018. However, if there were consistent improvements in the MTTR in the last quarter of the year, it may be possible to meet the benchmark set by the AIHC.

Industry	Current use of CMMS tool	Potential harnessed
Adcock Ingrams Critical Care (AICC)	*Development of maintenance management, preventive and breakdown maintenance	*Making maximum use of the maintenance management module *The synergy being created can facilitate the activation of the automation module *Progressing towards the practice of smart maintenance
	*Periodic analysis of operational details in CMMS database	
	*Benchmarking and identification of areas requiring performance improvements	
	*Improvements on MTTR (or JIT) and effects on machine reliability	
	*Develop synergy with finance, purchasing and store departments to ensure timely resourcing of spare parts for maintenance operations	

Table 5.
Summary of findings.

4.4.5 Summary of findings

Table 5 provides a summary of the functional use of the CMMS tool by AICC industry. Currently, the maintenance unit have activated and effectively using the different components of the maintenance management modules, as well as create synergy with other divisions of the industry. This will facilitate automation of operation and help in achieving the objectives of smart maintenance.

4.5 Discussion

The manufacturing company demonstrated a higher level of harnessing the potential inherent in a typical CMMS through the practice of comprehensive maintenance management, which includes planned, preventive, proactive and breakdown maintenance. The practice includes the analysis of the information in the operational database of the CMMS. These analyses enabled the maintenance unit to know the impact of the frequency of breakdowns, the length of time before repairs are completed and the effect of the duration of downtime of machines on the reliability and availability of machines in the production line. The analysis exposed the impact of the length of downtime on the productivity of the company. This could be described in the following understandable terms.

> *The average duration of downtime for 2017 was 250 hours per month, which is considerably higher than the benchmark, representing a level of 38.8% above target. This translates into 2.9 days' production lost per month and an average of 34.8 days per year, resulting in a revenue loss equivalent to 1 month per year.*

This realisation challenged the maintenance unit to improve on the MTTR. Success in the MTTR or JIT [3] is the product of appropriate maintenance planning, positive work ethics and the professional attitude of the workforce. This includes effective coordination between the maintenance unit, purchasing department, finance department and stores, the inventory control of stock and the timely availability of spare parts. It is important to note that the industry will have value for money through the effective management of an adequate stock of spare parts rather than purchasing on demand, which is more expensive [36]. Spares that are available reduce repair time and reduce the incidence of the 'fire-fighting approach' when sourcing spare parts. This enables the maintenance unit to strive towards achieving best practice, which suggests that 85% of repairs should be executed through planned maintenance and 15% through breakdown repairs [37]. These efforts culminate in the benchmarking of plant reliability with their sister industry. It is worth noting that plant reliability facilitates production planning, sales and marketing projection, achieving customers' satisfaction and profitability [37, 38].

The detailed analysis of the information in the operational database of plants in the production network enables maintenance units to identify area(s) requiring critical attention around which performance improvement strategies should be developed [39]. It provides intelligent information relating current performance against predetermined goals to decision-makers at all levels [19, 21]. This exercise challenges maintenance units to practise the art of continuous data collection, analysis and the interpretation of information to facilitate the development of appropriate improvement strategies [15]. Furthermore, it supports compliance activities, proposals for changes or requests for additional resources as it illuminates the link between strategies, performance and expected outcomes. It also achieves the objectives of smart maintenance [4, 19, 20].

5. Conclusions and recommendations

Although maintenance management and operations have significantly evolved from the use of manual to digital facilities, many of the operations of the maintenance units using CMMS tools are still operating at rudimentary levels. Generally, a typical CMMS tool has the capacity for planning and managing maintenance activities, resource management, analysis and interpretation, the contextual use of information in operational databases, the development of periodic reports, effective communication with stakeholders, performance and production improvements. However, the quality and quantity of the human capacity of the users of the system influences the potential of the CMMS tool that can be harnessed and put to productive use. Three case studies were used in this chapter to illustrate the potential capacity of a CMMS tool accessed by the maintenance units. Two of them refer to maintenance units in academic institutions and the third to a manufacturing company. In the academic institutions, the CMMS module activated was underutilised, as is evident from the poorly structured periodic reports. The reports were too technical, economical in details, best understood by those who prepared the reports and not suitable for effective communication with customers. The reports were silent about what the units were doing with work requests not completed at the time of reporting or the plans for addressing backlogs or deferred maintenance. The CMMS tool was being used for reactive maintenance instead of harnessing the vast potential of the CMMS tool for proactive maintenance, which is the backbone of smart maintenance.

In contrast, the manufacturing company uses the CMMS tool for planned and breakdown maintenance operations. It included the analysis of the information in the operational database, which enabled the unit to know the impact of the MTTR on the length of downtime, machine reliability and availability in the production network. This has challenged the maintenance unit to create synergy with the finance, purchasing and store departments, to facilitate the availability of spare parts and motivate the maintenance crew to timely identify and execute repairs. The activation and contextual use of the inter-related modules of the CMMS has enabled the maintenance unit to continuously improve on the MTTR, aiming at executing maintenance activities through planned maintenance rather than through a breakdown approach. It therefore improves on the practice of JIT, which is necessary for smart maintenance.

Although considerable effort is put in and investments are made to provide modern technology to aid maintenance operations, the potential uses of this tool are not adequately harnessed by many of the maintenance operatives. The maintenance units in the manufacturing company, in contrast to those in the academic institutions, through the analysis of the operational information in the CMMS database, have taken up some of the latent and rich potential inherent in a typical CMMS tool. Furthermore, they demonstrated the relationship of the 'work order management' tool with the 'assets management, preventive maintenance management, inventory control, report management and procurement management' tools. By doing so, they highlighted the collaborative relationships between the maintenance unit and other stakeholders, such as the finance, purchasing and store departments, in inventory control of stock and the timely availability of spare parts. Consequently, improvements in the MTTR translate into improved machine reliability, which facilitates production planning, sales and market projection, achieving customer satisfaction and profitability.

It is safe to conclude that the quality of the human capacity available to operate a typical CMMS tool influences the ability to harness the inherent potential of a CMMS. Therefore, this research recommends adequate resourcing and continuous development of the human capacity for the effective operation of the CMMS.

Author details

Edoghogho Ogbeifun*, Patrick Pasipatorwa and Jan-Harm C. Pretorius
Postgraduate School of Engineering Management, Faculty of Engineering and the Built Environment, University of Johannesburg, South Africa

*Address all correspondence to: edobuni@gmail.com

References

[1] Lopes I, Senra P, Vilarinho S, Sa VTeixeira C, Lopes J, Alves A, et al. Requirements specification of a computerized maintenance management system – A case study. Procedia CIRP. 2016;**52**:268-273. DOI: 10.1016/j.procir.2016.07.047

[2] Kostuk R. Computerised maintenance management systems. Studies & Proceedings of Polish Association for Knowledge Management. 2010;**35**:277-281

[3] Pheng LS, Arain FM, Fang JWY. Applying just-in-time principles in the delivery and management of airport terminal building. Built Environment Project and Asset Management. 2011;**1**(1):104-121. DOI: 10.1108/20441241111143812

[4] Uhlmann E, Franke D, Hohwieler E. Smart maintenance – Dynamic model-based instructions for service operations. Procedia CIRP. 2019;**81**:1417-1422. DOI: 10.1016/j-procir.2019.04.327

[5] Tsang AHC. Condition-based maintenance: Tools and decision making. Journal of Quality in Maintenance Engineering. 1995;**1**:3-17. DOI: 10.1108/13552519510096350

[6] Stamboliska Z, Rusiński E, Moczko P. Proactive Condition Monitoring of Low-Speed Machines. Cham: Springer International Publishing; 2015. pp. 9-34. DOI: 10.1007/978-3-319-10494-2_2

[7] Murthy DNP, Atrens A, Eccleston JA. Strategic maintenance management. Journal of Quality in Maintenance Engineering. 2002;**8**:287-305. DOI: 10.1108/13552510210448504

[8] Lavy S. Facility management practices in higher education buildings. Journal of Facilities Management. 2008;**6**(4):303-315. DOI: 10.1108/14725960810908163

[9] Gabriel G. Decentralising asset management in a university environment using web enabled technology. Facilities. 2003;**21**(10):233-243. DOI: 10.1108/02632770310493607

[10] Amadi-Echendu JE, de Wit FCP. Technology adoption: A study on post-implementation perceptions and acceptance of computerised maintenance management systems. 2015. Available from: https://repository.up.ac.za/bitstream/handle/2263/51452/AmadiEchendu_Technology_2015.pdf?sequence=1 [Accessed: 14 July 2020]

[11] Carder P. Knowledge-based FM: Managing performance at the workplace interface. Facilities. 1995;**13**(12):7-11. DOI: 10.1108/02632779510101033

[12] Chou JS, Chen HM, Hou CC, Lin CW. Visualised EVM system for assessing project performance. Automation in Construction. 2010;**19**:596-607. DOI: 10.1016/j.autcon.2010.02.006

[13] Horton D. Planning implementing and evaluating capacity development. International Service for National Agricultural Research (ISNAR). 2002;**50**:1-8

[14] Farazmand A. Innovation in strategic human resource management: Building capacity in the age of globalisation. Public Organization Review. 2004;**4**:3-24. DOI: 10.1023/B:PORJ.0000015649.54219.b7

[15] Melnyk SA, Bititci U, Platts KJ, Tobias J, Andersen B. Is performance measurement and management fit for the future? Management Accounting Research. 2014;**2**:173-186. DOI: 10.1016/j.mar.2013.07.007

[16] Uzarski DR, Grussing MN. Building condition assessment metrics: Best

practices. In: Amekudzi A, McNeil S, editors. Infrastructure Reporting and Asset Management. Reston, Virginia, USA: Infrastructure Reporting and Asset Management, American Society of Civil Engineers; 2008. pp. 147-152. DOI: 10.1061/9780784409589

[17] Hayes FX. A Proactive Solution. American School and University. USA: Endeavor Business Media, Janesville Ave, Fort Atkinson; 2006. pp. 310-312

[18] Myeda NE, Pitt M. Understanding the performance measurement system (PMS) for facilities management (FM) industry in Malaysia. In: Proceedings: PMA Conference. Cambridge UK; 11-12 July. 2012. pp. 1-10

[19] Amaratunga D, Baldry D. Moving from performance measurement to performance management. Facilities. 2002;**20**(5&6):217-223. DOI: 10.1108/02632770210426701

[20] Ogbeifun E. Performance management as a tool for performance improvement in the management of built facilities. Journal of Construction Project Management and Innovation. 2018;**8**(2):1852-1871. DOI: 10.36615/jcpmi.v8i2.162

[21] Buys F, Nkado RA. Survey of maintenance management systems in South African tertiary educational institutions. Construction Management and Economics. 2006;**24**:97-1005. DOI: 10.1080/01446190600851009

[22] Lindhard S, Larsen JK. Identifying the key process factors affecting project performance. Engineering Construction and Architectural Management. 2016;**23**(5):657-673. DOI: 10.1108/ECAM-08-2015-0123

[23] Loosemore M, Hsin YY. Customer-focused benchmarking for facilities management. Facilities. 2001;**19**(13&14):464-476. DOI: 10.1108/EUM0000000006204

[24] Attiany MC. Competitive advantage through benchmarking: Field study of industrial companies listed in Amman stock exchange. Journal of Business Studies Quarterly. 2014;**5**(4):41-51

[25] Wauters B. The added value of facilities management: Benchmarking work process. Facilities. 2005;**23**(3&4):142-151. DOI: 10.1108/02632770510578511

[26] Ogbeifun E. Developing performance measuring standards for assessing facilities management unit in higher education institution: a case study [thesis]. Johannesburg: University of Johannesburg; 2016

[27] Varcoe BJ. Business-driven facilities benchmarking. Facilities. 1996;**14**(3&4):42. DOI: 10.1108/02632779610112535

[28] Pasipatorwa P. Aligning the maintenance management system with the business strategy in order to achieve the strategic objectives of a manufacturing company: A case study [dissertation]. Johannesburg: University of Johannesburg; 2020

[29] Lateef OA, Khmidi MF, Idrus A. Building maintenance management in a Malaysian university campus: A case study. Australasian Journal of Construction Economics and Building. 2010;**10**(1&2):76-89. DOI: 10.5130/AJCEB.v10i1-2.1593

[30] Yin RK. Case Study Research and Applications: Design and Methods. 6th ed. Singapore: SAGE Publishing; 2014. Available from: https://lccn.loc.gov/2017040835

[31] Hallowell MR, Gambatese JA. Qualitative research: Application of the Delphi method to CEM research. Journal of Construction

Engineering and Management. 2010;**136**:1-9. DOI: 10.1061/(ASCE) CO.1943-7862.0000137

[32] Hsieh H-F, Shannon SE. Three approaches to qualitative content analysis. Qualitative Health Research. 2005;**2005, 15**(9):1277-1288. DOI: 10.1177/1049732305276687

[33] Bowen GA. Document analysis as a qualitative research method. Qualitative Research Journal. 2009;**9**(2):27-40

[34] Ogbeifun E. Facilities management in a multi-campus setting: a case study of the University of the Witwatersrand [dissertation]. Johannesburg: University of the Witwatersrand; 2011

[35] Operations Department. Central Technical Services. [Annual Report]. Johannesburg: University of Johannesburg; 2013

[36] Joachim A. Spare parts planning and control for maintenance operations [thesis]. Eindhoven: University of Technology; 2013

[37] Gulati R. Maintenance and Reliability Best Practices. 2nd ed. New York: Industrial Press; 2013

[38] Vishnu CR, Regikumar V. Reliability based maintenance strategy selection in process plants: A case study. In: Proceedings: Global Colloquium in Recent Advancement and Effectual Researches in Engineering, Science and Technology (RAEREST 2016). Vol. 25. Procedia Technology; 2016. pp. 1080-1087. DOI: 10.1016/j.protcy.2016.08.211

[39] Simatupang TM, Piboonrungroj P, Williams SJ. The emergence of value chain thinking. In: Proceedings of T-LOG. 2014. Available from: http://s3.amazonaws.com/academia.edu.documents/34563986/66.pgf?AWSAccessKeyId=AKIAJ56TQJRTWSMTNPEA&Expires=1469032947&Signature=Mh3eQaqStDrID47%2FDGFYC7dGDDI%3D&response-content-disposition-inline%3B%20filename%3DThe [Accessed: 30 June 2016]

Chapter 5

Multi-Strategy *MAX-MIN* Ant System for Solving Quota Traveling Salesman Problem with Passengers, Incomplete Ride and Collection Time

Bruno C.H. Silva, Islame F.C. Fernandes,
Marco C. Goldbarg and Elizabeth F.G. Goldbarg

Abstract

This study proposes a novel adaptation of *MAX-MIN* Ant System algorithm for the Quota Traveling Salesman Problem with Passengers, Incomplete Ride, and Collection Time. There are different types of decisions to solve this problem: satisfaction of the minimum quota, acceptance of ride requests, and minimization of travel costs under the viewpoint of the salesman. The Algorithmic components proposed regards vehicle capacity, travel time, passenger limitations, and a penalty for delivering a passenger deliverance out of the required destination. The ant-based algorithm incorporates different sources of heuristic information for the ants and memory-based principles. Computational results are reported, showing the effectiveness of this ant-based algorithm.

Keywords: Traveling Salesman, integer programming, transportation, shared mobility, Ant Colony Optimization

1. Introduction

The lives of ordinary consumers have changed almost beyond recognition in the past 20 years. First, with the introduction of high-speed internet access; but, more recently, with the arrival of mobile computing devices such as smartphones and tablets. According to data from the 2017 Gallup World Survey [1], 93 of adults in high-income economies have their cell phones, while 79% in developing economies. In India, 69% of adults have a cell phone, as well as 85% in Brazil and 93% in China [1]. Smartphones and the internet have created a novel digital ecosystem where the adoption of new paradigms is increasingly fast, and each innovation that appears and presents itself to the market can disrupt an entire segment.

In the transportation segment, a central theme is how the digital revolution has created opportunities to consider new models of delivering services under the paradigm of Mobility as a Service (MaaS) [2]. There is a growing interes%t in *MaaS* due to the notion of a sharing economy. Millennials own fewer vehicles than previous generations [3]. As evidenced by the ascension of on-demand mobility platforms, they are quickly adopting car sharing as a mainstream transportation solution. Investments in new travel patterns have become a priority to enable the transformation of opportunities in the urban mobility segment into new revenue streams.

This study deals with a novel optimization model that can improve the services provided by on-demand mobility platforms, called Quota Traveling Salesman Problem with Passengers, Incomplete Ride, and Collection Time (QTSP-PIC). In this problem, the salesman is the vehicle driver and can reduce travel costs by sharing expenses with passengers. He must respect the budget limitations and the maximum travel time of every passenger. Each passenger can be transported directly to the desired destination or an alternate destination. Lira et al. [4] suggest pro-environmental or money-saving concerns can induce users of a ride-sharing service to agree to fulfill their needs at an alternate destination.

The QTSP-PIC can model a wide variety of real-world applications. Cases related to sales and tourism are the most pertinent ones. The salesman must choose which cities to visit to reach a minimum sales quota, and the order to visit them to fulfill travel requests. In the tourism case, the salesman is a tourist that chooses the best tourist attractions to visit during a vacation trip and can use a ride-sharing system to reduce travel expenses. In both cases, the driver negotiates discounts with passengers transported to a destination similar to the desired one.

The QTSP-PIC was introduced by Silva et al. [5]. They presented a mathematical formulation and heuristics based on Ant Colony Optimization (ACO) [6]. To support the ant algorithms, they proposed a Ride-Matching Heuristic (RMH) and a local search with multiple neighborhood operators, called Multi-neighborhood Local Search (MnLS). They tested the performances of the ant algorithms on 144 instances up to 500 vertices. One of these algorithms, the Multi-Strategy Ant Colony System (MS-ACS), provided the best results. They concluded that their most promising algorithm could improve with learning techniques to choose the source of information regarding the instance type and the search space.

In this study, a *MAX-MIN* Ant System (*MM*AS) adaptation to the QTSP-PIC, called Multi-Strategy *MAX-MIN* Ant System (MS-*MM*AS), is discussed. *MM*AS improves the design of Ant System [6], the first ACO algorithm, with three important aspects: only the best ants are allowed to add pheromone during the pheromone trail update; use of a mechanism for limiting the strengths of the pheromone trails; and, incorporation of local search algorithms to improve the best solutions. Plenty of recent studies proved good effectiveness of the *MM*AS in correlated problems to QTSP-PIC [7–10]. However, none of these explored the Multi-Strategy (MS) concept.

In the traditional ant algorithms applied to Traveling Salesman Problem (TSP), ants use the arcs' cost as heuristic information [6]. The heuristic information adopted is called visibility. When solving the QTSP-PIC, different types of decisions must be considered: the accomplishment of the minimum quota, management of the ride requests, and minimization of travel costs. The MS idea is to use different mechanisms of visibility for the ants to improve diversification. Every ant decides which strategy to use at random with uniform distribution. The MS proposed in this study extends the original implementation proposed in [5]. MS-*MM*AS also incorporates RMH and MnLS and uses a memory-based technique proposed in [11] to avoid redundant work. In MS-*MM*AS, a hash table stores every solution constructed and used as initial solutions to a local algorithm. When the algorithm constructs a new solution, it starts the local search phase if the new solution is not in the hash table.

The benchmark for the tests consisted of 144 QTSP-PIC instances. It was proposed by Silva et al. [5]. Numerical results confirmed the effectiveness of the MS-*MM*AS by comparing it to other ACO variants proposed in [5].

The main contributions of this chapter are summarized in the following.

- The extension of the MS concept proposed in [5] with a roulette mechanism that orients the ants to choose their heuristic information based on the best quality solutions achieved;

- Improvement of the *MMAS* design with a memory based technique proposed in [11];

- Presentation of a novel *MMAS* variant that combines the improved MS concept and memory-based principles and assessment of its performance;

- Experiments on a set of QTSP-PIC instances ranging: 10 to 500 cities; and 30 to 10.000 travel requests. The results showed that the proposed *MMAS* variant is competitive regarding the other three ACO variants presented in [5] for the QTSP-PIC.

The remainder of this chapter is organized as follows. Section 2 presents the QTSP-PIC and its formulation. Section 3 presents the Ant Colony Optimization metaheuristic and the implementation design of the MS-*MMAS*. Section 4 presents experimental results. The performance of the proposed ant-based algorithm is discussed in Section 5. Conclusions and future research directions are outlined in Section 6.

2. Problem definition

The TSP can be formulated as a complete weighted directed graph $G = (N, A)$ where N is the set of vertices and $A = \{(i,j) \mid i,j \in N\}$ is the set of arcs. $C = [c_{ij}]$ is the arc-weight matrix such that c_{ij} is the cost of arc (i,j). The objective is to determine the shortest Hamiltonian cycle in G. Due to its applicability, many TSP variants deal with specific constraints [12]. Awerbuch et al. [13] presented several quota-driven variants. One of them, called Quota Traveling Salesman Problem (QTSP), is the basis for the problem investigated in this study. In the QTSP, there is a bonus associated with each vertex of G. The salesman has to collect a minimum quota of bonuses in the visited vertices. Thus the salesman needs to figure out which cities to visit to achieve the minimum quota. The goal is to find a minimum cost tour such that the sum of the bonuses collected in the visited vertices is at least the minimum quota.

The QTSP-PIC is a QTSP variant in which the salesman is the driver of a vehicle and can reduce travel costs by sharing expenses with passengers. There is a travel request, associated with each person demanding a ride, consisting of a pickup and a drop off point, a budget limit, a limit for the travel duration, and penalties associated with alternative drop-off points. There is a penalty associated with each point different from the destination demanded by each person. The salesman can accept or decline travel requests. This model combines elements of ride-sharing systems [14] with alternative destinations [4], and the selective pickup and delivery problem [15].

Let $G(N, A)$ be a connected graph, where N is the set of vertices and $A = \{(i,j) \mid i,j \in N\}$ is the set of arcs. Parameter q_i denotes the quota associated with vertex $i \in N$ and g_i the time required to collect the quota. c_{ij} and t_{ij} denote, respectively, the cost and time required to traverse edge $(i,j) \in A$. Let L be the set of passengers. List $l_i \subseteq L$ denotes the subset of passengers who depart from $i \in N$. Let $org(l)$ and $dst(l) \in N$ be the pickup and drop-off points requested by passenger l. The salesman departs from city $s = 1$, visits exactly once each city of subset $N' \subseteq N$ and returns to s. The quota collected by the salesman must be at least K units. Along the trip, the salesman may choose which travel requests to fulfill. The travel costs are shared with vehicle occupants. The number of vehicle occupants, or passengers, cannot exceed R. Each passenger $l \in L$ imposes a budget limit w_l and a trip's maximum duration b_l. Let h_{lj} be the penalty to deliver passenger $l \in L$ at city $j \in N$, $j \neq dst(l)$. The value of variable h_{lj} is computed in the final cost of the tour if passenger l is delivered to city j. If $j = dst(l)$, then $h_{lj} = 0$. The objective of the QTSP-PIC is to find a Hamiltonian cycle $\Psi = (N', A')$ such that the ride-sharing cost and eventual penalties are minimized, and the quota constraint is satisfied.

The QTSP is NP-hard [13]. It is a particular case of the QTSP-PIC, in which the list of persons demanding a ride is empty and the time spent to collect the bonus in each vertex is zero. Thus, QTSP-PIC also belongs to the NP-hard class.

Silva et al. [5] presented an integer non-linear programming model for the QTSP-PIC. They defined a solution as $S = (N', Q', L', H')$, where N' is a list of vertices that represents a cycle, Ψ, such that the minimum quota restriction, K, is satisfied; Q' is a binary list in which the i-th element is 1 if the salesman collects the bonus from city i, $i \in N'$; L' is a binary list in which the l-th element is 1 if the salesman accepts the l-th travel request; and H' is a list of integers in which the l-th element is the index of the city where the l-th passenger leaves the car. If $L'[l] = 0$, then $H'[l] = 0$. The cost of solution S, denoted by $S.cost$, is calculated by Eq. (1).

$$S.cost = \sum_{i,j \in N'} \frac{c_{ij}}{1 + \sum_{l \in L'} v_{ij}^{l}} + \sum_{l \in L'} h_{lH'_l} \tag{1}$$

3. Ant Colony Optimization

In the Ant Colony Optimization, artificial ants build and share information about the quality of solutions achieved with a communication scheme similar to what occurs with some real ants species. Deneubourg et al. [16] investigated the behavior of Argentine ants and performed some experiments, where there were two bridges between the nest and a food source. He observed that the ants initially walked on the two bridges at random, depositing pheromone in the paths. Over time, due to random fluctuations, the pheromone concentration of one bridge was higher than the other. Thus, more ants were attracted to that route. Finally, the whole colony ended up converging towards the same route. The behavior of artificial ants preserves four notions of the natural behavior of ants:

- Pheromone deposit on the traveled trail;
- Predilection for trails with pheromone concentration;
- Concentration of the amount of pheromone in shorter trails;
- Communication between ants through the pheromone deposit.

Pheromone is a chemical structure of communication [17]. According to Dorigo et al. [18], pheromone enables the process of stigmergy and self-organization in which simple agents perform complex and objective-oriented behaviors. Stigmergy is a particular form of indirect communication used by social insects to coordinate their activities [18].

Considering the context of ant algorithms applied to the TSP, when moving through the graph G, artificial ants tend to follow paths with higher pheromone deposits rates. As ants tend to deposit pheromone along the path they follow, as more ants choose the same path, the pheromone rate tends to increase in these paths. This cooperation mechanism induces artificial ants to find good solutions, as it works as a shared memory that is continuously updated and can be consulted by every ant in the colony [19].

The base-line of the Ant Colony Optimization is the algorithm Ant System [6]. In the TSP application, N is the set of vertices to visit. Ants construct solutions iteratively. Every iteration, the ant chooses the next vertex based on heuristic information, η, and pheromone trails, τ. Initially, pheromone trails have the same amount of pheromone, τ_0, computed by Eq. (2), where n is the number of vertices in N and $Cost(D)$ is the value of the TSP tour built by a greedy heuristic.

$$\tau_{ij}^{0} = (n \times Cost(D))^{-1} \tag{2}$$

The k-th ant iteratively adds new vertices to the solution. The algorithm uses Eq. (3) to compute the probability of the k-th ant to move from vertex i to j at the t-th iteration, where $\eta_{ij} = \frac{1}{c_{ij}}$ is the heuristic factor, $\tau_{ij}(t)$ is the pheromone in arc (i,j) in the t-th iteration, and Λ^k is the list of vertices not visited by the k-th ant. Coefficients α and β weight the influence of the pheromone and heuristic information, respectively. They are user-defined parameters. If $\alpha = 0$, the probability computed by Eq. (3) depends only on the heuristic information. So, the ant algorithm behaves like a greedy method. If $\beta = 0$, ants tend to select paths with higher pheromone levels. It may lead the algorithm to early stagnation. So, balancing the values of α and β is critical to guarantee a suitable search strategy [5].

$$\Upsilon_{ij}^{k}(t) = \frac{[\tau_{ij}(t)]^{\alpha} \cdot [\eta_{ij}]^{\beta}}{\sum_{w \in \Lambda^k} \left[[\tau_{iw}(t)]^{\alpha} \cdot [\eta_{iw}]^{\beta} \right]}, \qquad j \in \Lambda^k \tag{3}$$

Eqs. (4) and (5) show the formulas used to update pheromone trails, where ρ is the evaporation coefficient, $Cost(W^k)$ is the cost of the route W^k built by the k-th ant, and $\Delta\tau_{ij}^{k}$ is the pheromone deposited on arc (i,j) by the k-th ant, computed by expression (6).

$$\tau_{ij} = (1-\rho) \times \tau_{ij} + \rho \times \Delta\tau_{ij}, \qquad \rho \in [0,1] \tag{4}$$

$$\Delta\tau_{ij} = \sum_{k=1}^{m} \Delta\tau_{ij}^{k} \tag{5}$$

$$\Delta\tau_{ij} = \begin{cases} \dfrac{1}{Cost(W^k)}, & \text{if arc } (i,j) \in W^k. \\ 0, & \text{otherwise.} \end{cases} \tag{6}$$

The ant algorithms proposed after AS improved its implementation design with elitist pheromone update strategies and local search algorithms to improve solutions [6]. Two well-known variants of AS are the Ant Colony System (ACS) [20] and *MAX-MIN* Ant System [21]. Silva et al. [5] presented AS and ACS adaptations for the QTSP-PIC. Section 3.1 presents the *MAX-MIN* Ant System algorithm.

3.1 Multi-strategy *MAX-MIN* ant system

MMAS uses Eq. (3) to compute the probability of an ant to move from vertex i to j. Besides, it incorporates improvements to avoid search stagnation and a pheromone update rule that limits pheromone concentration rates. Eq. (7) presents the pheromone update rule. Limits τ_{max} and τ_{min} prevent stagnation of pheromone values.

$$\tau_{ij} = max\left\{\tau_{min}, min\left\{\tau_{max}, (1-\rho)\times\tau_{ij} + \rho\times\Delta\tau_{ij}^{best}\right\}\right\}, \qquad \rho\in[0,1] \qquad (7)$$

$$\Delta\tau_{ij}^{best} = \begin{cases} \dfrac{1}{Cost\left(W^{best}\right)}, & \text{if arc}(i,j)\in W^{best}. \\ 0, & \text{otherwise.} \end{cases} \qquad (8)$$

There are three possibilities for the best route (W^{best}) considered in the algorithm: the best route in the current iteration, the best route found so far, and the best route since the last time pheromone trails were reinitiated. In the *MMAS* original design [21], these routes were chosen alternately. The initial value of pheromone trails was t_{max}. If the algorithm reached stagnation, i.e., the best current route remained the same for several iterations, the pheromone value reinitialized to t_{max}. Assigning t_{max} to pheromone trails produces a small variability among pheromone levels at the start of the search [21].

The implementation of the *MMAS* for the QTSP-PIC extends the original proposal [21] with the following adaptions:

- Ants start at vertex s;
- Ants include vertices in the route up to reach the minimum quota;
- Solution S^k, built by the k-th ant, is computed by assigning passengers to route W^k with the *RMH* algorithm [5];
- Use of the MS concept.

The ants in the *MMAS*, use arc costs to compute heuristic information. In the MS-*MMAS*, ants use four sources for this task, listed in the following.

- Cost oriented: uses c_{ij} as heuristic information, such that $\eta_{ij} = \frac{1}{c_{ij}}$;
- Time oriented: uses t_{ij} as heuristic information, such that $\eta_{ij} = \frac{1}{t_{ij}}$. This heuristic information guides ants to vertices that lead to travel time savings.
- Quota oriented: q_j is used as heuristic information, $\eta_{ij} = \frac{q_j}{c_{ij}}$. This heuristic information guides ants to go to vertices that lead to the maximization of the quota collected.
- Passenger oriented: the heuristic information is $\eta_{ij} = \frac{|L_j|}{c_{ij}}$. This strategy orients ants to maximize the number of travel requests fulfilled.

In the MS concept proposed in [5], every ant decides which strategy to use at random with uniform distribution. A roulette wheel selection improves this concept. The proportion of the wheel assigned to each heuristic information is directly related to the quality of solutions achieved. So, ants learn, at each iteration, the best heuristic information. At the final iterations, ants tend to use the heuristic information that proved to be most promising.

Algorithm 1 presents the pseudo-code of the MS-*MMAS*. It has the following parameters: maximum number of iterations ($maxIter$), number of ants ($m \in \mathcal{Z}_{>0}$), pheromone coefficient ($\alpha \in \mathcal{R}_{>0}$), heuristic coefficient ($\beta \in \mathcal{R}_{>0}$), evaporation factor ($\rho \in [0, 1]$), and pheromone limits (τ_{max}, $\tau_{min} \in \mathcal{R}_{>0}$). It also has the following parameters and variables:

- N: set of vertices;
- ξ: index of the heuristic information source;
- W^k: route built by the k-th ant;
- S^k: solution produced after applying the *RMH* heuristic [5] to route W^k;
- W^i: the best route built in the i-th iteration;
- S^i: the best solution produced in the i-th iteration;
- W^*: the best route found so far;
- S^*: the best solution found so far;
- W^{best}: route used as input to the pheromone updating procedure;
- Π: hash table that stores every solution S^i constructed and used as initial solution to the local search algorithm;

Algorithm 1: MS-$MMAS(maxIter, m, \alpha, \beta, \rho\ \tau_{max}, \tau_{min})$

```
Π ← ∅
Initialize pheromone trails
For k = 1 to m.
    W^k[2] ← random_city(N\{s})
For i = 1 to maxIter
    For k = 1 to m
        ξ ← chose_heuristic_information()
        W^k ← build_route(α, β, ξ)
        S^k ← assign_passengers(W^k)
        Update(W^i,S^i)
    If S^i ∉ Π
        S^i ← MnLS(S^i)
        Store(Π, S^i)
    Update(W*,S*)
    W^best ← alternate(maxIter, i, W^i, W*)
    Pheromone_update(W^best,ρ, τ_max, τ_min)
Return S*
```

The algorithm sets τ_{max} as the initial value of pheromone trails (step 2). Since ants begin at vertex s, the second vertex is selected randomly with uniform distribution (steps 3 and 4). The k-th ant decides which heuristic information, ξ, to use (step 7) and builds a route (step 8). The algorithm uses the *RMH* heuristic to assign passengers to W^k, completing a solution (step 9). The algorithm updates W^i and S^i (step 10). The MnLS algorithm is applied to S^i (step 12) if the solution $S^i \notin \Pi$. After the local search, the algorithm stores S^i in the hash table Π. At the next iteration, the current S^i is the starting solution of the local search if it is not in Π. This procedure prevents redundant work. The algorithm updates the best route and the best solution found so far, W^* and S^* (step 14). Similar to the original design of *MM*AS, W^i is assigned to W^{best} at the first 25% iterations or if i ranges from [50%,75%] of $maxIter$. W^* is assigned to W^{best} if i ranges from [25%,50%] of $maxIter$ or if it is greater than or equal to 75% of $maxIter$ (step 15). This procedure improves diversification by shifting the emphasis over the search space. W^{best} is used to update pheromones (step 16). Finally, the algorithm returns S^*.

4. Experiments and results

This section presents the methodology for the experiments and results from the experiments. Section 4.1 presents the methodology. Section 4.2 presents the parameters used in the MS-*MM*AS algorithm. Section 4.3 presents the results.

4.1 Methodology

The experiments were executed on a computer with an Intel Core i5, 2.6 GHz processor, Windows 10, 64-bit, and 6GB RAM memory. The algorithms were implemented in C ++ lan and compiled with GNU g++ version 4.8.4. The benchmark set proposed in [5] was used to test the effectiveness of the MS-*MM*AS. The sizes of those instances range from 10 to 500 vertices. Small instances have up to 40 vertices, medium up to 100, and large more than 100 vertices. The instances are available for download at https://github.com/brunocastrohs/QTSP-PIC.

The best, average results, and average processing times (in seconds) are reported from 20 independent executions of the MS-*MM*AS. Experiments are conducted to report the distance between the best-known solutions and the best results provided by the MS-*MM*AS. The variability in which the MS-*MM*AS achieved the best-known solutions stated in the benchmark set is also calculated. With these experiments, it is possible to conclude if the MS-*MM*AS algorithm was able to find the best-known solution of each instance and with what variability this happens.

The Friedman test [23] with the Nemenyi post-hoc procedure [24] are applied, with a significance level 0.05, to conclude about significant differences among the results of the MS-*MM*AS and the other three ACO variants proposed in this [5]. The instances were grouped according to their sizes (number of vertices) for the Friedman test. There are eight groups of symmetric (asymmetric) instances, each of them contains nine instances, called $g<n>$, where $<n>$ stands for the size.

4.2 Parameter tuning

The IRACE software was used, presented by [22], to tune the parameters of the MS-*MM*AS algorithm. 20 symmetric and 20 asymmetric instances were submitted to adjust the parameters. Those instances were selected at random. The IRACE uses the *maxExperiments* and *maxTime* parameters as stopping criteria. This parameters were set as follows: $maxExperiments = 10^3$; and, $maxTime = \infty$.

For the asymmetric instance set, the parameters were defined as follows: $maxIter = 31$; $m = 51$; $\alpha = 3.08$; $\beta = 10.31$; $\rho = 0.52$; $\tau_{max} = 0.8$; and $\tau_{min} = 0.2$. For the symmetric instance set, the parameters were: $maxIter = 29$; $m = 57$; $\alpha = 2.92$; $\beta = 9.53$; $\rho = 0.67$; $\tau_{max} = 0.7$; and $\tau_{min} = 0.2$.

4.3 Results

In this section, the results of the MS-*MM*AS are tested and compared to those produced by the other three ACO variants proposed in [5]: AS, ACS, and MS-ACS.

Table 1 presents the comparison between the ant algorithms. The best results obtained by MS-*MM*AS were compared with those achieved by each ant algorithm proposed in [5]. The results are in the $X \times Y$ format, where X and Y stand for the number of instances in which the ant algorithm X found the best solution and the number of instances in which the ant algorithm Y found the best solution, respectively.

Table 1 shows that the MS-*MM*AS was the algorithm that reported the best solution for most instances. This algorithm performed best than other ACO variants due to its enhanced pheromone update procedures. The MS implementation with roulette wheel selection proved to be effective at finding the best heuristic information used by the ants during the run. **Table 1** also shows that the MS-*MM*AS provides results with better quality than the MS-ACS in the most symmetric cases. The MS-ACS was superior to the MS-*MM*AS in seventeen asymmetric cases and fourteen symmetric instances. Was observed that the pseudo-random action choice rule of MS-ACS [20], which allows for a greedier solution construction, proved to be a good algorithmic strategy for solving large instances.

Tables 2 and **3** shows the ranks of the ant algorithms based on the Friedman test [23] with the Nemenyi [24] post-hoc test. The first column of this Tables presents the subsets of instances grouped according to their sizes. The other columns of this Tables present the p-values of the Friedman test and the ranks from the Nemenyi post-hoc test. In the post-hoc test, the order ranks from a to c. The c rank indicates that the algorithm achieved the worst performance in comparison to the others. The a rank indicates the opposite. If the performances of two or more algorithms are similar, the test assigns the same rank for them. In this experiment, the significance level was assigned with 0.05.

The p-values presented in **Tables 2** and **3** show that the performance of the ant algorithms was not similar, i.e., the null hypothesis [24] is rejected in all cases. In these Tables, can be observed that MS-*MM*AS ranks higher than AS and ACS for all subsets. The ranks of MS-ACS and MS-*MM*AS were the same in the most cases. This implies that the performance of only these two algorithms where similar, i. e., the relative distance between the results achieved by these two algorithms are short.

To analyze the variability of the results provided by each ant algorithm compared to the best results so far for the benchmark set, three metrics regarding the

	Asymmetric			Symmetric		
	AS	**ACS**	**MS-ACS**	**AS**	**ACS**	**MS-ACS**
MS-*MM*AS	68 x 0	68 x 1	45 x 17	66 x 1	68 x 2	48 x 14

Table 1.
Comparison between the ant algorithms.

			Asymmetric		
Subset	**p-value**	**AS**	**ACS**	**MS-ACS**	**MS-*MMAS***
*g*10	0.003159	b	b	a	a
*g*20	0.000040	b	c	a	a
*g*30	0.000017	c	c	a	a
*g*40	0.000024	b	c	a	a
*g*50	0.000048	c	c	b	a
*g*100	0.000045	b	c	a	a
*g*200	0.000031	b	c	a	a
*g*500	0.000037	c	b	a	a

Table 2.
Results of Friedman's test and Nemenyi post-hoc test over asymmetric instances set.

			Symmetric		
Subset	**p-value**	**AS**	**ACS**	**MS-ACS**	**MS-*MMAS***
*g*10	0.003543	b	b	a	a
*g*20	0.000205	b	c	a	a
*g*30	0.000045	c	c	b	a
*g*40	0.000059	b	c	a	a
*g*50	0.000035	b	b	a	a
*g*100	0.000024	b	b	a	a
*g*200	0.000045	c	b	a	a
*g*500	0.000098	c	b	a	a

Table 3.
Results of Friedman's test and Nemenyi post-hoc test over symmetric instances set.

		Asymmetric		
Metric	**AS**	**ACS**	**MS-ACS**	**MS-*MMAS***
ν	4.30%	2.56%	6.8%	11.84%
Φ	0.2075333	0.2773624	0.0541799	0.0054741
Ω	0.2835401	0.4023659	0.1754952	0.5854892

Table 4.
Variability of the ants algorithms for asymmetric instances.

		Symmetric		
Metric	**AS**	**ACS**	**MS-ACS**	**MS-*MMAS***
ν	2.01%	0.69%	8.75%	9.05%
Φ	0.2169756	0.2285948	0.0547890	0.0113889
Ω	0.3017957	0.3620682	0.1656022	0.6432748

Table 5.
Variability of the ants algorithms for symmetric instances.

	Asymmetric				Symmetric			
Instance	**Best**	**Average**	**Time**	**Percentage**	**Best**	**Average**	**Time**	**Percentage**
A-10-3	478.42	863.13	0.15	100%	545.92	996.34	0.15	25%
A-10-4	523.57	1069.33	0.14	80%	460.00	838.84	0.18	20%
A-10-5	482.60	690.08	0.14	5%	371.93	658.97	0.19	10%
A-20-3	519.67	936.47	0.35	5%	679.75	1363.59	0.32	20%
A-20-4	458.10	1145.88	0.38	0%	346.30	661.82	0.43	35%
A-20-5	398.75	669.82	0.35	10%	351.50	1006.24	0.48	5%
A-30-3	618.33	1180.70	0.48	10%	574.33	1469.68	0.50	5%
A-30-4	401.20	805.32	1.28	5%	654.80	1202.15	0.40	5%
A-30-5	475.83	1033.91	0.58	10%	464.05	911.56	0.66	5%
A-40-3	692.00	1060.67	2.83	5%	718.25	1399.04	0.72	10%
A-40-4	658.95	1088.41	2.52	5%	570.98	961.13	2.95	5%
A-40-5	460.90	900.51	3.11	5%	441.22	836.52	2.89	5%
B-10-3	729.50	925.35	0.13	5%	834.67	1485.47	0.20	20%
B-10-4	306.90	421.35	0.13	15%	493.58	757.45	0.16	10%
B-10-5	434.75	835.01	0.18	55%	726.35	1160.97	0.23	5%
B-20-3	805.42	1251.39	0.28	10%	950.00	1666.22	0.45	5%
B-20-4	848.62	1366.69	0.35	5%	822.82	1386.67	0.49	5%
B-20-5	895.17	1275.78	0.28	70%	660.22	1215.12	0.35	5%
B-30-3	747.75	1316.96	1.31	5%	718.67	1358.11	0.93	5%
B-30-4	723.27	1301.57	1.51	5%	650.35	1272.86	0.69	5%
B-30-5	700.75	1205.96	1.39	5%	504.68	1091.11	0.89	5%
B-40-3	964.42	1574.00	2.02	0%	889.83	1682.76	1.97	5%
B-40-4	1195.62	2134.73	1.20	5%	743.82	1508.16	2.09	0%
B-40-5	819.28	1537.71	1.11	10%	749.82	1351.64	0.85	5%
C-10-3	359.25	604.70	0.17	55%	597.83	697.51	0.05	0%
C-10-4	307.10	514.66	0.20	5%	408.45	514.65	0.15	85%
C-10-5	566.58	783.35	0.17	10%	409.60	846.51	0.23	10%
C-20-3	650.25	978.84	0.46	10%	629.92	1063.99	0.36	0%
C-20-4	563.78	938.50	0.72	5%	441.65	1019.96	1.06	10%
C-20-5	739.22	1056.77	1.02	0%	711.87	1095.84	0.63	5%
C-30-3	837.58	1198.09	1.05	5%	830.17	1221.65	0.61	10%
C-30-4	754.10	1144.99	2.47	5%	745.92	1096.55	1.16	5%
C-30-5	560.18	998.28	2.29	10%	490.80	931.81	2.21	5%
C-40-3	1008.00	1541.54	2.59	5%	607.00	946.57	3.45	10%
C-40-4	695.30	1172.76	2.06	10%	699.80	1136.17	2.25	0%
C-40-5	623.33	1097.22	3.24	10%	475.67	898.80	7.82	0%

Table 6.
Results of the MS-MMAS executions for small instances.

	Asymmetric				Symmetric			
Instance	**Best**	**Average**	**Time**	**Percentage**	**Best**	**Average**	**Time**	**Percentage**
A-50-3	1058.33	2128.68	2.31	5%	1000.33	1943.13	2.91	5%
A-50-4	774.93	1473.57	4.48	5%	783.40	1345.19	3.53	5%
A-50-5	673.42	1314.54	4.16	5%	583.08	1008.33	2.59	0%
A-100-3	1431.42	2046.96	53.44	5%	1514.08	2292.08	15.66	20%
A-100-4	1456.47	2705.07	23.12	5%	1165.90	1595.91	37.11	5%
A-100-5	1106.17	1778.38	40.57	10%	980.28	1366.09	58.20	5%
A-200-3	2806.75	3610.22	424.60	0%	2793.33	3272.00	257.42	0%
A-200-4	2388.88	3196.15	381.07	0%	2199.45	2807.06	79.32	10%
A-200-5	1753.00	2286.08	1380.26	10%	2086.82	3237.85	55.37	10%
A-500-3	6878.42	7165.39	1641.92	33%	6331.75	7679.65	732.94	10%
A-500-4	5572.42	5637.26	4939.84	0%	5030.48	5080.66	913.82	0%
A-500-5	4389.92	4539.44	16990.77	50%	4610.95	4696.91	73.97	33%
B-50-3	1338.42	2356.24	5.82	10%	966.42	1770.88	4.60	5%
B-50-4	951.87	1757.61	5.00	5%	772.67	1490.01	1.47	5%
B-50-5	1083.18	1943.22	3.91	5%	692.42	1342.78	4.74	5%
B-100-3	1781.00	3115.78	30.96	5%	1803.33	3258.90	15.11	5%
B-100-4	1409.65	2467.02	61.76	5%	1648.58	3360.93	11.00	5%
B-100-5	1361.20	2734.24	39.89	5%	1018.37	1536.34	99.02	5%
B-200-3	3302.83	4840.94	317.37	0%	3016.67	4267.17	56.62	0%
B-200-4	2536.80	3477.13	681.75	0%	2326.97	3139.57	81.34	10%
B-200-5	2127.88	2814.24	906.74	0%	1893.67	2506.64	102.33	10%
B-500-3	6994.84	7203.61	3857.77	0%	6433.92	6475.67	126.51	0%
B-500-4	5419.87	5730.97	24183.87	100%	5191.77	5276.98	267.72	0%
B-500-5	4546.28	4643.03	21118.98	0%	4379.43	4494.11	164.08	33%
C-50-3	1201.92	1801.68	3.12	5%	829.75	1482.86	2.82	5%
C-50-4	937.25	1651.11	7.96	5%	901.40	1605.24	6.16	10%
C-50-5	609.60	1127.99	16.58	5%	766.48	1366.58	7.43	5%
C-100-3	1496.58	2001.76	47.62	0%	1364.00	1819.03	14.21	10%
C-100-4	1352.85	2458.82	32.32	5%	1099.00	1445.29	23.98	0%
C-100-5	1022.70	1466.42	127.11	0%	991.12	1472.23	35.21	5%
C-200-3	2629.00	3197.13	478.90	10%	2510.50	3171.66	65.07	10%
C-200-4	2184.85	2648.38	710.70	10%	2141.40	2741.11	52.59	0%
C-200-5	1881.03	2296.53	486.62	0%	1713.17	2127.84	126.85	10%
C-500-3	6528.08	6618.16	2484.52	0%	6023.42	6087.14	138.12	33%
C-500-4	5139.54	5298.21	8188.46	0%	4942.28	4958.64	65.19	33%
C-500-5	4286.45	4278.49	6061.78	0%	4167.67	4178.10	302.79	0%

Table 7.
Results of the MS-MMAS executions for medium and large instances.

results produced by the experiments were adopted. The first metric, ν, shows the percentages relative to the number of times an ant algorithm found the best-known solution along 20 independent executions. The second metric, Φ, is the relative distance between the cost of the best-known solution χ^* and the best solution χ^{min} of each ant algorithm. The third metric, Ω, is the relative distance between χ^* and the average solution χ^a of each ant algorithm. To calculate Φ, the Eq. (9) was used. Ω is calculated using the formula (10). The average values of ν, Φ and Ω are reported in **Tables 4** and **5**.

$$\Phi = \frac{\chi^{min}}{\chi^*} - 1 \quad (9)$$

$$\Omega = \frac{\chi^a}{\chi^*} - 1 \quad (10)$$

It can be observed from **Tables 4** and **5** that the MS-*MMAS* is the best one concerning the ν and Φ metrics. The MS-ACS is the best algorithm concerning the Ω metric. **Tables 6** and **7** show the data regarding the results of MS-*MMAS* reported in **Tables 4** and **5**. The results of the other ACO variants can be seen in [5].

Tables 8 and **9** present the average processing time (in seconds) spent by each heuristic. Instances are grouped by the number of vertices. From these tables, it can be conclude that the MS-*MMAS* was the ant algorithm that demanded more

n	AS	ACS	MS-ACS	MS-*MMAS*
10	0.05	0.06	0.12	0.15
20	0.10	0.13	0.26	0.46
30	0.20	0.30	1.92	1.37
40	0.34	0.48	2.29	2.29
50	0.41	2.20	5.77	5.92
100	6.68	28.85	32.69	50.75
200	31.81	270.94	409.72	640.89
500	41.72	3477.13	3545.20	9940.87

Table 8.
Average time spent by the ant algorithms for the set of asymmetric instances.

n	AS	ACS	MS-ACS	MS-*MMAS*
10	0.05	0.06	0.12	0.17
20	0.10	0.13	0.27	0.51
30	0.18	0.24	0.49	0.89
40	0.28	0.38	0.73	2.78
50	0.40	0.56	5.41	4.02
100	8.13	13.51	29.17	34.93
200	22.94	51.92	112.58	97.43
500	40.53	75.72	127.83	309.46

Table 9.
Average time spent by the ant algorithms for the set of symmetric instances.

processing time. **Tables 6** and **7** (1) present detailed results concerning the average time required by the MS-*MM*AS. Data regarding the time consumption of the other ACO variants can be seen in [5].

5. Discussion

The purpose of this study was to adapt *MM*AS to the QTSP-PIC and compare its performance with the ACO variants proposed in [5]. As expected, MS-*MM*AS proved to be competitive regarding the other ACO variants proposed to solve QTSP-PIC. Similarities and differences that were observed in the results are discussed in section 5.1. The limitations of the study are discussed in Section 5.2.

5.1 Comparison between ACO algorithms

The ACO algorithms proposed by Silva et al. [5] showed to be a viable method for solving QTSP-PIC. Yet, the performance of AS and ACS algorithms, when compared to MS-*MM*AS, was rather poor for the benchmark set studied. MS-*MM*AS improved the results achieved by AS in 134 instances. Compared to ACS, MS-*MM*AS performed better in 136 instances. The results achieved by MS-*MM*AS improved those produced by MS-ACS in 93 instances. It is interesting to note that the MS-ACS algorithm performed slightly better on the large instances than the MS-*MM*AS. The Friedman test and Nemenyi post-hoc ranked these two ACO algorithms with the same scale for the most instance groups, which means that difference between the results achieved by the MS-ACS and MS-*MM*AS was significantly small.

These observations are also supported by the variability results of each ACO algorithm. Metric ν showed that MS-*MM*AS was the algorithm that achieved the best know solutions of the benchmark set in most cases. Metric Φ showed that the MS-*MM*AS performed slightly better overall on the benchmark set than the MS-ACS. A possible explanation for this is that the MS-ACS variation might converge to a local minimum faster than the MS-*MM*AS.

Results presented in this study showed that the MS-*MM*AS algorithm is better suited than the other three ACO variants proposed in [5] to solve QTSP-PIC. This suggests a positive impact of the implementation design proposed in this study and a contribution to the *MAX-MIN* Ant System state of the art.

5.2 Limitations

Due to limited time, parallel computing techniques could not be tested to improve the performance of MS-*MM*AS. A previous study done by Skinderowicz [10] investigated the potential effectiveness of a GPU-based parallel *MAX-MIN* Ant System in solving the TSP. In this study, the most promising *MM*AS variant was able to generate over 1 million candidate solutions per second when solving a large instance of the TSP benchmark set. Other techniques can improve the MS-*MM*AS design and could not be tested due to the lack of time:

- A rank-based pheromone updating rule [25];
- The application of the pseudo-random action choice rule proposed in [21];
- Hybridization with other meta-heuristics [26–28].

6. Conclusions

This work dealt with a recently proposed variant of the Traveling Salesman Problem named The Quota Traveling Salesman Problem with Passengers, Incomplete Ride, and Collection Time. In this problem, the salesman uses a flexible ride-sharing system to minimize travel costs while visiting some vertices to satisfy a pre-established quota. He must respect the budget limitations and the maximum travel time of every passenger. Each passenger can be transported directly to the desired destination or an alternate destination. The alternative destination idea suggests that when sharing a ride, pro-environmental or money-saving concerns can induce persons to agree to fulfill their needs at a similar destination. Operational constraints regarding vehicle capacity and travel time were also considered.

The Multi-Strategy *MAX-MIN* Ant System, a variant from the Ant Colony Optimization (ACO) family of algorithms, was presented. This algorithm uses the MS concept improved with roulette wheel selection and memory-based principles to avoid redundant executions of the local search algorithm. The results of MS-*MM*AS were compared with those produced by the ACO algorithms presented in [5]. To support MS-*MM*AS, the ride-matching heuristic and the local search heuristic based on multiple neighborhood operators proposed by [5] were reused.

The computational experiments reported in this study comprised one hundred forty-four instances. The experimental results show that the proposed ant algorithm variant could update the best-known solutions for this benchmark set according to the statistical results. The comparison results with three other ACO variants proposed in [5] showed that MS-*MM*AS improved the best results of MS-ACS for ninety-three instances, and a significant superiority of MS-*MM*AS over AS and ACS.

The presented work may be extended in multiple directions. First, it would be interesting to investigate if the application of the pseudo-random action choice rule [20] could improve the MS-*MM*AS results. Another further promising idea is the use of pheromone update rule based on ants ranking [25]. Extension of the MS-*MM*AS implementation design with parallel computing techniques [10] and hybridization with other meta-heuristics [26–28] is other interesting opportunity for the future research.

Abbreviations

MaaS	Mobility as a Service
QTSP-PIC	Quota Traveling Salesman Problem with Passengers, Incomplete Ride and Collection Time
ACO	Ant Colony Optimization
RMH	Ride-Matching Heuristic
MnLS	Multi-neighborhood Local Search
MS-ACS	Multi-Strategy Ant Colony System
*MM*AS	*MAX-MIN* Ant System
MS-*MM*AS	Multi-Strategy *MAX-MIN* Ant System
MS	Multi-Strategy
TSP	Traveling Salesman Problem
QTSP	Quota Traveling Salesman Problem
AS	Ant System
ACS	Ant Colony System

Author details

Bruno C.H. Silva[1,2]*, Islame F.C. Fernandes[2], Marco C. Goldbarg[2] and Elizabeth F.G. Goldbarg[2]

1 Federal University of Ceará No. 1, Fortaleza, Brazil

2 Federal University of Rio Grande do Norte No. 2, Natal, Brazil

*Address all correspondence to: bruno@crateus.ufc.br

References

[1] Mobile Tech Spurs Financial Inclusion in Developing Nations. shorturl.at/bjyOZ [Accessed in: 2020-08-08]

[2] Hensher D., Ho C., Mulley C., Nelson J., Smith G., Wong Y. (2020). Understanding Mobility as a Service (MaaS): Past, Present and Future. h ttps://doi.org/10.1016/C2019-0-00508-0 [Accessed in: 2020-08-08]

[3] Sin C. Ho, W.Y. Szeto, Yong-Hong Kuo, Janny M.Y. Leung, Matthew Petering, Terence W.H. Tou, A survey of dial-a-ride problems: Literature review and recent developments, Transportation Research Part B: Methodological, Volume 111, 2018, Pages 395–421, ISSN 0191–2615.https://doi.org/10.1016/j.trb.2018.02.001 [Accessed in: 2020-08-02]

[4] de Lira, V. M., Perego, R., Renso, C., Rinzivillo, S., Times, V. C. (2018). Boosting ride sharing with alternative destinations. IEEE Transactions on Intelligent Transportation Systems, 19 (7), 2290–2300. https://doi.org/10.1109/TITS.2018.2836395 [Accessed in: 2020-08-02]

[5] Bruno C.H. Silva, Islame F.C. Fernandes, Marco C. Goldbarg, Elizabeth F.G. Goldbarg. Quota travelling salesman problem with passengers, incomplete ride and collection time optimization by ant-based algorithms. Computers & Operations Research, Volume 120, 2020, 104950, ISSN 0305–0548. https://doi.org/10.1016/j.cor.2020.104950. [Accessed in: 2020-08-02]

[6] M. Dorigo, M. Birattari and T. Stutzle, "Ant colony optimization," in IEEE Computational Intelligence Magazine, vol. 1, no. 4, pp. 28–39, Nov. 2006, https://doi.org/10.1109/MCI.2006.329691. [Accessed in: 2020-08-04]

[7] J. Pedro Schmitt, F. Baldo and R. Stubs Parpinelli. A MAX-MIN Ant System with Short-Term Memory Applied to the Dynamic and Asymmetric Traveling Salesman Problem. 2018 7th Brazilian Conference on Intelligent Systems (BRACIS), Sao Paulo, 2018, pp. 1–6, https://doi.org/10.1109/BRACIS.2018.00009. [Accessed in: 2020-08-04]

[8] J. Pedro Schmitt, R. Stubs Parpinelli and F. Baldo. (2019). Analysis of Max-Min Ant System with Local Search Applied to the Asymmetric and Dynamic Travelling Salesman Problem with Moving Vehicle. In International Symposium on Experimental Algorithms (pp. 202–218). Springer, Cham.https://doi.org/10.1007/978-3-030-34029-2$_$14[Accessed in: 2020-08-04]

[9] Yang, K., You, X., Liu, S. et al. A novel ant colony optimization based on game for traveling salesman problem. Appl Intell (2020). https://doi.org/10.1007/s10489-020-01799-w [Accessed in: 2020-08-04]

[10] R. Skinderowicz. Implementing a GPU-based parallel MAX–MIN Ant System, Future Generation Computer Systems,Volume 106, 2020, Pages 277–295, ISSN 0167-739X. https://doi.org/10.1016/j.future.2020.01.011 [Accessed in: 2020-08-04]

[11] S.L. Martins, P.M. Pardalos, M.G.C. Resende, and C.C. Ribeiro. Greedy randomized adaptive search procedures for the steiner problem in graphs. In P. M. Pardalos, S. Rajasekaran, and J. Rolim, editors, Randomization methods in algorithmic design, volume 43 of DIMACS Series on Discrete Mathematics and Theoretical Computer Science, pages 133–145. American Mathematical Society, 1999. https://bookstore.ams.org/dimacs-43. [Accessed in: 2020-08-04]

[12] K. Ilavarasi and K. S. Joseph. Variants of travelling salesman problem: A survey. International Conference on Information Communication and Embedded Systems (ICICES2014), Chennai, 2014, pp. 1–7. https://doi.org/10.1109/ICICES.2014.7033850. [Accessed in: 2020-08-04]

[13] Awerbuch, B., Azar, Y., Blum, A., Vempala, S. New approximation guarantees for minimum-weight k-trees and prize-collecting salesmen. SIAM Journal on computing, 28(1), 254–262, 1998. https://epubs.siam.org/doi/abs/10.1137/S009753979528826X. [Accessed in: 2020-08-04]

[14] Niels Agatz, Alan Erera, Martin Savelsbergh, Xing Wang. Optimization for dynamic ride-sharing: A review, European Journal of Operational Research, Volume 223, Issue 2, 2012, Pages 295–303. https://doi.org/10.1016/j.ejor.2012.05.028 [Accessed in: 2020-08-04]

[15] Schonberger, J., Kopfer, H., Mattfeld, D. C. A combined approach to solve the pickup and delivery selection problem. In Operations Research Proceedings 2003 (pp. 150–155). Springer, Berlin, Heidelberg. https://doi.org/10.1007/978-3-642-55537-4$_$24 [Accessed in: 2020-08-04]

[16] Deneubourg, J. L., Aron, S., Goss, S., Pasteels, J. M. The self-organizing exploratory pattern of the argentine ant. Journal of insect behavior, 3(2), 159–168, 1990. https://doi.org/10.1007/BF01417909 [Accessed in: 2020-08-04]

[17] Hart, Adam; Jackson, Duncan E. U-turns on ant pheromone trails. Current Biology, v. 16, n. 2, p. R42-R43, 2006. https://doi.org/10.1016/j.cub.2006.01.015 [Accessed in: 2020-08-04]

[18] Marco Dorigo, Eric Bonabeau, Guy Theraulaz, Ant algorithms and stigmergy, Future Generation Computer Systems, Volume 16, Issue 8, 2000, Pages 851–871, ISSN 0167-739X. https://doi.org/10.1016/S0167-739X(00)00042-X [Accessed in: 2020-08-04]

[19] Skinderowicz R. Ant Colony System with Selective Pheromone Memory for TSP. Lecture Notes in Computer Science, vol 7654. Springer, Berlin, Heidelberg. 2012. https://doi.org/10.1007/978-3-642-34707-8_49 [Accessed in: 2020-08-04]

[20] M. Dorigo and L. M. Gambardella. Ant colony system: a cooperative learning approach to the traveling salesman problem. in IEEE Transactions on Evolutionary Computation, vol. 1, no. 1, pp. 53–66, April 1997. https://doi.org/doi: 10.1109/4235.585892 [Accessed in: 2020-08-04]

[21] Thomas Stutzle, Holger H. Hoos. MAX–MIN Ant System. Future Generation Computer Systems. Volume 16, Issue 8, 2000, Pages 889–914, ISSN 0167-739X. https://doi.org/10.1016/S0167-739X(00)00043-1 [Accessed in: 2020-08-04]

[22] Manuel Lopez-Ibanez, Jeremie Dubois-Lacoste, Leslie Perez Caceres, Mauro Birattari, Thomas Stutzle. The irace package: Iterated racing for automatic algorithm configuration. Operations Research Perspectives, Volume 3, 2016, Pages 43–58, ISSN 2214–7160. https://doi.org/10.1016/j.orp.2016.09.002 [Accessed in: 2020-08-04]

[23] O'Gorman, Thomas. A comparison of the F-test, Friedman's test, and several aligned rank tests for the analysis of randomized complete blocks. Journal of Agriculture Biology and Environmental Statistics. 6. 367–378, 2001. https://doi.org/10.1198/108571101317096578 [Accessed in: 2020-08-04]

[24] Alan C. Elliott, Linda S. Hynan. A SAS macro implementation of a multiple comparison post hoc test for a Kruskal–Wallis analysis, Computer

Methods and Programs in Biomedicine, Volume 102, Issue 1, 2011, Pages 75–80, ISSN 0169–2607. https://doi.org/10.1016/j.cmpb.2010.11.002 [Accessed in: 2020-08-04]

[25] Q. Song, Q. Zhao, S. Wang, Q. Liu and X. Chen. Dynamic Path Planning for Unmanned Vehicles Based on Fuzzy Logic and Improved Ant Colony Optimization. IEEE Access, vol. 8, pp. 62107–62115, 2020. https://doi.org/10.1109/ACCESS.2020.2984695 [Accessed in: 2020-08-10]

[26] M. Xia. An ant colony algorithm Hybridized with Iterated Local Search for the QAP. 2009 Asia-Pacific Conference on Computational Intelligence and Industrial Applications (PACIIA), Wuhan, pp. 80–83, 2009. https://doi.org/10.1109/PACIIA.2009.5406542 [Accessed in: 2020-08-27]

[27] Li, X., Tian, P., Leung, S. An ant colony optimization metaheuristic hybridized with tabu search for open vehicle routing problems. J Oper Res Soc 60, 1012–1025 (2009). https://doi.org/10.1057/palgrave.jors.2602644 [Accessed in: 2020-08-27]

[28] P.S. Shelokar, Patrick Siarry, V.K. Jayaraman, B.D. Kulkarni. Particle swarm and ant colony algorithms hybridized for improved continuous optimization. Applied Mathematics and Computation, Volume 188, Issue 1, P. 129–142, 2007. https://doi.org/10.1016/j.amc.2006.09.098 [Accessed in: 2020-08-27]

Chapter 6

System of Data Transfer from and to Social and Economic Processes via Creative Economy Networks Created Based on Cultural Heritage Administration Processes and Vice Versa

Jozef Stašák and Jaroslav Mazůrek

Abstract

In general, a functionality and performance of any social or economic process is conditioned by an appropriate information support, while several parts of social and economic process might be involved into macroprocess structure and functionality and quantified via two independent linguistic sets. On one hand, the first linguistic set contains elements, which provide transfer of signals coming from external environment to macroprocesses and are denoted as sensors. On the other hand, the second linguistic set contains elements, which provide transfer of signals from macroprocesses to external environment denoted as effectors. However, a content of the above-mentioned linguistic sets is being transferred microeconomy and creative economy business processes (CE Processes) as well, while appropriate business strategy KPI indicators and parameters for setting of adequate business process metrics items, which could enable fulfillment of business strategy goals and aims, are being generated However, the data transfer from creative economy to macroeconomic process (MAC Processes) is important as well, while the microeconomy business processes (MIC Processes) play a role of go-between elements for both directions of transfer too. On the other hand, the chapter deals with the DTS System structure, functionality description as well as conceptual, design and implementation model too, where appropriate networks play a role of principle importance.

Keywords: data transfer, creative economy networks, cultural heritage administration

1. Introduction

In general, a functionality and performance of any social or economic process is conditioned by an appropriate information support, while several parts of social and economic process might be involved into macroprocess structure and functionality and quantified via two independent linguistic sets. On one hand, the first linguistic set contains elements, which provide transfer of signals coming from external environment to macroprocesses and are denoted as **sensors.** On the other hand, the second linguistic set contains elements, which provide transfer of signals from macroprocesses to external environment denoted as **effectors**. However, a content of the above-mentioned linguistic sets is being transferred **microeconomy** and **creative economy** business processes (CE Processes) as well, while appropriate business strategy KPI indicators and parameters for setting of adequate business process metrics items, which could enable fulfillment of business strategy goals and aims, are being generated However, the data transfer from creative economy to macroeconomic process (MAC Processes) is important as well, while the microeconomy business processes (MIC Processes) play a role of go-between elements for both directions of transfer too. On the other hand, the chapter deals with the DTS System[1] structure, functionality description as well as conceptual, design and implementation model too, where appropriate networks play a role of principle importance.

The presented chapter is divided into nine sub-chapters, where the first two of then deal with economic system and economic object and contain a brief description of macroprocesses (MAC Process), microprocesses (MIC Processes) and creative economy (CE) processes, where the process denoted as Cultural Heritage Creation and Management plays a role of principle importance (see also Section 4) and the DTS System plays a role of the core process and is being discussed within Sections 5 and 6. However, the DTS System provides a bi-directional data transfer as well, while the first transfer direction is getting started from CE processes to MAC and the second one starts from MAC and is finished at CE Processes as well (see also Sections 7 and 8), while the MIC Processes play a role of the data transfer mediator. The chapter is being closed by Section 9, which deals with DTS System implementation aspects, while an appropriate economic network seems to be a principal facility applied for those purposes (see also Section 9).

2. Economic system and economic object

The creative economy creates an integral part of standardized economy represented by micro and macroeconomic processes implemented and operated within adequate economic system and object, while any firm or company is considered to be the economic object. On the other hand, any economic object might be identified with managed, management system and information system while any system is represented by external and internal structure and between both structure types exists a zone denoted as a **grey zone**, where a set of sensors and effectors is located. The sensors provide transfer of signals from external environment to the investigated system internal structure, and the effectors provide signal transfer from the investigated system internal structure to external environment [1].

The Data Transfer System (hereinafter as DTS System) *external structure* is represented by set of social and economic processes surrounding any *creative economy system*,[2] while an appropriate creative economy system consists of pre-defined business processes closely related to cultural heritage creation and management in most cases, while the DTS internal structure also is represented by microprocesses, which play a role of intermediate element between creative processes and macroprocesses.

---

[1] DTS System–Data Transfer System.

[2] The creative economy system creates an integral part of the DTS system internal structure.

3. Macroprocesses, microprocesses and creative economy processes

When considering the cultural heritage creation process, the microeconomic business processes represent those types of processes, which enable producing of products denoted as cultural heritage artifacts. That production is realized within firm or company, which has its own business strategy. With respect to Balanced Scorecard Methodology [2], the strategy includes five perspectives: (a) economic perspective, (b) perspective of internal business processes, (c) customer's perspective, (d) education and growth perspective, and (e) technological perspective. The perspectives related to (a), (b), (d), and (e) are closely related to **production of cultural heritage artefacts**, the customer's perspective and financial perspectives are partially concerned with **cultural heritage management** and with **creative economics** as well. When looking at creative economics from financial perspective, its integral part is created by allocated resources[3] and generated outputs (assets), which might of material, financial, or intellectual nature too [3]. However, the microeconomic business processes represent an intermediate facility at transfer of data and information from creative economics BP to macroeconomic busines processes as well, while the same is concerned with data or information transfer from macroeconomic busines processes to creative economics business processes too.

4. Cultural heritage creation and management

4.1 Creative business process structure and functionality

The business processes denoted as **Cultural Heritage Creation** and **Cultural Heritage Management** create an integral part of any Cultural and Historical Process (hereinafter known as CHP Process). The CHP process seems to be a core business process (BP), when considering the data transfer from creative BP to micro and macro business processes.[4] However, the term cultural and historical process consists of two relatively individual main processes denoted as **culture** and **history** as well, while the cultural heritage seems to be an important output related to both above-mentioned processes. When looking at literary resources, you might find many semantic interpretations related to those terms [3].

In general, the business processes denoted as **Cultural Heritage Creation** and **Cultural Heritage Management** are affected by three types of economic processes: (a) **macroeconomic processes,** (b) **microeconomic processes**, and (c) **creative economics processes**. A need of the cultural heritage creation and management is being derived based on *macroeconomic* processes and might be interpreted as a *public order or demand* and it creates basis for appropriate resources procurement. On the other hand, the *microeconomic processes* might initiate a process of the cultural heritage creation and that processes is running within actual firm or company and are represented by its own vertical and horizontal structure, metrics and information support as well, while those aspects determinate its functionality and performance. The third group of business processes is closely related to **creative economics.** However, the culture heritage creation and management data and information are being transferred to appropriate microprocesses and macroprocesses subsequently [3].

---

[3] Those resources represent a decisive input import by macroeconomic processes, first. They might be of material, financial, or intellectual nature.

[4] The terms process and business process have the same semantic meaning within that chapter.

On one hand, the Cultural Heritage Creation and Cultural Heritage Management processes are running within different cultural institutions and organizations, while museums and galleries play a role of principal importance there and should be managed. There might be applied conventional and advanced approaches to Cultural Heritage Management. There might be applied various advanced approaches, while one of them can be based on Knowledge Management. This approach has been applied, when preparing the exhibition *Uchovávanie sveta* (Preserving the World) installed in the Slovak National Gallery (SNG) from 16 December 2015 to 28 February 2016. The project was the first larger attempt to exploit digitized art collections systematically and massively in CEDVU – Central Evidence of Visual Art Items – the outcome of the national project *Digital Gallery* funded by European Union Operational program Information Society

The success of the exhibition revealed that digitized art can be efficiently and effectively combined with presentations of original artefacts and bring synergy effects highly appreciated by its audience. The paper represents an attempt to make relevant conclusions from this experiment that might help to organizers to build similar innovative events in the future [4].

4.2 A taxonomy of performance dimensions and levels of analysis in the creative industries

A taxonomy of performance closely related to Creative Economy identifies three fundamental forms of capital observable in the financial assets, access to liquidity or monetary income of its owner. On the other hand, the cultural capital manifests itself as long lasting dispositions for ownership of academic knowledge, skills, cultural goods as well, while the social capital is defined as the resources accumulated through belonging to durable network of institutionalized relationships. The symbolic capital transcends economic, cultural, and social capital as a form of prestige bestowed upon its owner based on recognition of her legitimate competence and authority [5–9].

The above-mentioned four categories are closely related to performance in the creative industries. All creative industries display a significant ration of "symbolic content" to functional usage, which also varies from one to the next. As a result of that, the creative outcomes derive a large part of their value from subjective experiences that rely heavily on using symbols to manipulate emotions and perceptions [10–12]. Cultural goods are non-material goods directed to public of consumers, for whom they generally serve as an aesthetic or expressive rather than clearly utilitarian function [6, 11, 13–16].

Economic, cultural, and social capitals are all reflected in the following three core dimensions of performance in the creative industries: commercial performance, artistic merit, and social impact. The first two core dimensions, commercial performance and artistic merit represent key components of creative industry performance research [11, 17, 18]. However, those components usually have long and short term nature Whilst the former directly reflects the notion of national capital, the latter is close in its definition to that of "symbolic cultural capital" as the capacity to define and legitimize cultural and artistic values, standards and styles" [6, 19].

This taxonomy is hereby applied to all relevant levels of performance analysis in the creative industries. They create two distinct categories within existing literature. The first is being focused on creative production processes and investigates roles and contributions of the individual worker, the creative project team and the creative group assigned to this process. The second deals with creative outcome, its distribution and consumption. It examines the creative project, the creative organization, and the creative network of organizations from one-off event (e.g. an annual festival) to semi-permanent regional cluster and industry [6]. However, the above-mentioned description creates a good basis for further research with DTS System as well, while many of those aspects are being applied and developed within Section 7 of that chapter.

4.3 Creative business process linguistic modelling: structure and functionality

The Business Process Linguistic Modelling (hereinafter known as the BPLM approach) is working with the use of semantic networks and so-called reference databases, while is based on two principle issues [20]:

Issue no. 1 Linguistic Representation of BP Functions

A structure and functionality of any BP might be described via TNL text, which consists of common logical sentences, while any logical sentence consists of text strings [Ts (I, j)] i = 1… .n, (serial number of logical sentence contained within TNL text) and j = 1… .m1 (serial number of the text string contained within logical sentence).

Each text strings Ts (I, j) has its own semantic meaning assigned via [Tsem (i, j)], while formulas (1)–(3) might be postulated

$$\{\text{Word}\ (i,j)\} = \{[\text{Ts}\ (i,j)], [\text{Tsem}\ (i,j)]\} \tag{1}$$

$$\{\text{Words}\ (i,j)\} = \Pi\ \{\text{Word}\ (i,j)\} \tag{2}$$

$$\{\text{Words}\ (i,j)\} \subset \{\text{TNLtext}\ (i,j)\} \tag{3}$$

Where

The {Word (i, j)} set represents linguistic representation facility denoted as the linguistic set, while its content is created by two subsets:

- [Ts (i, j)] – which contains text strings being created as a result of the first stage related to TNL text content semantic analysis, however those strings have not yet adequate semantic meaning and they might be of a text or numeric nature, e.g. [Ts (i, j)] = [1500 centigrade, IA poise, 25 grams]

- However, they get it after assignment of adequate text string selected from semantic dictionary quantified via [Tsem (i, j)] subset as well. while the [Tsem (i, j)] subset content is represented so that Tsem (i, j)] = [glass melt temperature, glass melt viscosity, glass melt quantity].

- After unification of both subsets with respect to formula (1) a final content of the {Word (i, j)} = {[glass melt temperature, 1500 centigrade], {[glass melt temperature, IA poise], {[glass melt quantity, 25 grams]}

- A content of [Tsem (i, j)] is created by linguistic variable items and a content of [Ts (i, j)] is created by linguistic variable values. Therefore, the {Word (i, j)} is denoted as the linguistic set.

However, we need an appropriate set such words, to describe any business process (Pe) structure functionality, while the {Pe (i, j)} set might be applied for those purposes. Because, the words quantified via sets represented by formulas (1)–(3) and they have a linguistic character, they will be considered to be linguistic sets and they will be applied business process (Pe) structure functionality. It means, the {Pe (i, j)} might be approximated via {Words (i, j)} set and formula (4) might be postulated

$$\{Pe\ (i,j)\} \approx \{Words\ (i,j)\} \tag{4}$$

{Pe (i, j)} – is a linguistic set, which quantifies a business process closely related to linguistic variable items and values contained in the {Words (i, j)} linguistic sets.

On the other hand, any business horizontal structure is being created by set functions, which generate pre-defined outputs based on appropriate inputs, while formula might be postulated

$$\{Pe\ (i,\ j)\} = \{[F1\ (i,\ j)],\ [F2(i,\ j)] \ldots .[Fn\ (i,\ j)]\} = \prod_{k=1}^{m}[F_k(i,\ j)] \tag{5}$$

However, those functions may be described via {Words (i, j)} linguistic sets as well, while formula (6) can be postulated

$$\forall [Fk\ (i,j)] \exists \{Wordsk\ (i,j)\} \Rightarrow [Fk\ (i,j)] = \{Wordsk\ (i,j)\} \tag{6}$$

It means, any business process (BP) function can be approximated via set of words, which creates an integral part of TNL text. This is the first important principle of BP modeling linguistic approach. However, there is the second important issue as well Any BP function set consists of three principal subsets [Object (i, j)], [Action (i, j)] and [Result (i, j)] see also formula (7)

$$\{[Fk\ (i,j)]\} = \{[Object\ (i,j)], [Action\ (i,j)], [Result\ (i,j)]\} \tag{7}$$

and the [Action (i, j)] set elements "are responsible for" generation of pre-defined BP outputs based on appropriate BP inputs. The [Object (i, j)] subset elements provide interconnection to BP input set and the [Result (i, j)] subset elements provide interconnection to BP output set.

Issue No. 2 Linguistic Representation of BP Outputs and Inputs

In general, no BP can generate required pre-defined outputs without appropriate inputs. As a result of that, they must be defined and quantified very precisely. Because of considering the BP modeling linguistic approach, they have to be postulated via linguistic sets as well, denoted as {Petx (i, j')}, where i = 1 ... n and has the same meaning like in the case of BP linguistic sets and j' is a serial number of linguistic subset {Petx (i, j')}[5] consists of (see also formula (8))

$$\{Petx\ (i,j')\} = \{[Petx\ (i,1)], [Petx\ (i,2)], [Petx\ (i,m2)]\} \tag{8}$$

When applying business process {Pe (i, j)}[6] linguistic set to {Petx (i, j')} linguistic set, with respect to formulas (9)–(11), we can get adequate BP functionality results in form of {Res1 (i, j")} set

$$\{Petx\ (i,j')\} \otimes \{Pe\ (i,j)\} = \{Res1\ (i,j")\} \tag{9}$$

where j" =1 ... m_3

However, the {Res1 (i, j")} linguistic set represents two type of business process outputs: primary and secondary outputs, while the {Tbex (i, j"') linguistic set elements represent BP functionality primary products and {Retx (i, j"") linguistic

[5] The {Petx (i, j') linguistic set contains elements closely related to BP inputs.

[6] The {Pe (i, j)} – linguistic set contains elements closely related to BP transformation operands and BP metrics.

set elements represent BP functionality secondary products and formula (10) might be postulated

$$\{\text{Res2 (i, j")}\} = \{\text{Tbex (i, j"')} \otimes \{\text{Retx (i, j"")} \tag{10}$$

and

$$\{\text{Res2 (i, j")}\} = \{\text{Res1 (i, j")}\} \tag{11}$$

Formulas (9)–(11) create basis of Principle Businesses Process Linguistic Modeling Equation (hereinafter known as PBPL Equation [21]. This equation has an endless number of solutions and any of them is closely related to the actual problem area solution.

4.4 Business process transformation operands and metrics

4.4.1 Business process transformation operands

The transformation operands play an important role when providing conversion of process inputs into pre-defined outputs and might be quantified adequate linguistic sets {[TOP1 (i, j_1)]}, {[TOP2 (i, j_2)]} [22]. A linguistic set denoted as {[TOP1 (i, j_1)]} is a set, the elements of which contain data concerned with the *transformation functions* and [TOP2 (i, j_2)] is a set, the elements of which contain data concerned with the *transformation tools.* However, the transformation functions contain the following elements: (a) inspiration, (b) definition, (c) research, (d) production, (e) approval, and (f) realization as well, while the transformation tools are represented by: (a) words, images and sound effect, (b) drawings, (c) conceptual models and (d) implantation models [3].

4.4.2 Business process metrics

A business metric is any type of measurement used to gauge some quantifiable component of a company's performance, such as return on investment (ROI), employee and customer churn rates, revenues, and EBITDA. Business metrics are part of the broad arena of business intelligence, which comprises a wide variety of applications and technologies for gathering, storing, analyzing, and providing access to data to help enterprise users make better business decisions.

The core parts of metrics include: (a) measurement units, (b) reporting period (c) reporting frequency, (d) the current value of the metric with the latest data, (e) previous values of metrics and (f) trend – this is the change in value over time when comparing it with the actual value to previous values[7] [24].

On one hand, the business provided in any firm or company is identified by a set of business processes, which are running there. On the other hand, any business process (BP) is represented by its own vertical and horizontal structure, while both structure types might be quantified via adequate linguistic sets and the {Pe (i, j)} linguistic set consists of two subsets {Pe_1 (i, j_1)} and {Pe_2 (i, j_2)}. The {Pe_1 (i, j_1)} linguistic set consists of further subordinated set {[TOP1 (i, j_1)]} concerned with BP transformation operands and {[TOP2 (i, j_2)]} one is concerned with business process transformation tools.

However, the {Pe_2 (i, j_2)} linguistic set is concerned with business process metrics issues as well, while two types of BP metrics might be postulated:

---

[7] IT Manager's Guide Metrics TechExcel [23].

(a) external metrics and (b) internal metrics. When quantifying both metrics types via linguistic sets formula (12) might be postulated

$$\{Pe_2\ (i, j_2)\} = \{[EM\ (i, j_3)], [IM\ (i, j_4)]\} \tag{12}$$

Where

[EM (i, j_3)] – is a linguistic set, which represents BP external metrics

[IM (i, j_4)] – is a linguistic set, which represents BP internal metrics

Both the above-mentioned linguistic sets seem to be independent and they might be postulated as individual linguistic sets {[EM (i, j_3)]} and {[IM (i, j_4)]}. The {[EM (i, j_3)]} linguistic set represents a BP external metrics, which is concerned with BP inputs [IMP ($_I$, j_{3a})], and BP outputs [OUTP (i, j_{3b})]}. The {[IM (i, j_4)]} linguistic set represents a BP internal metrics, which is concerned with BP production devices [DEV (i, j_6)], BP production tools [TOOL (i, j_7)], and BP human resources [HR (i, j_8)] [22] With respect to the above-mentioned issues the following formulas might be postulated

$$\{[EM\ (i, j_3)]\} = \{[IMP\ (_I, j_{3a})], [OUTP\ (i, j_{3b})]\} \tag{13}$$

$$\{[IM\ (i, j_4)]\} = \{[DEV\ (i, j_6)], [TOOL\ (i, j_7)], [HR\ (i, j_8)]\} \tag{14}$$

However, that representation of business process metrics will be applied within Section 7 of that chapter as well.

5. DTS-system: general overview

A System of Data Transfer from Creative Economy business processes to Micro and Macro processes and vice versa based on Cultural Heritage Processes (hereinafter known as Data Transfer System—DTS-System) represents a complex system, which should provide an appropriate information and knowledge-based support for business processes running within Creative Economy System (CE System), System of Microprocesses (MIC System) and System of Macroprocesses (MAC System), which create an integral part of social and economic processes as well, while a Culture Heritage Creation and Management plays a role of principle importance in the above-mentioned DTS System. On the other hand, the DTS System should provide a bi-directional data transfer: (a) Data Transfer from Creative Economy business processes to Micro and Macro processes and (b) Data Transfer from Macroprocesses (MA Processes) to Creative Economy business processes (CE Processes), while the MIC System plays a role of intermediator agent as well. The entire DTS system should be implemented and operated via Economic Network System (EN System), which consist of three EN subsystems.

6. DTS-system: structure and functionality

6.1 External and internal structure

The DTS system is represented by its own *internal* and *external* structure and both structure types are separated via adequate *grey zone* (see also Section 2). The DTS System external and internal structure, incl. adequate grey zone is shown **Figure 1** and create basis for design of DTS System conceptual, logical and implementation model.

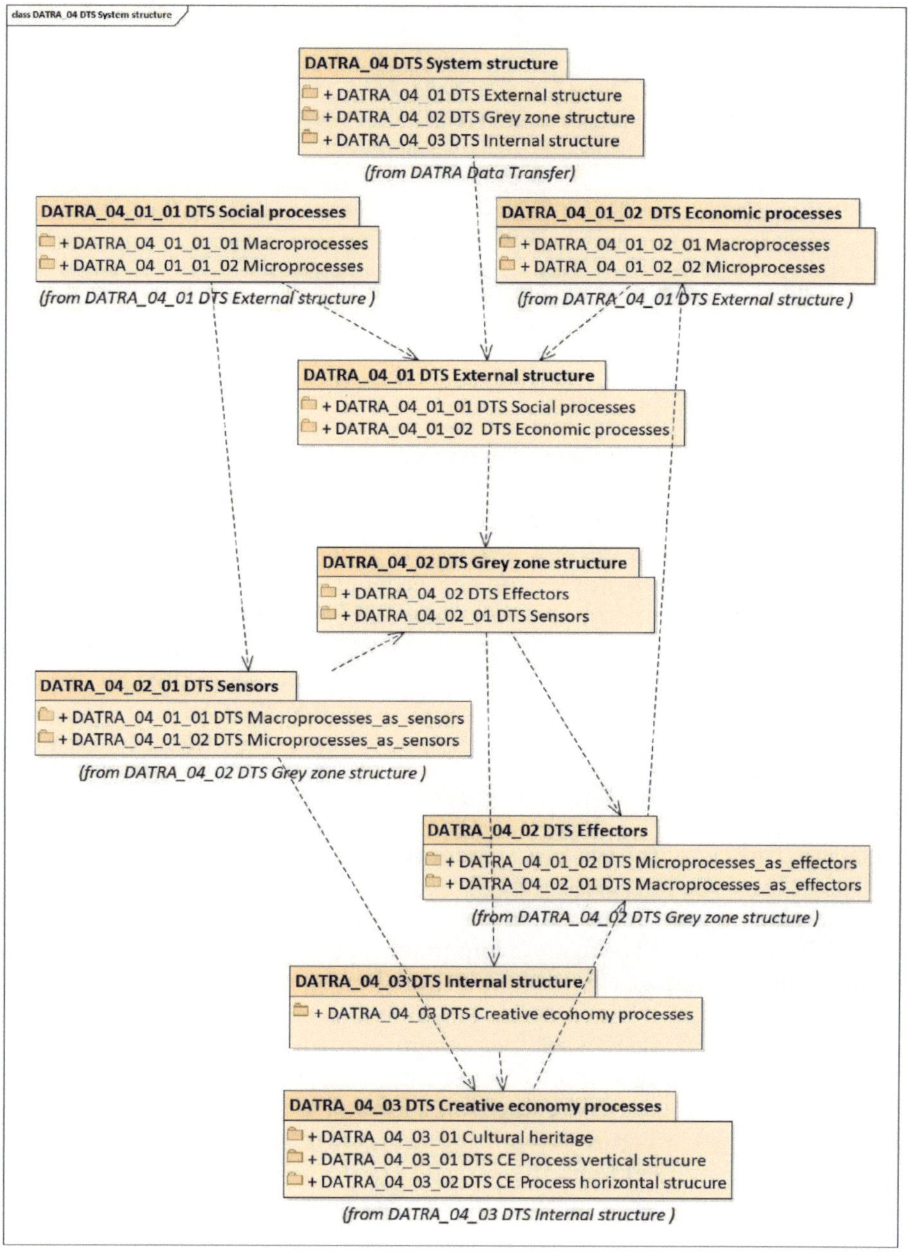

Figure 1.
Layout of the DTS system external and internal structure. Source: The Authors.

The DTS system is represented by its own *internal* and *external* structure and both structure types are separated via adequate *grey zone* (see also Section 2). The DTS System external structure is created by social and economic processes, while both of them generate data or information for macroprocesses (MAC Processes), which provide further data or information for microprocesses (MIC Processes) and creative economy processes (CE Processes). However, the MAC Processes contain elements, which enable data transfer from external environment to MAC processes (sensors) and from MAC processes to external environment (effectors) as well. The same is concerned to MIC and CE Processes).

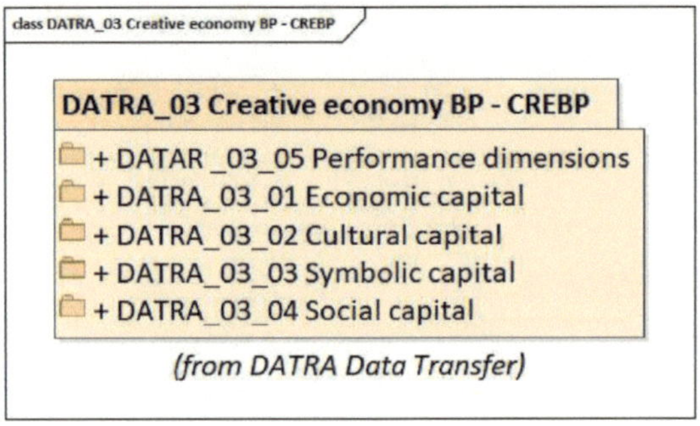

Figure 2.
CEPS system linguistic set objectives. Source: The Authors.

6.2 The creative economy process system structure

In general, any system of business processes might have its own vertical and horizontal structure, while the same is concerned with wit business processes, which create an integral part of the creative economy system. With respect to the fact, that the creative economy process system (CEPS) described within that contribution contains a set of business processes closely related to cultural heritage production and management, that system vertical structure is created by two business process: (a) Cultural heritage creation - production (b) (CHCP) and Cultural heritage management (CHMP). The CEPS system horizontal structure is created by elements shown in **Figure 2**.

6.3 The creative economy process system content

The previous section deals with CEPS system horizontal and vertical structure elements, however there is one more important aspect, which is closely related to content of culture heritage creation and culture heritage management processes, which play a role of principle importance within Creative Economy System to be investigated. However, there is another component, which create the CEPS system content as well, while any of those components has its own structure elements and those elements are denoted as the linguistic sets[8] and are concerned with appropriate objectives shown in **Figure 3**, while the structure of appropriate subordinated linguistic sets is shown in **Figure 4**.

In general, any process or business process horizontal structure is created by appropriate business process functions (BPFs), which might be quantified via adequate linguistic sets, which enable quantifying the BPFs information and knowledge-based support closely related to a given business process. When looking at **Figure 2** you can see the creative economy business process functions DATRA_03_01 up to DATRA_03_05 and the above-mentioned linguistic sets postulated as {[Performance dimensions (i. j)]}, {[Economic capital (i. j)]}, {[Cultural capital (i. j)]}, {[Symbolic capital (i. j)]}, {[Social capital (i. j)]}.

In **Figure 2**, there are specified BPFs and appropriate linguistic sets closely related to those BPFs information and knowledge-based support, however in **Figure 3** the reader might see the specification in more details as well.

[8] The problems of linguistic sets mentioned in that section are explained within Section 4.3.

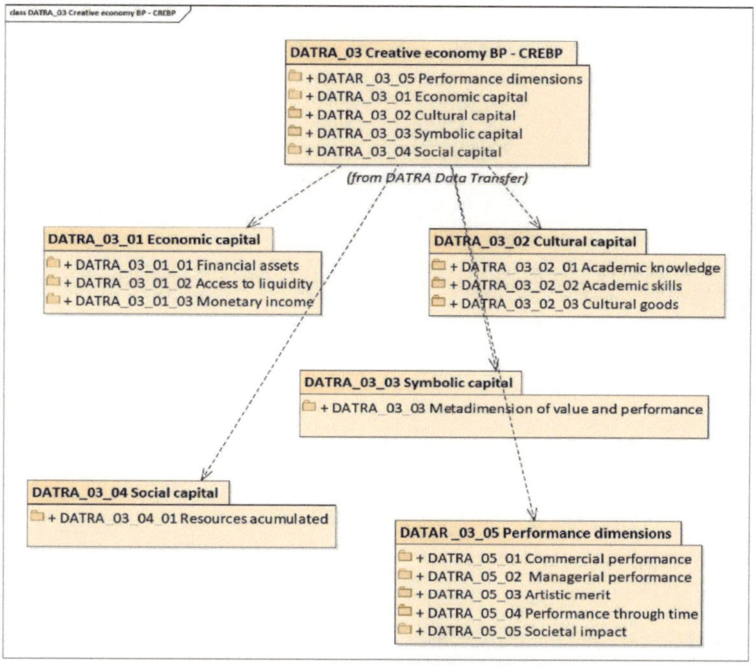

Figure 3.
CEPS system linguistic set objective subsets. Source: The Authors.

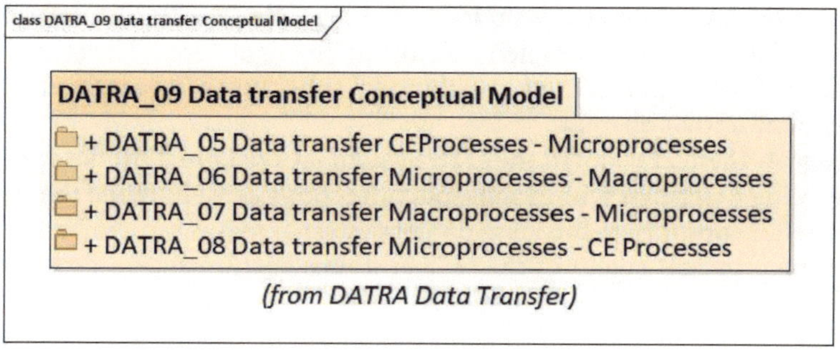

Figure 4.
The DTS System – Conceptual Model. Source: The Authors.

7. Conceptual model

7.1 General overview

The DTS System – Conceptual Model is a standardized system denoted as DATRA_09 Data Transfer Conceptual Model, which should provide a bi-directional data transfer. On one hand, from Creative Economy Business Processes (CEP Processes) to Microprocesses (MIC Processes) and (MIC Processes) and from Microprocess to Macroprocesses (MAC Processes) and on the other hand, we have to consider the data transfer from macroprocesses, throughout microprocesses up to CEP Processes as well, while DTS System – Conceptual Model consists of four subsystems as shown in **Figure 4**. Moreover, the individual subsystems include sets of adequate components will be written within further sections of that chapter.

However, the DTS System contains a set of adequate business processes as postulated within **Figure 4** as well, while a business process modeling linguistic

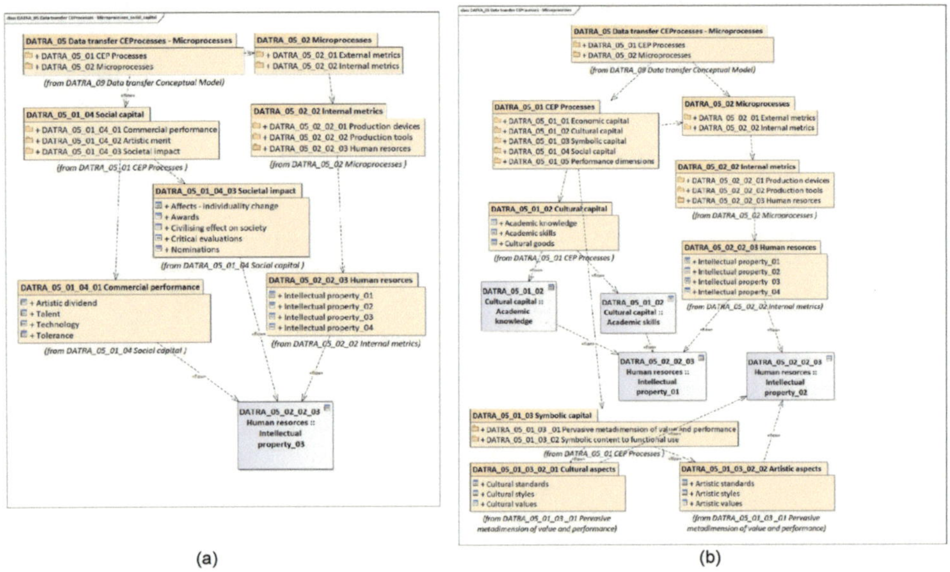

Figure 5.
(a) The principal layout of Data transfer between CE Process and MIC Process internal metrics items Part 1 (cultural and symbolic capital) and intellectual capital part 2. Source: The Authors. (b) The principal layout of Data transfer between CE Process and MIC Process internal metrics items Part 2 (social capital) and intellectual capital part 3. Source: The Authors.

approach (hereinafter known as BPLM Approach) should be applied to quantify the data transfer from CEP Processes up to MAC Processes. The BPLM Approach is based on specialized type of sets denoted as the linguistic sets.[9] The structure and functionality of data transfer conceptual model subsystems is described within further sections of that chapter

In **Figure 4**, there are shown the DTS System – Conceptual Model subsystems, which are being described within subsequent sections as well, while the data transfer related to an appropriate business process internal metrics is shown in **Figure 5a** and **b** and the principal layout of the DTS System structure and bi-directional functionality is shown in **Figures 6** and 7.

7.2 The subsystem DATRA_05 – CE processes versus microprocesses: components and modules

When investigating the subsystem DATRA_05 structure and functionality, we must consider CE Processes and MIC Process structure and metrics, on the other hand, the MIC Process functions might be defined generally regardless their content, while CE Processes and MIC Process metrics plays a role of principal importance within that subsystem functionality description.

7.2.1 The linguistic sets closely related to CE Processes and MIC Process internal and external metrics

Now, we shall postulate the linguistic sets closely related to internal and external metrics[10] of CE Processes and MIC Process, while the CE Process external metrics items are represented by the following linguistic sets: access to liquidity {ALI(i, j)]},

---

[9] The linguistic set structure and functionality is being discussed within Section 4.3.

[10] The problems of BP internal and external metrics are explained in Section 4.3.

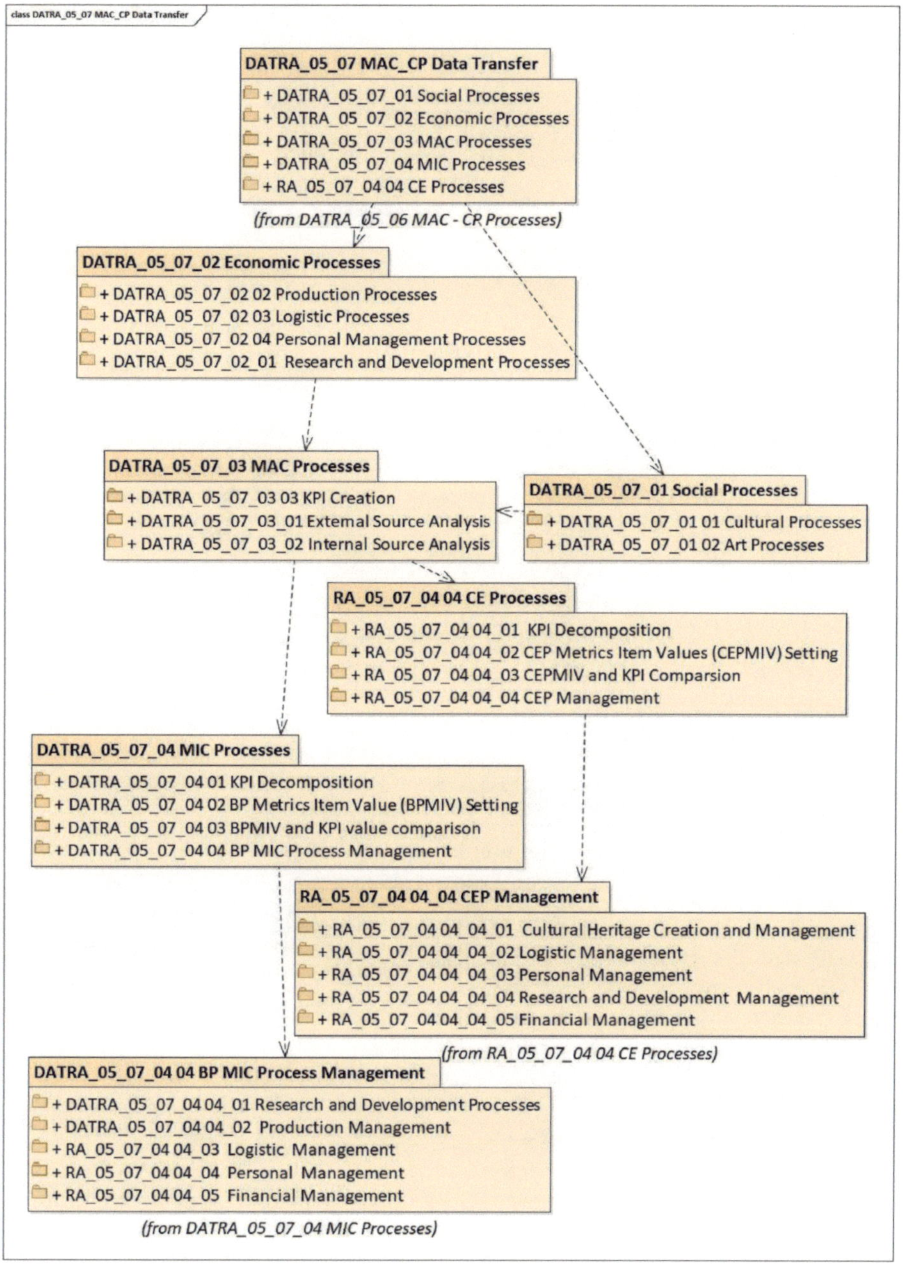

Figure 6.
The principle layout of DTS2 System Conceptual Model. Source: The Authors.

financial assets, {FAT(i, j)]}, monetary income {MOI(i, j)]} and the CE Process internal metrics items are represented by those linguistic sets cultural capital {[CULC (i, j)]}, symbolic capital {[SYMC (i, j)]}, social capital {[SOCC (i, j)]}, and performance dimensions {[PEDC (i, j)]. On the other hand, the MIC Process external metrics items are postulated as follows: material input {[MI (i, j)]}, material input cost {[MICS (i, j)]},, output material {[OMP (i, j)]}, products and output material product assets {[OMPA (i, j)]}, while MIC Process external metrics items are represented by production device {[PDEV (i, j)]} linguistic set, by production tool{[PTOOL (i, j)]} linguistic set, and human resources {[HR (i, j)]} linguistic set.

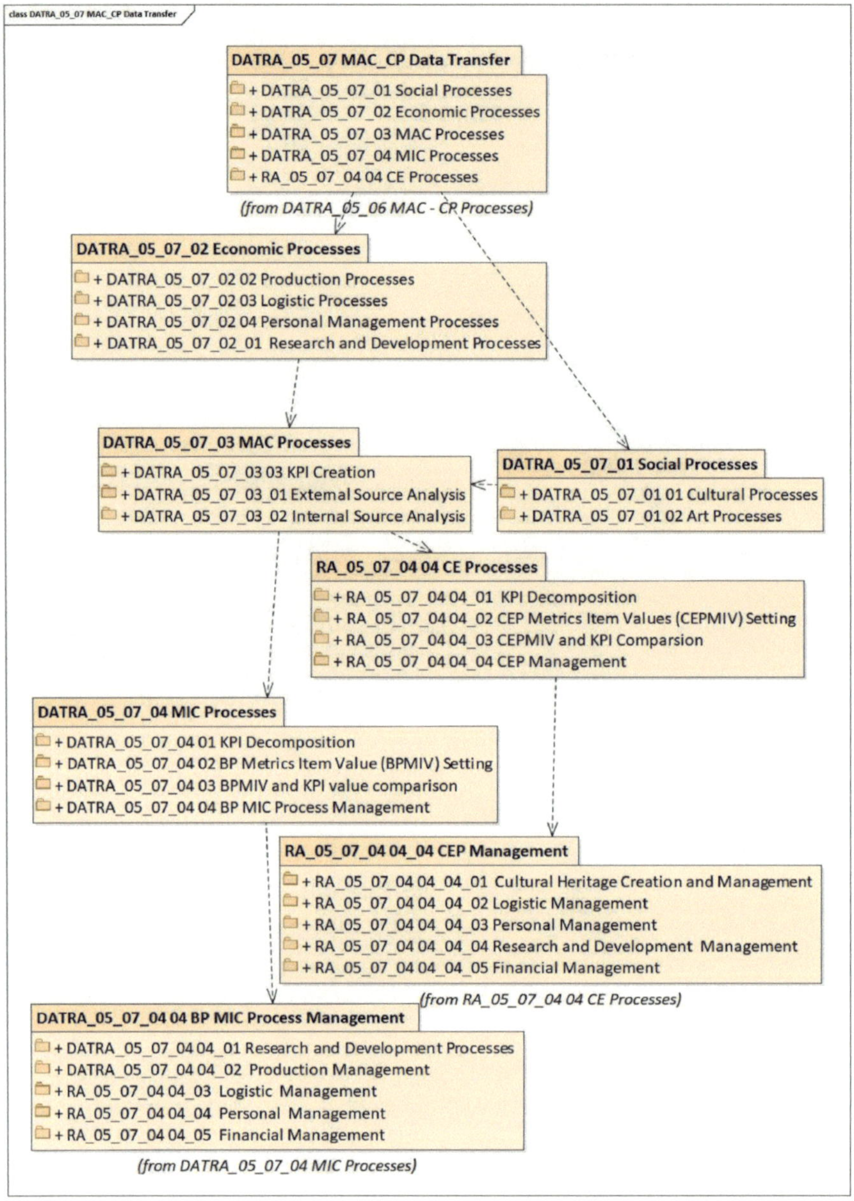

Figure 7.
Data transfer from MAC to MIC and CE processes. Source: The Authors.

The human resources {[HR (i, j)]} linguistic set contains three subordinated sets: (a) intellectual capital 1 – (ICA1 (i, j)), (b) intellectual capital 2 – (ICA2 (i, j)) and (c) intellectual capital 3 – (ICA3 (i, j)) linguistic set.

7.3 Data transfer between CE Process and MIC Process external metrics items

7.3.1 Economic and cultural capital versus MIC Process external metrics item

Now, we shall investigate the data transfer between BPFs denoted as economic capital {[EC (i, j)]} and cultural capital {[CULC (i, j)]} within CE Process items: access to liquidity {ALI(I, j)]}, financial assets FAT(I, j)]}, monetary income

{MOI(I, j)]}, {[CULG(I, j)]}, and MIC Process external metrics items postulated as: material input {[MI (i, j)]}, material input cost {[MICS (i, j)]},, output material {[OMP (i, j)]}, products and output material product assets {[OMPA (i, j)]}.

The data transfer between Economic and cultural capital within CE Process and MIC Process external metrics items is running in the following steps:

- The CE Process linguistic set{[EC (i, j)]: [ALI(I, j)]} content determines a possibility of the cultural heritage production or acquisition initiation and is closely related {[MICS (i, j)]} set within MIC Process.
- The CE Process linguistic set {[EC (i, j)]: [FAT (I, j), MOI (I, j)]]} content indicates financial aspects of cultural heritage artefact production or acquisition financial contribution within creative economy activities.
- The CE Process linguistic set {[CULC (i, j)]: [CULG (i, j)]} content indicates of cultural heritage artefact production or acquisition material contribution within creative economy activities.

7.4 Data transfer between CE Process and MIC Process internal metrics items

7.4.1 Cultural capital, symbolic and social capital incl. performance dimensions versus MIC Process internal metrics items

Now, we shall investigate the data transfer between BPFs, the internal metrics of which is represented by linguistic sets denoted as cultural capital {[CULC (i, j)]}, symbolic capital {[SYMC (i, j)]}, social capital {[SOCC (i, j)]}, and performance dimensions {[PEDC (i, j)]}within CE Process items and MIC Process internal metrics items postulated as follows:

$$\{[\mathrm{IM}\ (i,j)]\} = \{[\mathrm{DEV}\ (i,j)], [\mathrm{TOOL}\ (i,j)], [\mathrm{HR}\ (i,j)]\} \tag{15}$$

$$\{[\mathrm{HR}\ (i,j)]\} = \{[\mathrm{INP1}(i,j)], [\mathrm{INP2}(i,j)], [\mathrm{INP3}(i,j)]\} \tag{16}$$

7.4.2 Description of structure related cultural capital, symbolic and social capital incl. performance dimensions structure

The **cultural capital linguistic set {[CULC (i, j)]}** consists of two subordinated sets [CULC1 (i, j)], the content of which is concerned with cultural goods represented by [CULGOODS (i, j)] and closely related to external metrics subset {[OMP (i, j)]}, while the [CULC2 (i, j)] contains further subordinated sets concerned to academic knowledge (ACKN (i, j)), and academic skills (ACSK (i, j)), while the [CULC2 (i, j)] subset content is closely related to intellectual capital 1 – (ICA1(i, j)) linguistic subset existing within MIC processes.

The **symbolic capital linguistic set {[SYMC (i, j)]}** consists of two subordinated sets: (CULTAP (i, j)) – cultural aspects and (ARTAP (i, j)) – artistic aspects, while the (CULST (i, j)) linguistic set includes further subordinated sets postulated as follows: (a) cultural standards (CULTSTA (i, j)), (b) cultural styles (CULTSTY (i, j)), and cultural values (CULTVAL (i, j)).

However, the (ARTAP (i, j)) – artistic aspect linguistic set includes further subordinated sets postulated as follows: (a) artistic standards (ARTTSTA (i, j)), (b) artistic styles (ARTTSTY (i, j)), and artistic values (ARTVAL (i, j)) as well, while the {[SYMC (i, j)]} subset content is closely related to intellectual capital 2 – (ICA2(i, j)) linguistic subset existing within MIC processes.

The principal layout of Data transfer between CE Process and MIC Process internal metrics items Part 1 (cultural and symbolic capital) and intellectual capital part 2 is shown in **Figure 5a**.

The **social capital linguistic set {[SOCC (i, j)]}** consists of three subordinated sets: (a) commercial performance (COMPER (i, j)), (b) artistic merit (ARTMER (i, j)), and societal impact (SOCIMP (i, j)) linguistic set, while all the above-mentioned linguistic sets contain

Further subordinated linguistic sets postulated as follows:

- (COMPER (i, j)
    - Artistic dividend (ARTDI (i, j))
    - Talent (TALEN (i, j))
    - Technology (TECHN (i, j))
    - Tolerance (TOLER (i, j))
- (ARTMER (i, j))
- (SOCIMP (i, j
    - Affects individuality change (ICAF (i, j))
    - Awards (AVAR (i, j))
    - Civilizing effect on society (CEOS (i, j))
    - Critical evaluations (CEVA (i, j))
    - Nominations (NOMI (i, j))

and the {[SYMC (i, j)]} subset content is closely related to intellectual capital 3 – (ICA3 (i, j)) linguistic subset existing within MIC processes.

The **performance management linguistic set {[PERFM (i, j)]}** consists of three subordinated sets: (a) performance through time (PERFT (i, j)) and (b) managerial performance (MANPERF (i, j)), while all the above-mentioned linguistic sets contain further subordinated linguistic sets postulated as follows:

- (PERFT (i, j))
    - Artistic merit cumulative values through time and space (AMECVTS (i, j))
    - CEP[11] cumulative values through time and space (CEPCVTS (i, j))
- (MANPERF (i, j)),
    - Creative production process (CRP (i, j))

[11] CEP – Creative economy process.

- Outcome consumption (OUC (i, j))
- Outcome distribution (OUD (i, j))

while the (PERFT (i, j) subset content is closely related to intellectual capital 3 – (ICA3 (i, j)) linguistic subset and (MANPERF (i, j)) subset content is closely related to material input {[MI (i, j)]}, material input cost {[MICS (i, j)]},, output material {[OMP (i, j)]}, products and output material product assets {[OMPA (i, j)]} linguistic sets existing within MIC processes.

7.5 Cultural heritage administration versus creative economy process functionality

7.5.1 Cultural heritage administration process external and internal metrics linguistic sets

When considering the System of Creative Economy two principle aspects should be considered and respected. The first one is closely related to a huge set of appropriate business processes running within that system. However, the second aspect is concerned with the fact, that system should respect adequate system theory principles as well, while any system vertical structure is being created by: (a) subsystems, (b) components and (c) modules and (d) sub-modules actually. With respect to that theory, we can define two principle subsystems related to creative economy system. The first subsystem is closely related to art creative activities and the second one is concerned with scientific, research and development activities. We shall discuss the art creative activities, where cultural heritage administration plays a role of principle importance and creates a target subsystem of the Creative Economy system discussed within that section, while the subsystem concerned with scientific research and development will be omitted. The cultural heritage administration relatively individual system seems to be the principle subsystem of the creative system and the core business process running within that system. As a result of that two main business processes might be defined: (a) CHC Cultural heritage artifact creation and CHM Cultural heritage artifact management. Furthermore, we shall discuss a set of problems concerned with structure and metrics related to both above-mentioned main business processes.

7.5.2 CHC Cultural heritage artifact creation

The CHC Cultural heritage artifact creation main business process horizontal structure might contain a lot of different business process functions (BPFs), however the linguistic sets representing its external and internal metrics might be defined as follows: (a) cultural heritage artifact creation economic data {[CHACED (i, j)]}, (b) cultural heritage artifact data {[CHAD (i, j)]}, and (c) cultural heritage artifact production data {[CHAPD (i, j)]}.

7.5.3 CHC Cultural heritage artifact management

On the other hand, the CHM Cultural heritage artifact management main business process has its own vertical structure represented by three subprocesses: (a) CHM-01 Cultural heritage artifact acquisition, (b) CHM-02 Cultural heritage artifact processing, and (c) CHM-03 Cultural heritage artifact delivery, while the CHM- 01 business process external and internal metrics is represented by the following linguistic sets: (a) cultural heritage artifact technical data {[CHATD (i, j)]}, (b) cultural heritage artifact commercial data {[CHACD (i, j)]}, (c) cultural heritage artifact operational data {[CHAOD (i, j)]}, and (d) cultural heritage artifact economic data {[CHAED (i, j)]}. The CHM- 02 business process external and internal metrics is represented by the following linguistic sets: (a) Cultural heritage artifact archival processing data {[CHAAPD (i, j)], (b) Cultural heritage artifact economic processing data {[CHAEPD (i, j)]}, (c) Cultural heritage artifact objective processing data {[CHAOPD (i, j)]}.

7.5.4 CHC Cultural heritage artifact delivery

The CHM- 03 business process external and internal metrics is represented by the following linguistic sets: (a) cultural heritage delivered artifact data {[CHADD (i, j)]}, (b) cultural heritage delivered searcher's data {[CHASD (i, j)].

Remark:

It should be noted that all the linguistic mentioned in that contribution have a rather complicated hierarchic structure from objective point of view, while that structure cannot be discussed in more details because the contribution limited number of pages.

7.5.5 Data transfer among Cultural Heritage Administration subprocesses and CEP processes

The business process denoted as Cultural Heritage Administration seems to be the core process and consists of two: main process as mentioned within previous section. However, the metrics linguistic sets related to both main processes are described within previous section as well, while the algorithms of data transfer among Cultural Heritage Administration subprocesses and CEP processes are described within following subsections with respect to the CEP Process functions and their metrics. We shall describe the transfer of data contained within linguistic sets {[CHAED (i, j)]}, {[CHAD (i, j)]}, and{[CHAPD (i, j)]} closely related with CHC Cultural heritage artifact creation business process and CEP Process subordinated business processes generalized as follows:

The {[CHAED (i, j)]}, and{[CHAPD (i, j)]} LS content is being transferred to **{[EC (i, j)]:** {ALI(I, j)]} LS and {[EC (i, j)]: {FAT(I, j)]} LS and {[EC (i, j)]: {MOI (i, j)]} LS content

The {[CHAD (i, j)]} and {[CHAPD (i, j)]} LS content is being transferred to **{[CULC (i, j)]: [CULG (i, j)]}** LS content

The {[CHAED (i, j)]}, {[CHAD (i, j)]}, and {[CHAPD (i, j)]} linguistic set (LS) content is being transferred to **{[SYMC (i, j)]:** [AASP (i, j)], [ARTSTA (i, j)], [ARTSTY (i, j)], [ARTSTV (i, j)]} linguistic set content

The {[CHAED (i, j)]}, {[CHAD (i, j)]}, and {[CHAPD (i, j)]} linguistic set (LS) content is being transferred to{[**SOCC (i, j)**]} (COMPER (i, j) (ARTDI (i, j)), (TECHN (i, j)), (TALEN (i, j)), (TOLER (i, j))

The {[CHAED (i, j)]}, {[CHAD (i, j)]}, and {[CHAPD (i, j)]} linguistic set (LS) content is being transferred to**{[PERFM (i, j)]}** (PERFT (i, j)) (AMECVTS (i, j)) and {[PERFM (i, j)]} (PERFT (i, j)) (CEP (i, j)) LS content

The {[CHAED (i, j)]}, {[CHAPD (i, j)]}, and {[CHAD (i, j)]} LS content is being transferred to **{[PERFM (i, j)]}** (PERFT (i, j)) (AMECVTS (i, j)) and {[PERFM (i, j)]} (PERFT (i, j)) (CEP (i, j)) LS content.

7.6 MIC Process external and internal metrics items

In general, any business process is represented by an appropriate vertical and horizontal structure and external and internal metrics approximated by adequate linguistic sets[12] as well. While the **BP external metrics** is created by linguistic sets shown in **Figure 3**.

On the other hand, the **MIC BP internal metrics** is created via subordinated linguistic sets as production devices {MIC: IM:[DEV (i, j)]}, production tools {MIC: [IM:TOOL (i, j)], and human resources {MIC:IM:HR (i, j)]} as well, while the {[MIC:IM:DEV (i, j)]} linguistic set (LS) contains one subordinated LS denoted as device production output performance {MIC:IM:[DPOP (i, j)]}, tool production output performance {MI:[IM:TPOP (i, j)]}, and the {[IM:HR (i, j)]} contains four subordinated linguistic sets.

7.7 MAC Process metrics items

The MAC Process metrics[13] item is represented by hierarchic structure, while at the first hierarchic level is created by three subordinated linguistic sets: (a) goods market {MAC:[GM (i, j)]}, (b) financial market {MAC:[FM (i, j)]}, and labour market {MAC:[LM (i, j)]}, while the second hierarchic level has the following structure {MAC:[GM(GDP (i, j)]}, {MAC:[GM(EA (i, j)]}, {MAC: [LM(WA (i, j)]}, {MAC: [FM(CA (i, j)]}, {MAC: [FM(SA (i, j)]}, {MAC: [LM(WA (i, j)]}, MAC: [LM (PR (i, j)]}, MAC: [LM(WAPRE[14] (i, j)]

7.8 The data transfer from microprocesses to macroprocesses

7.8.1 MIC Process external metrics versus MAC Processes linguistic sets

The microprocess metrics linguistic sets cover MIC Process external metrics, while the external metric set contain the following subsets: (a) Material input cost items {[MICI (i, j)]}, (b) Material input items {[MII (i, j)]}, (c) Output material product assets {[OMPA (i, j)]} and (d) Output material product assets {[OMP (i, j)]}

The data transfer from microprocesses external metrics to macroprocesses is running in the following steps:

- Material input cost {[MII (i, j)]} and {[MICI (i, j)]}, linguistic set content is being transferred to {MAC: [GM (EA (i, j)]},]} linguistic set

- Output material product assets {[OMPA (i, j)]} and Output material product assets {[OMP (i, j)]} linguistic set content is being transferred to {MAC:[GM (EA (i, j)]}, linguistic set

- Output material product assets {[OMPA (i, j)]} and Output material product assets {[OMP (i, j)]} linguistic set content is being transferred to {MAC:[GM (GDP (i, j)]}, linguistic set

- Output material product assets {[OMPA (i, j)]} and Output material product assets {[OMP (i, j)]} linguistic set content is being transferred to {MAC:[FM (GDP (i, j)]}, linguistic set MAC: [FM(CA (i, j)]} and {MAC: [FM(SA (i, j)]}.

---

[12] The problems concerned with linguistic sets, which create basis of Business Process Linguistic Modelling approach are discussed within Section 4.3.

[13] MAC Process metrics does not contain the external and internal part.

[14] Abbreviation explanations: MAC—macroprocess, GM—goods market, GDP—gross domestic product, EA—economic agents, FM—financial market CA—checking account, SA—saving account, LM—labor market, WA—wages, PR—prices, WAPRE—wages-prices equilibrium.

- Output material product assets {[OMPA (i, j)]} and Output material product assets {[OMP (i, j)]} linguistic set content is being transferred to {MAC: [LM (WA (i, j)]}, MAC: [LM(PR (i, j)]}, MAC: [LM(WAPRE.

The data transfer from microprocesses internal metrics to macroprocesses is running in the following steps:

- The {MIC: IM:[DEV (i, j)]} and {MIC:[IM:TOOL (i, j)]} linguistic content is being transferred to MAC: [FM(CA (i, j)]}, {MAC: [FM(SA (i, j)]}, and {MAC: [GM(EA (i, j)]} linguistic sets

- The {MIC: IM:[DEV (i, j)]} and {MIC:[IM:TOOL (i, j)]} linguistic content is being transferred to {MAC: [LM(WA (i, j)]}, MAC: [LM(PR (i, j)]}, MAC: [LM (WAPRE.

- The {MIC: IM:[HR (i, j)]} linguistic content is being transferred to {MAC: [LM (WA (i, j)]}, MAC: [LM(PR (i, j)]}, MAC: [LM(WAPRE.

- After having completed the previous step, the extracted data and data segments or fragments are transformed to data structure, which might be accepted by application, which enables generating adequate creative economics strategies.

8. Data Transfer System MAC versus CE Processes (DTS2 System)

8.1 DTS2 System structure and functionality

As mentioned above, the DTS Systems operates within two directions. It provides data transfer from CE Processes to MAC Processes, while the MIC Processes play a role of mediator, when running the data transfer between CE and MAC Processes and that type of transfer is covered by DTS1 Subsystem, the data transfer between MAC and CE processes is covered by DTS2 Subsystem. We shall describe a structure and functionality of DTS2 Subsystem in that section.

In general, the creative economy processes (CE Processes) are closely related to social and economic processes, which represent external environment and provide transfer of signals or data to set of macroprocesses, which represent an outgoing point for data transfer concerned to social and economic requirements related to CE and MIC processes. On one hand, when considering the social processes, a set of cultural and art processes play a role of principle importance. On the other hand, when considering the economic processes the research and development processes create basis from creative economy point of view, while the logistics, and personal management processes play a role of principle importance for the research and development processes and for production processes as well. Both of the above-mentioned processes (social and economic) provide transfer of data to MAC processes, which are necessary from external and internal source analysis and they play an important role, when generating values closely related to KPI indicators postulated at national economy level.

However, the KPI indicator values postulated at national economy level are being transferred to MIC processes, where their decomposition related to appropriate MIC process busines levels (strategic, tactic and operational) and play an important role at BPMIV setting and evaluation of performance related to adequate processes running within MIC processes set. However, the KPI indicators play an important role MIC and CE process management as well, while cultural heritage creation and management seems to be the core process from creative economy processes point of view. The principal layout of data transfer from MAC to MIC and CE processes is shown in **Figure 6**.

8.2 DTS2 System Conceptual Model

As mentioned above (see also Section 7.1), the DTS System – Conceptual Model is a standardized system denoted as DATRA_09 Data Transfer Conceptual Model, which should provide a bi-directional data transfer. On one hand, from Creative Economy Business Processes (CEP Processes) to Microprocesses (MIC Processes) and (MIC Processes) and from Microprocess to Macroprocesses (MAC Processes) and on the other hand, we have to consider the data transfer from macroprocesses, throughout microprocesses up to CEP Processes as well, while DTS System – Conceptual Model consists of four subsystems as shown in **Figure 4**. However, the social and economic processes create significant elements of DTS System external environment and generate appropriate data important for functionality of MAC Processes, which provide services for KPI indicator generation [22] within an appropriate firm or company business strategy creation [25, 26].

However, there is data and information generated as a result of text [27, 28] and image sematic analysis [29, 30], as well, while some of those algorithms might be applied, when providing analysis of the firm or company external and internal resources before getting started business strategy creation and KPI generation too. Subsequently, the KPI indicators are being decomposed [22] and assigned to adequate MIC Processes organization units and their metrics [24] item values are being set simultaneously. After having completed the above-mentioned action the management of MIC and CEP Processes might be started, where logistic, personal financial and research and development management play a role of principle importance as well, while the production processes are significant for MIC process management and cultural heritage creation and management are significant from CEP Processes point of view.

The principle layout of DTS2 System Conceptual Model is shown in **Figure 7**.

9. Economic networks structure and functionality

9.1 Economic network structure

The above-mentioned DTS System implemented via so called economic network, which seems to be relatively independent system and consist subsystems covering data transfer within CE, Micro and Macroprocesses (EN-S1 Subsystem) and Macro, Micro and CE Processes (EN-S2 Subsystem) as well.

9.2 Functionality of economic network EN-S1 and EN-S2 subsystems

The EN-S1 Subsystem is being covered by creative **economy processes**, which consists of two sub-processes: (a) Cultural heritage administration and (b) Research and development administration. However, the BP denoted as Cultural heritage administration is described within previous sections as too, while the Research and development administration is not discussed within that contribution. On the other hand, there are MIC and MAC processes and **MIC Processes** include production and non-production domain and plays a role of go-between elements, when providing data transfer from creative economy processes to macroprocesses

However, **the MAC Processes** create an integral part of social processes and represent the go-between elements, when preparing goals and aims related to creative economy process functionality and performance as well, while they play a role of go between elements social, between MIC and CE Processes as well.

The **EN-S2 Subsystem** is being covered by **EN_S2_01 Macroprocesses** component, which consists of the following modules: (a) Normative act processing, (b) Document semantic analysis, (c) Datamining execution, (d) KPI creation and (d) KPI execution.

However, the **EN-S2 Subsystem** is being covered by **EN_S2_02 Microprocesses** component, which consists of the following modules: (a) MIC Process – KPI Setting and (b) MIC Process – Metrics Indicator Setting as well, while the latest component is denoted as EN-S2-03 CE Process, which consists of the following modules: (a) CE[15] Process KPI Setting and (b) CE Process – Metrics Indicator Setting, and CHA[16] Process – Metrics Indicator Setting.

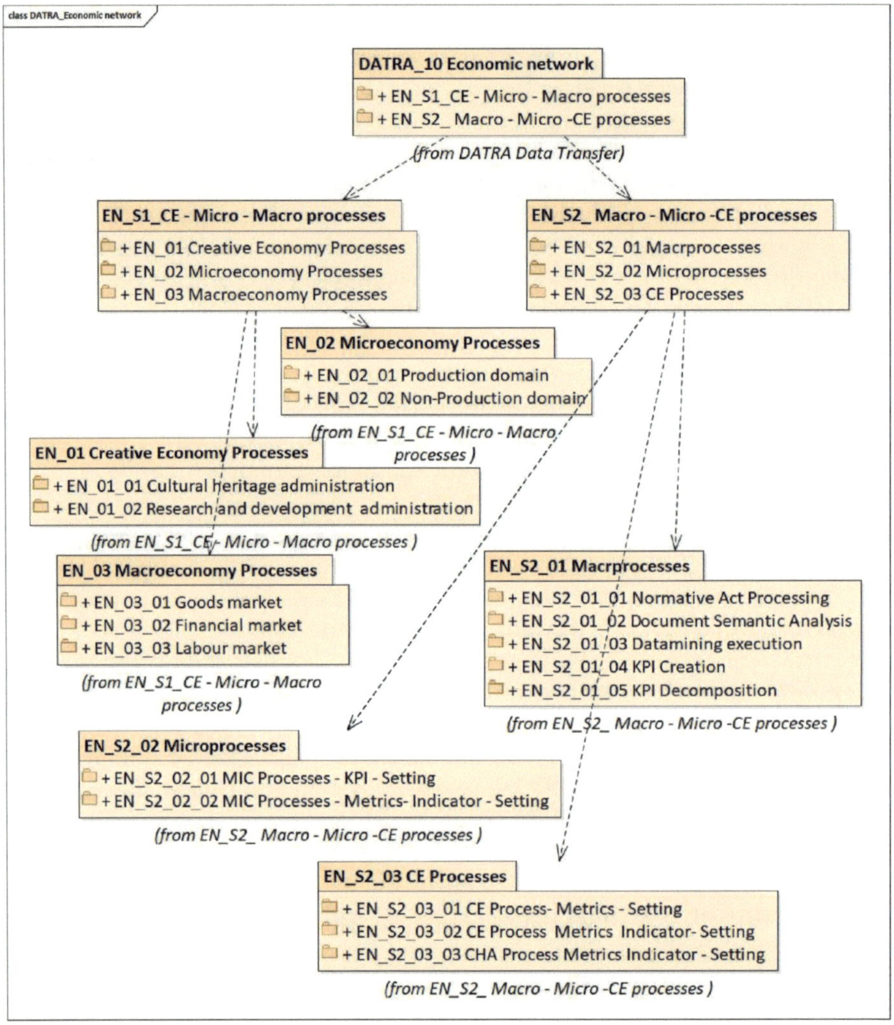

Figure 8.
The principal layout of Economic Network structure. Source: The Authors.

---

[15] CE—Creative Economy.

[16] CHA—Cultural Heritage Administration.

The EN-S1 subsystem structure and functionality is being discussed within previous sections and the EN-S2 subsystem structure and functionality will be discussed within individual and independent contribution.

The principal layout of Economic Network structure is shown in **Figure 8**.

10. Design and implementation aspects

The Economic Network system (EN System) plays a role of information and knowledge-based supporting facility for CE, MIC, and MAC Processes. However, the EN System provides services related to KPI creation and decomposition, which is important for business strategy design within MIC and CE processes as well and is being designed as an information system and as a knowledge-based system too. On one hand, when considering the information system, the linguistic sets are implemented as standardized database tables, on the other hand, considering the knowledge based system the linguistic sets create basis of adequate reference databases and play a role semantic network elements. When considering the user's communication with the information system an appropriate database management system for those purposes should be applied and when considering the user's communication with an adequate knowledge-based system an appropriate inference system is used for those purposes. The above-mentioned aspects should be respected within EN System design and implementation, while a selected GraphDB [31] system will be applied for EN System design and implementation and an adequate application program should be created.

11. Discussion

At present, the creative economy seems to a new branch, which provides interconnection between cultural standardized economy sphere and includes different activities related to culture and cultural heritage creation and management. However, providing those activities requires intellectual, material, and financial support as well, while all of them are closely related to macroeconomy and microeconomy objectives. On the other hand, the creative economics processes are the business processes too. However, they have adequate vertical and horizontal structure require an appropriate information, knowledge-based and organization support as well in order to assure their proper and efficient functionality, while their information support and metrics elements are closely related to each other. As a result of that the data or information transfer plays a role of principal importance and is running in two stages: (a) from creative economy to microeconomy BPs and (b) from microeconomy to macroeconomy BPs. Of course, the data transfer from DTS System external area represented by social and macroprocesses to CE Processes play an important role as well, while a set of MIC Processes represent an appropriate go-between element and an integral part of sensors and effectors, acting within grey zone, which exists between BP external and internal environment.

The above-mentioned DTS System implemented via appropriate economic network, which seems to be relatively independent system and consist subsystems covering data transfer within CE, Micro and Macroprocesses (EN-S1 Subsystem) and Macro, Micro and CE Processes (EN-S2 Subsystem) as well.

The above-mentioned stages are being implemented and operated via appropriate economic networks, the (a) phase activities are covered by economic network I and the (b) phase are covered by economic network II. Both above-mentioned economic network subsystems should be implemented and operated via adequate

Datawarehouse system, where the economic network subsystems EN-S1 and EN-S2 could be implemented via Datamart I and Datamart II and both have their own ETL systems, which provide extraction of data from appropriate linguistic sets (E), the data transformation (T) to pre-defined structure of multidimensional tables and data loading (L) to the above DataMart I and Datamart II. On the other hand, there is an appropriate OLAP system, which should enable providing data selection and analysis with respect to authorized user requirements. Moreover, that system could contain a software, which enables providing a semantic analysis of EU documents and extracting data, which are inserted into a software, which prepares supporting data concerned to business strategy of the firm or company, which deals with creative economy activities.

12. Conclusion

At present, the creative economy seems to a new branch, which provides interconnection between cultural standardized economy sphere and includes different activities related to culture and cultural heritage creation and management. However, providing those activities requires intellectual, material, and financial support as well, while all of them are closely related to macroeconomy and microeconomy objectives. On the other hand, the creative economics processes are the business processes too. However, they have adequate vertical and horizontal structure require an appropriate information, knowledge-based and organization support as well in order to assure their proper and efficient functionality, while their information support and metrics elements are closely related to each other. As a result of that the data or information transfer plays a role of principal importance and is running in two stages: (a) from creative economy to microeconomy BPs and (b) from microeconomy to macroeconomy BPs. Of course, the data transfer from DTS System external area represented by social and macroprocesses to CE Processes play an important role as well, while a set of MIC Processes represent an appropriate go-between element and an integral part of sensors and effectors, acting within grey zone, which exists between BP external and internal environment. The above-mentioned DTS System implemented via appropriate economic network, which seems to be relatively independent system and consist subsystems covering data transfer within CE, Micro and Macroprocesses (EN-S1 Subsystem) and Macro, Micro and CE Processes (EN-S2 Subsystem) as well The above-mentioned stages are being implemented and operated via appropriate economic networks, the (a) phase activities are covered by economic network I and the (b) phase are covered by economic network II. Both above-mentioned economic network subsystems should be implemented and operated via adequate Datawarehouse system, where the economic network subsystems EN-S1 and EN-S2 could be implemented via Datamart I and Datamart II and both have their own ETL systems, which provide extraction of data from appropriate linguistic sets (E), the data transformation (T) to pre-defined structure of multidimensional tables and data loading (L) to the above DataMart I and Datamart II. On the other hand, there is an appropriate OLAP system, which should enable providing data selection and analysis with respect to authorized user requirements. Moreover, that system could contain a software, which enables providing a semantic analysis of EU documents and extracting data, which are inserted into a software, which prepares supporting data concerned to business strategy of the firm or company, which deals with creative economy activities.

Author details

Jozef Stašák[1] and Jaroslav Mazůrek[2]*

1 Institute of Technology and Business, Okružní 10, 370 01 České Budějovice, Czech Republic

2 Department of Mediamatics and Cultural Heritage, Faculty of Humanities, University of Žilina, Žilina, Slovak Republic

*Address all correspondence to: jaroslav.mazurek@fhv.uniza.sk

References

[1] Stašák J. 2010: Princípy modelovania ekonomických objektov EKONOM. Bratislava; 2010

[2] Grell M, Stašák J. 2011 Balanced scorecard – Semantic approach, Aspects of Methodology In: Ekonomika a manažment: Vedecký časopis Fakulty podnikového manažmentu Ekonomickej univerzity v Bratislave. - Bratislava: Fakulta podnikového manažmentu Ekonomickej univerzity v Bratislave. 2011;8(2):34-46. (Slovak) ISSN 1336-3301

[3] Mazůrek, J., Stašák, J., Škorvagová, E., Mazůreková, M. 2020 Mediamatika a kultúrne dedičstvo ako súčasť spoločenských procesov (Mediamatics as an integral Part of social Processes) EDIS vydavateľské centrum Žilinskej univerzity, Žilina 2020 (printed)

[4] Bohumelová, M., Hvorecký, J: 2016 Museums and Galleries: From Traditional to Learning organizations. In: 14th International Conference on Emerging eLearning Technologies and Applications (ICETA), 7 pp., Starý Smokovec, 2016, ISBN: 978-1-5090-4699-7

[5] Bordieu P. Le capital social: Notes provisoires. Actes de la recherche en sciences sociales. 1980;**31**(1):2-3

[6] Halida, A.L 2010 Taxonomy of Performance Dimensions and Levels of Analysis in the Creative Industries [cit. 15.8.2020] available at https://www.academia.edu/12833133/Performance_in_the_Creative_Industries

[7] FloridA R. The Rise of Creative Class. New York: Basic Books; 2002

[8] Howkins, J. 2001 The Creative Economy: how people make money from ideas. 2 vydanie. London: Penguin. 304 s. ISBN: 978-0141977034

[9] Jensen, R. (1999). The Dream Society: How the Coming Shift From Information to Imagination Will Transform Your Business. McGraw-Hill Professional.

[10] Hesmondhalgh D. The cultural industries. London UK: Sage (2nd edition) systems. American Journal of Sociology. 2007;77:639-659

[11] Hirsch, P. Hirsch, P.M. 1972 Processing fads and fashions. An organization-set analysis of cultural industry M. 2000 Cultural industries revisited. Organization Science 11 (3): 356-361

[12] Pine BJ, Gilmore JH. Welcome to experience economy. July-August: Harvard Business Review; 1998. pp. 97-105

[13] Thornton PH, Jones C, Kury K. Institutional logics and institutional change in organizations: Transformation in accounting, architecture, and publishing. Research in the Sociology of Organizations. 2005;**23**:125-170

[14] Lash S, Urry J. Economies of Signs and Space. London: Sage; 1994

[15] Me, 2009 MEASURING THE ECONOMIC CONTRIBUTION OF CULTURAL INDUSTRIES A review and assessment of current methodological approaches, http://uis.unesco.org/sites/default/files/documents/measuring-the-economic-contribution-of-cultural-industries-a-review-and-assessment-of-current-methodological-approaches-en_1.pdf

[16] Rikalović, G. and H. Mikić (2011). "Razvojna uloga kreativnog sektroa i njegova pozicija u politici zaposljavanja".Contemporary Trends in European Economy: Implication for Serbia. Novi Sad: High Business

[17] Caves RE. Creative industries: Contracts between art and commerce. Cambridge, MA: Harward University Press; 2000

[18] Lampel J, Shamsie J, Lant TK, editors. The Business of culture: Strategic perspectives on entertainment and media. Mahwah, NJ: Lawrence Elbraum Associates; 2006

[19] Townley B. The role of competing rationalities in institutional change. Academy of Management Journal. 2002; **45**:163-179

[20] Stašák, J. 2015 Business Process Linguistic Modelling – Philosophy & Principles Computer Science and Information Technology 3(5): 198-213, 2015 DOI: 10.13189/csit.2015.030505

[21] Stašák J., Vaníčková R., Grell M. 2015 Business Process Modelling Linguistic Approach – Problems of Business Strategy Design Universal Journal of Management, 2015 3, (7).

[22] Stašák, J., Schmidt, P, 2018 Key Performance Indicators versus Business Process Metrics International Journal of Advanced Operations Management on 29/June/2018

[23] IT Manager's Guide Metrics TechExcel. http://techexcel.com/resources/whitepapers/Metrics.pdf

[24] Stašák J. 2017 Contribution to Linguistic Modeling of Business Process Structure, Functionality and Performance Metrics *Journal of Scientific Research & Reports 13(3): 1-16, 2017; Article no.JSRR.30233 ISSN: 2320-0227*

[25] Stašák, J., 2013 Strategy creation versus analysis of business subject internal capabilities linguistic approach. In: Proceeding of the Conferences Present Day Trends of Innovations p. 212-223, Dubnica nad Váhom, 2013, ISBN 978-80-89400-59-1

[26] Stašák, J., Schmidt, P. 2019 Semantic Technology and Linguistic Modeling in Business Strategy Design and Evaluation International Journal of Business Information Systems on 21/Jun/2019

[27] Stašák, J, 2004a.: A Contribution to Semantic Text Analysis In: Electronic Computers and Informatics ECI 2004, The University of Technology Košice, Department of Computers and Informatics of FEI, 22-24. 9.2004 Košice – Herľany, SR, p.132-144, ISBN 80-8073-150-0

[28] Stašák, J, 2011: How Image and Text Semantic Analysis Systems may be applied for Educational and Teaching Purposes – Acta Technologica Dubnicae, 2011, č.1, s.1-18, ISSN 1338-3965

[29] Stašák J. 2004b Application of Fuzy Sets Apparatus in Image Semantic Analysis In: Proceeding of 5th International Caprpathian Control Conference Faculty of Mechanical Engineering and Robotics AGH-UST Krakow Zakopane, Poland, p. 255-260, ISBN 83-89772-00-0 ICCC 2004 Zakopane, Poland, May 25-28,2004

[30] Stašák, J., 2004c A Contribution to Image Semantic Analysis In: Informace na dlani, Albertina Income Praha, s.r.o., Praha 2004, ISSN: 1214-1429, electronic version, Infoforum Praha 2004

[31] Robinson, I., Webber, J., Eifrem, E. 2013 Graph Databases O'Reilly Media, Inc., 1005 Gravenstein Highway North, Sebastopol, CA 95472 (147699300 61761_441d82b38d2194b0018ef 8fce03b2b5f.pdf)

Chapter 7

Conceptualization, Definition and Assessment of Internal Logistics through Different Approaches Using Artificial Intelligence

Orlem Pinheiro de Lima, Sandro Breval Santiago, Carlos Manuel Rodríguez Taboada, Jorge Laureano Moya Rodríguez, Maida Bárbara Reyes Rodríguez, Márcia Ribeiro Maduro, Paulo César Diniz de Araújo, Nilson José de Oliveira Junior, André Luiz Nunes Zogahib and José Carlos da Silva Lima

Abstract

The aim of this chapter is to develop a new concept of internal logistics, its components parts and how to evaluate it. To quantify the level of performance of the internal logistics of a company is an important issue to gain competitiveness. There are few papers now at days that analyze how to quantify this issue. In recent years, it has been developed numerous applications of Fuzzy logic and Neural Networks to solve diverse problems of Engineering. Fuzzy logic is a mathematical tool that emulates the method used for humans for managing and processing information and Neural Networks are computing systems inspired by the biological neural networks that constitute human brains. Such systems "learn" to perform tasks by considering examples, generally without being programmed with task-specific rules. This chapter offers a new definition of internal logistics and shows the procedure to evaluate its level in a company. This procedure for assessing the internal logistics was developed through an Excel tab, a fuzzy inference system and a neural network. To validate this procedure, it was applied to 93 companies in the Industrial Pole of Manaus. Results obtained by different approaches are very similar, demonstrating the validity of the procedure developed.

Keywords: internal logistics, measurement, neural networks, fuzzy logic, performance, industrial companies

1. Introduction

The advent of globalization promotes organizations with the persistent pursuit of competitiveness, forcing businesses moving competitive advantages. Still, organizations that are in the forefront of their sectors and considered successful are those that actually trying to develop its core competencies to offer a standard of excellence in goods and services and are concerned with its strategy and with the workforce. The market organizations have demanded a set of features that include efficiency, effectiveness, dynamism, creativity, agility, flexibility and having holistic vision, to be competitive and having defined their strategies, seeking business sustainability.

In [1] are highlighted the indicators that assess the efficiency of internal activities and processes, and logistics performance indicators are suggested. They are classified into the following categories: stock management, cost, productivity, quality and customer service. However, it is necessary to develop a performance evaluation form for the supply chain, using external and internal indicators together to evaluate the performance of the entire chain, not only internal indicators of logistics, so that, working together, companies manage to achieve the best return business of supply chain.

In [2], it is considered the designation of logistics as "logistics management." Also it is cited that this concept can be included into customer service, traffic and transportation means, storage, selection of local to manufacture and store, inventory control, order processing, acquiring, transportation of materials, distribution, supply of parts, packaging, returning goods and order volume forecast and that an organization must provide products and services to customers according to their needs and requirements as efficiently as possible.

Sometimes logistics is related to the marketing, as the strategic process to managing the acquisition, transfer, storage, parts, and finished products, together with the flow of information and its marketing channels to maximize profit with cost-cutting. There is no single definition for conceptualizing logistics, which be accepted by all researchers in the field. The important thing is that companies know that it is present in the business world and that professionals must understand their target that "is to make available products and services where they are needed, when they are desired" [3].

The traditional logistics refers to activities such as packaging, transportation, loading, unloading and storage, etc. The modern logistics reaffirms the concept of integrated logistics management and its implementation. It is important to outstanding that modern logistics must be understood as the medium during the acquisition, production, and operation of the whole process to delivery to the final consumer [4].

In contemporary organizational environment, logistics appears as a strategic concept, because of not only materials management and physical distribution, but also for providing values of time and place for customers, for becoming an element that stands for organizations, with agility, flexibility and integration of internal and external channels. Several authors describe that the concept of logistics can be separated into three basic items: food (suppliers), plants (internal) and distribution (customers). This represents a group that is often defined as highly empirical, resulting in negative effects that directly influence the outcome of the final performance of organizations [5, 6].

Despite the importance of internal logistics, it has not been fully understood, particularly in manufacturing industry [7–9]. However, it constitutes a large part of the total cost for businesses [10]; average logistics costs represent between 10% and 30% of total sales volume in a typical production company [11]. This chapter is composed by eight parts: An introduction, a Literature Review, a section of materials and methods where it is explained all the procedure developed, a section of results analysis and finally the conclusions, the acknowledgments, the conflict of interest and the references.

2. Literature review

2.1 Internal logistics concepts

Several authors state that an internal logistics system well designed and correctly used increases the efficiency of an organization [12, 13].

To summarize, several aspects of logistics performance are very important for a company, and among others refer to delivery, quality, robustness, information, and cost and customer service. However, it is also important to consider which is the combination of high efficiency, performance, and effectiveness [14]. For [15] the performance efficient logistics activities alone are not enough. To create competitiveness for the company, it is essential that the right kind of logistics activities to be prioritized and through the right performance variables. However, as discussed by [16], there is a lack of standardized ways of dealing with the internal logistics requirements influencing the overall logistics performance. In general, logistics managers are trying to use measurements to help design and manage more effective and efficient logistic systems for the client. Identifying the value of internal logistics and its critical performance criteria can be a way to help this development [17].

Design and improve the internal logistics system involves decision making at different levels, such as strategic, tactical and operational issues. As such, it involves long-term planning (strategic) and aspects of planning and control (management) of short and medium term [18]. An internal logistics system that works well requires involvement and understanding of the system at all levels. Logistics professionals should be empowered with the necessary experience in essential and critical functions for their own company and fully understand how they affect the entire value chain [19].

Internal Logistics handles all the management of the internal supply process, storage, transportation and distribution of goods within the organization, that is, to meet its domestic demands as support for manufacturing [20]. According to [6], the cycle of support to manufacturing activities is directly related to internal logistics, i.e., planning and production control. Thus, the logistical support to the production aims mainly to establish and maintain an economic and orderly flow of materials and stocks in process in order to meet the schedules of the production sector. The logistics support production has the operational responsibility for the following activities: handling and storage of products, materials, components and semi-finished parts. With the changes in the business environment, logistics service concepts have evolved and various issues were added in operational logistics tasks such as packaging, outsourced inventory management, bar code, and information systems. These operational logistics tasks were considered, and called as "internal logistics," and these activities should "interact with other functional areas" [21]. Internal logistics thus involves logistics activities within the walls of an organization, e.g. internal transport, materials handling, storage and packaging [15].

Other more recent studies indicate that the internal logistics has been the attempt to organize and optimize the internal activities with the cost reduction objective for organizations in different segments. However, organizational issues such as the lack of a strategic vision that become in difficulties need to be addressed. The transfer of knowledge and technology used in the manufacturing industry could be of great benefit concluded by [22].

According to [23], the end consumer determines the success or failure of supply chains. Thus, an important part of logistics performance is linked to customer service and to be able to respond to their needs and requirements. When it comes to internal logistics as a system, both the client and the service provider are the same at the organization itself. So to see the internal logistics as a system, both the service provider and the customer are the same company. As such, the customer's needs and requirements can be translated for internal purposes. Therefore, the performance of internal logistics is under the control of the company, and can provide a more direct indication of the effects of the relationship involving structure and logistics [16].

Given this approach to internal logistics, was noticed a gap both in theory and in practice referred to this issue. Many discussions in the general theoretical field of logistics has been developed. Companies need support with tools and models or methods that make it possible to identify, organize and help to define and shape to analyze them; it is evident in the daily graded citations mentioned above about the lack of studies in this area. Therefore, this chapter suggests a way to define and evaluate the internal logistics.

Based on the readings of selected articles and the development of a pattern was possible to define a more comprehensive concept of internal logistics:

Internal logistics is planning, execution and control of the physical flow and internal information of the company, seeking to optimize the resources, processes and services with the highest possible profit.

According to the standpoint of logistics as a picture or an approach that consists of several parts and aspects, it is often described as a system, which is the perspective used in this work to analyze the internal logistics as part of the system. According to [24], the logistics system is always open and in a state of exchange with its environment. However, the limits of the system and subsystems and components included vary depending on different perspectives.

The system studied in this chapter is the internal logistics system, where the system boundaries are the physical limits of the company under study. Then internal logistics comprises logistics activities within the walls of an organization, such as internal transport, materials handling, storage and packaging [15].

There are three different angles from which, logistics operations can be seen: processes, resources and organization. All these aspects can be seen as parts of the logistics system, as the flow of goods and information to be made through a series of stages called activities and processes [25]. In addition, resources refer to all means, equipment and personnel needed to run the process. Finally, the organization includes all planning and control procedures necessary to implement and manage the system.

Several authors argue that an internal logistics well designed and properly used increases the efficiency of an organization [26]. The project of internal logistics system is therefore an aspect that strongly influences the competitiveness of the system and is therefore related to the objective of this chapter.

Projecting and improve internal logistics system comprises decision making at different levels, such as strategic, tactical and operational levels. As such, it includes long-term planning (strategic) and aspects of planning and control (management) of short and medium term [18]. Internal logistics system that works well requires participation and understanding of the system at all levels. Logistics professionals must be equipped with the necessary expertise in critical and essential functions for their own company and fully understand how they affect the entire value chain. Supply chains are often faced with the situation where they have to accept some degree of uncertainty, however, must develop a strategy that allows them to adjust supply to demand [27]. In general, it can be affirmed that a strategy is about how to make the planning, which is very different from doing [18]. To conduct portfolio analysis were consolidated by central themes articles and one can see some evidence as to the possible parts of internal logistics. For a better demonstration of sets of items was prepared to **Table 1** resulting in a preliminary view of the parts.

Articles key points	IT	LT	PCP	PCM	ST	IM	SP	RC	WP	LA	HL	IM	PP
Supply networks	X	X					X						
Decision making	X	X		X	X	X							X
Delivery		X	X					X					
Health care logistics	X	X		X			X				X		
Logistics performance	X	X	X			X	X					X	
Inbound logistics	X	X			X	X					X	X	X
Cross-docking		X			X						X		X
Automated material handling systems	X		X		X						X	X	
Identifying business value using the RFID e-Valuation framework	X	X											
Cost-cutting	X	X	X		X	X	X			X	X		X
Customer and supplier logistics	X	X	X									X	
Inventory and transportation decisions	X	X			X	X							
Intralogistics operations	X	X								X			
EDI in logistics	X	X				X		X				X	
Integration between logistics and Assembly lines	X	X	X		X	X				X			X
Optimization transport		X	X								X	X	X
Routing and inventory	X	X	X		X	X	X				X	X	X
Smart logistics	X	X	X				X		X				
Manufacturing strategy	X	X	X						X				
Consolidation loads	X	X					X					X	

IT = information technology; LT = logistics techniques; PCP = planning and control production; PCM = planning and control manufacturing; ST = storage; IM = inventory management; SP = supply; RC = receipt; WIP = working-in-progress; LA = layout; HL = handling; IM = internal transport; PP = picking and packing.
Source: Prepared by authors (2020).

Table 1.
Key points to compose the concept and components parts.

A well-formulated strategy helps to use all the resources of an organization and create value based on its internal competition and shortcomings in relation to the external environment. However, it is of great importance that the logistic function and logistics strategy are integrated and aligned with other functions and strategies of the organization, to create competitiveness [15].

Companies that emphasize logistics periodically reorganize its logistics functions in their attempt to find and keep the best design in the business environment which is rapidly changing [28]. The overall response capacity could be achieved through a greater sharing of information between partner organizations and a careful selection of suppliers by the purchaser. **Table 2** provides a summary of the components used in different studies applied to internal logistics.

The novelty of this chapter is related to a new definition of internal logistics which implies a description of its component parts according to this new definition, and the procedure for evaluating its level in anyone company or factory.

Internal logistics components	Source
Physical flow	
Receipt	[29–33]
Warehouse	[25, 34–39]
Supply	[40–42].
Movement	[32, 34, 43–47]
Working in process	[19, 48–53]
Internal transport	[39, 54–58]
Picking/packing	[32,59–65]
Information flow	
Information technology	[40, 55, 66–71]
Planning and material control	[54, 68, 72–75]
Planning and production control	[74, 76–79]
Customer service	[40, 80–84]
Order processing	[48, 68, 73, 85–87]
Inventory management	[32, 34, 41, 66, 68, 88–91]
Source: Prepared by authors (2020).	

Table 2.
Theoretical background of the internal logistics components.

3. Materials and methods

3.1 Approach developed for evaluating the components parts of internal logistics

There were identified a few systematic attempts, proposals and techniques that improve the manufacturing system and the internal logistics and their related performance. They have to be able to assess the dynamics of production and the corresponding improvement and taking into account environmental issues. For developing the objective of the chapter there were carried out the following steps:

- To identify the composition of the industrial pole of Manaus according to the different industries that compound it.
- To identify in literature, the component parts and definitions of internal logistics.
- To develop a new definition of internal logistics and its component parts.
- To confront and discuss the new definition and its component parts with industry professionals through surveys and interviews.
- When there were defined the component parts, it was elaborated a survey of then questions for assessing each one of the component part using a Likert scale of five points. This survey and questionnaire was also discussed with engineers and researchers that deal with internal logistics and supply chain.

- The survey was applied to the different industries of the industrial pole of Manaus.

- With all these information, then there were analyzed and decided which tools would be used for assessing the internal logistics according to its new definition and according to its components parts.

- It was stablished an Internal Logistics Index for evaluating its level in any company or factory

- For this porpoise there were used an Excel tab, the fuzzy logic, and the neural networks.

All this procedure is explained in detail below.

The Industrial Pole of Manaus has more than 565 companies of small, medium and large size, involving seven subsectors of different branches of activities, which can be seen in **Figure 1**. The emphasis of the companies to be researched will focus in the two-wheel sector that is the 16.77% of the PIM billing. The research will be developed in companies of medium, large and small size.

To assess the internal logistics, a survey was conducted to different companies of the Industrial Pole of Manaus. For data had statistical significance, it was analyzed what size the sample should have.

The "right" sample magnitude for a specific application depends on many factors, such as costs, administrative aspects, level of precision, level of reliability, variability within the population or subpopulation of interest and specimen method.

These factors interact in multifaceted ways. Although a consideration of all the variations is beyond the scope of this chapter, the remainder of this epigraph covers a situation that commonly occurs with simple random samples: How to find the minimum sample magnitude that offers the desired precision.

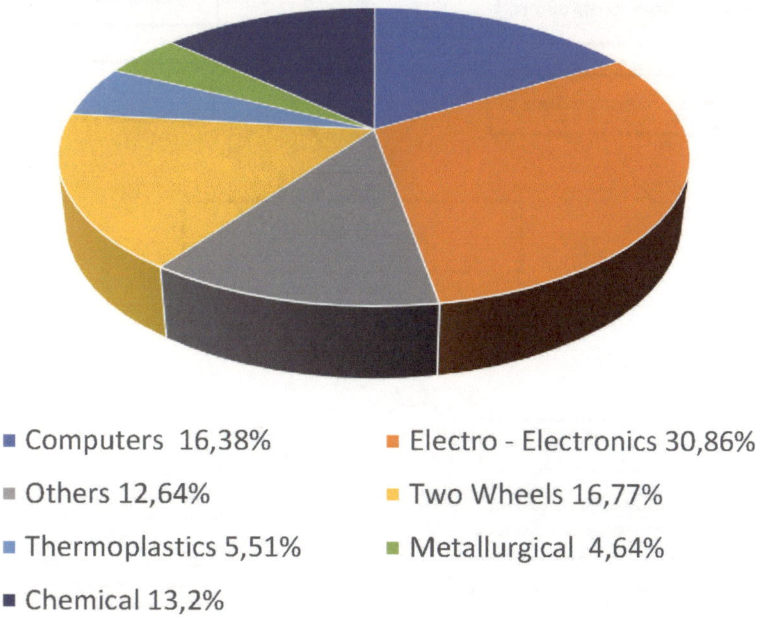

Figure 1.
Share of activities of sub-sectors in sales of the industrial pole of Manaus in the period from January to February 2015 (calculated based on sells in dollars). Source: Suframa – Industrial indicators (2015).

For demonstrating that a process has been improved, it is necessary to measure the process competence before and after improvements are implemented. This permits to measure the process improvement (e.g., defect reduction or productivity growth) and translate the effects into a projected financial result – something that corporate leaders can understand and appreciate. Determining sample dimension is a vital topic because samples that are too large may waste time, resources and money, while samples that are too small may lead to inaccurate results.

In the case of the industrial pole of Manaus, it is composed for 565 companies, and using the formulation expressed in [92], the number of companies to be considered for a good statistical representation has to be more than 60 companies.

Analyzing the sectors of the Industrial Pole of Manaus, it was possible to identify the components to assess the internal logistics. They were redesigned through interactions with business professionals from different companies in order to obtain the greatest possible standardization of component parts of internal logistics. **Figure 2** shows these parts. From this picture can be observed that there are component parts that have to do with the physical flow and other with the information flow.

Each component part of the figure above was evaluated by 10 properties or pertinent questions reflecting the respective training component behavior for performance, supported by Likert scale of 1 to 5, with 1 indicating little or no adhesion and 5 full adhesions between the question versus practice where each part can reach a maximum of 50 points is that the resulting properties of 10 x 5 points, and a total of 130 questions as a result of the 13 component parts of 10 questions each.

The questionnaire applied in enterprises, medium and large, the following segments: electronics, appliances, components and two wheels. In March 2015, 539

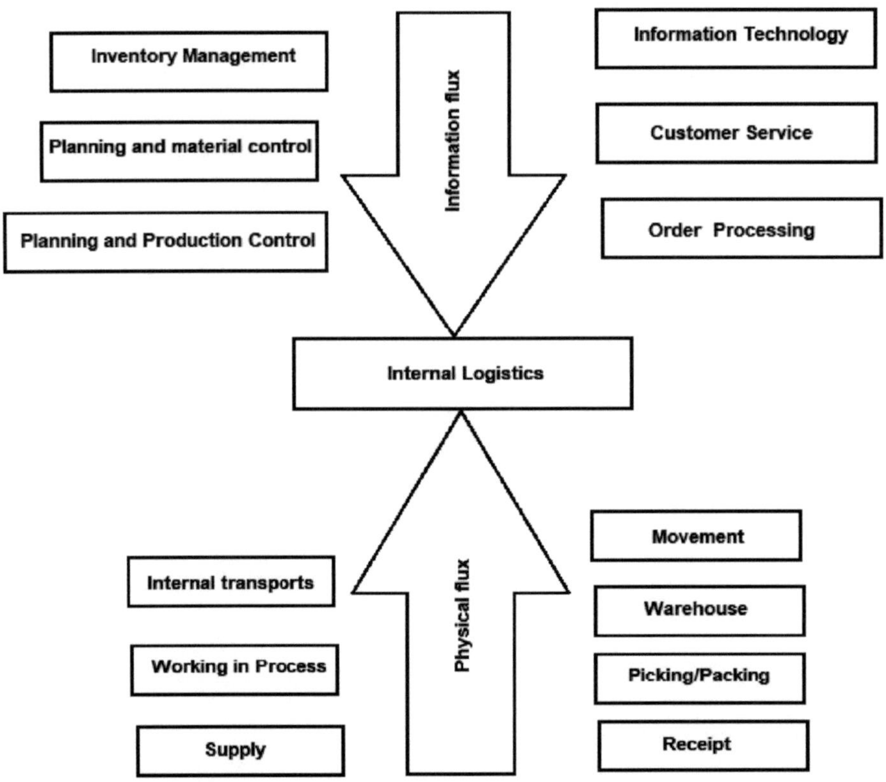

Figure 2.
Component parts of the internal logistics. Source: Authors (2020).

invitations were sent to participate in the study. A total of 327 responses received, being considered only 140 (25.97%) fit and consistent for research. The sample characteristics are shown in **Table 3**.

3.2 Assessment of the weight of the component parts of internal logistics by companies

To evaluate the weight of each component part of the Internal Logistics were sent a survey to 93 companies to analyze them and to attribute a weight of importance in a Likert scale of 1–5 where 1 was minor and five very important according to the particularity and priority that represents the component parts for the aforementioned companies. In **Table 4** there are offered the results of three of the companies investigated.

It was found that depending on the company and its respective sector, the priorities and the degree of importance may be subject to change and therefore affect the performance of internal logistics index.

The maximum score that each company can get is 65 points, which is the result of the multiplication of the 13 items by the maximum value of each item according to the Likert scale. It is noted for example that the company 1 attributed a very low note for the items: Storage, WIP and internal transport, while companies 2 and 3 attributed notes 5, 5 and 4 respectively for these same items, therefore, it follows which depending on the sector and type of production, whether continuous or discrete, the degree of importance may change. An arithmetic mean of the 3 companies was also developed in this tabulation and it was appreciated that from the maximum possible score of 65 points, company 1 scored 35 points, followed by 61 points by the company 2 and finally the company 3 with 59 points, and the arithmetic mean was 51.67 points.

3.3 Evaluation of the internal logistics index by companies using excel solver

Based on the literature investigated was developed the structure of diagnostic model of the component parts of the internal logistics, its filling, testing and subsequent validation. They were developed 10 questions to assess each property and was conducted a survey in different companies. These questions were developed based on the literature review, the survey results according to the criteria of specialists of logistics management, and consulting and business managers. It was developed an Excel tab to evaluate the performance of each of the component parts of the internal logistics as well as the Internal Logistics Index of a company.

Industrial sector	%	Size dcounter
Domestic appliances	15	100–250
Electronics	35	100–250
Components	40	250–500
Towels	9	500–1000
Others	1	100–250

Source: Authors based on survey (2016).

Table 3.
Firms demographics: Industry and size.

	Component part	Assigned weight by each company				
		Company 1	Company 2	Company 3	Arithmetic mean	Company 3 Weight
Component parts of the Internal Logistics	Receipt	3	5	4	4.00	6.8%
	Handling and movement	2	4	3	3.00	5.1%
	Picking/packing	4	4	4	4.00	6.8%
	Storage	1	5	5	3.67	8.5%
	Stocks management	2	5	5	4.00	8.5%
	Supplying	5	5	5	5.00	8.5%
	PMC- planning and material control	2	5	5	4.00	8.5%
	PPC - planning and production control	2	5	5	4.00	8.5%
	WIP- working in process	1	5	5	3.67	8.5%
	Order processing	4	4	5	4.33	8.5%
	Internal transports	1	4	4	3.00	6.8%
	Customer support	5	5	5	5.00	8.5%
	I. T. information technology	3	5	4	4.00	6.8%
	Internal Logistic Index	35	61	59	51.67	100%

Source: Authors.

Table 4.
Answers from the companies on the degree of importance of the elements of internal logistics.

The Excel Tab developed to calculate the Internal Logistic Index was based on the following equations:

$$ILI = \sum_{i=1}^{13}\left[\left(\frac{Z_i}{100}\right).W_i\right] \quad (1)$$

where ILI = General index of the performance of the Internal Logistics; W_i = Weight attributed to each component part i; i = each of the properties analyzed; Z_i = value reached in % for the property i based on the sum of all values given to each parameter of the corresponding property of the Likert scale from 1 to 5 and divided by the maximum possible value to reach in% i.e.:

$$Z_i = \sum_{j=1}^{10}\left(\frac{P_j.L_j}{50}\right).100 \quad (2)$$

where P_j = Each of the parameters that assess the Z_i property (always it going to assume the value 1 in the above expression); L_j = Value assigned to the parameter P_j at the Likert scale from 1 a 5.

3.4 Assessment of internal logistics through fuzzy logic

3.4.1 The fuzzy logic and internal logistics

Assessing the Internal Logistic Index of a Company is a very complex task due in some case to the lack of information and in other cases to the excess of information for decision-making. This leads to difficulty in defining, measuring and monitoring of objectives and targets to set rates compliance associated with measuring the performance of the Internal Logistics [93]. In response to these challenges of business management there have been emerged theories, approaches and methodologies (flexibility, resilience, etc.) using tools such as fuzzy logic for reliable solutions that adapt easily to changing parameters of imprecision [94].

In addition to the treatment of imprecise environments, another emerging challenge is to achieve that the measurement of organizational performance transcends the traditional financial approach and to be conducted throughout with suitable means to new generations of applications in the management of internal logistics.

3.4.2 Method of fuzzy inference

A fuzzy inference method allows deriving conclusions (a fuzzy value) from a set of if-then rules and a set of input values to the system, by applying composition ratios. The two inference methods commonly used are the Mamdani introduced by Mamdani and Assilian [95] and the TSK (Takagi-Sugeno-Kang) proposed by Takagi and Sugeno [96].

The main difference between these methods is the consequent type of the fuzzy rule. The systems Mamdani type use fuzzy sets as consistent rule and TSK used linear functions of the input variables with discrete data outputs. In this research the type Mamdani inference system (**Figure 3**) with outputs continuous values is used.

To facilitate the modeling of the problem in fuzzy logic, it was used the Fuzzy Logic Toolbox ™ of MATLAB software. The steps for formulating the model of fuzzy inference of Mamdani type were [97, 98]:

3.4.3 Selection of indicators

Performance measurement of the internal logistics can be based on the selection and definition of indicators used to evaluate the efficiency and effectiveness of its operations. Indicators should have a holistic approach and facilitate the implementation of initiatives for improvement. Indicators selected for the proposed fuzzy model to measure the performance of the Internal Logistics of the company studied are described in **Table 5**. The components were grouped into larger groups as shown therein. A letter from A to B. defines each group.

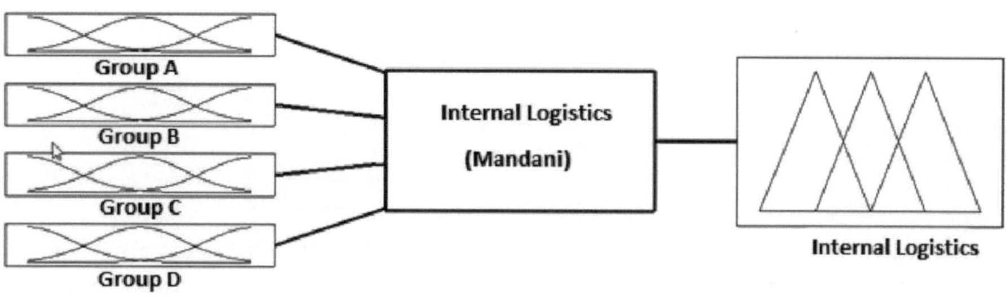

Figure 3.
Fuzzy system for asses the internal logistics. Source: Authors (2020).

A	B	C	D
Receipt	WIP- working in process	PPC - planning and production control	Picking/packing
Storage	Handling and movement	PMC- planning and material control	
Stocks management	Supplying	Customer support	
I. T. information technology	Internal transports	Order processing	

Source: Authors.

Table 5.
Component parts of internal logistics.

Component parts/groups	A	B	C	D	LI
1	M	B	M	B	B
2	M	G	G	B	M
3	G	B	M	B	B
4	G	G	G	B	M
5	B	G	G	B	B
6	G	B	G	G	M
7	B	G	B	M	B
8	M	G	B	M	M
9	B	M	M	M	M
10	M	B	M	M	M
11	G	M	M	M	M
12	M	B	G	B	M
13	G	G	M	G	G
14	M	B	G	G	M
15	G	M	B	M	M
16	B	B	M	B	B
17	M	M	G	G	G
18	G	M	B	M	M
19	B	B	M	G	M
20	M	G	B	G	M
21	G	M	B	G	M
22	B	M	B	G	M
23	B	G	B	M	B
24	B	M	G	R	M

Source: Authors (2020).
M = medium, G = good, B = bad.

Table 6.
Fuzzy rules.

Each component part of the previous group is evaluated using 10 pertinent questions that reflect the behavior of the respective part.

3.4.4 Development of fuzzy rules

The model has 24 rules, which were created from the experience of official logistics industry specialists and numerical data from surveys and they are offered in **Table 6**.

The model has four inputs that are the four groups described in **Table 5** and an output that is the Internal Logistics Index. The parameters of pertinence functions associated with each variable were also specified. There were adjusted all inference functions and the defuzzification method used. The rules of an inference engine of a fuzzy system has to be made by experts, or learned by the system, in this case using neural networks to strengthen future decision-making. For making the rules in this problem were used criteria of 20 specialists in the field of Internal logistics.

3.5 Assessment of internal logistics using neural networks

One problem with the method applied in the previous section is that the user of Excel tab has to assign a weight to each component part of the internal logistics based on in his own experience, which naturally influences the overall index of internal logistics of a company. Attempting to avoid subjectivity in determining this rate, it was looked to the technique of artificial neural networks. To analyze the Internal Logistics of an industrial company was used the Internal Logistics Index (ILI), evaluated between 0 and 100%. This index is calculated based on the values assigned to each of the internal logistics properties between 0 and 50 according to the 10 parameters of evaluation of each property in the Likert scale of 1 to 5.

There were selected the same 10 companies of the Industrial Pole of Manaus for their study and analysis, all of them belonging to the productive sector.

Property	Company number and value of the performance of each property									
	E1	E2	E3	E4	E5	E6	E7	E8	E9	E10
Receipt	30	35	36	50	42	47	32	18	50	22
Handling and movement	20	44	23	50	35	45	34	15	39	19
Picking/packing	50	22	32	48	26	34	45	21	40	23
Storage	40	33	41	43	12	50	23	32	35	34
Stocks management	30	11	25	21	18	16	15	18	18	35
Supplying	24	44	18	50	22	24	18	43	25	23
PMC- planning and material control	33	50	15	39	19	35	22	32	35	41
PP - planning and production control	28	33	21	40	23	32	18	19	28	19
WIP- working in process	16	22	32	35	34	18	34	42	47	32
Order processing	41	11	18	18	35	33	45	35	45	34
Internal transports	33	33	43	25	23	22	35	26	34	45
Customer support	23	45	32	35	41	43	25	12	50	23
I. T. information technology	33	44	19	28	19	21	18	18	16	15

Source: Authors.

Table 7.
Different properties that compose the Internal Logistic Index of a company and their values for the 10 companies evaluated.

	Company number and Internal Logistic Index possible according to the performance value of each property									
	E1	E2	E3	E4	E5	E6	E7	E8	E9	E10
ILI	65	75	70	67	78	60	70	65	78	65

Source: Authors.

Table 8.
Possible internal logistics indexes (ILI) for each company (targets) *for training the ANN.*

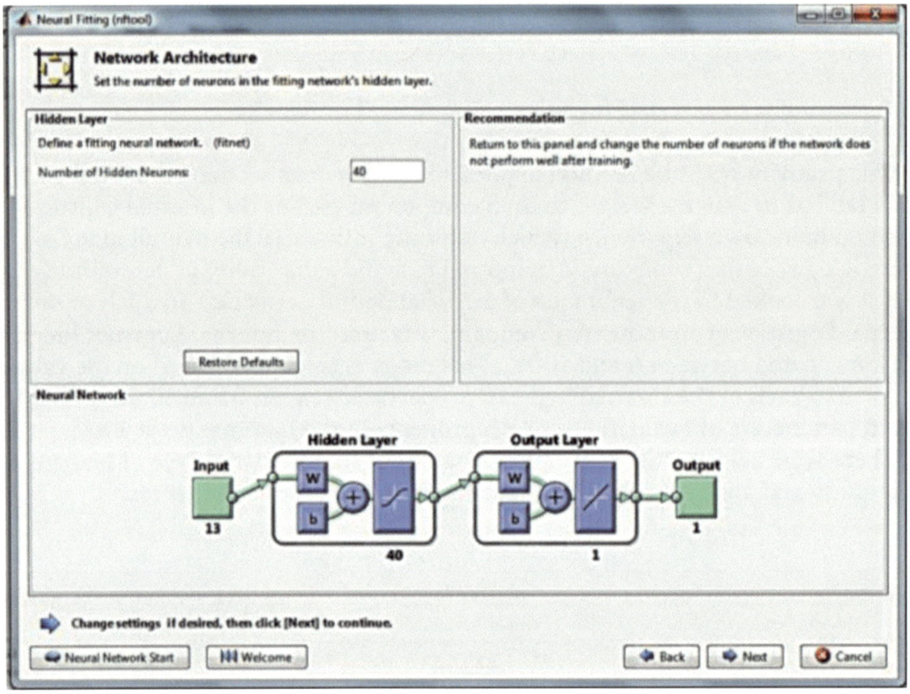

Figure 4.
Artificial neural network implemented in MATLAB. Source: Authors (from MATLAB).

It was proposed to the ANN to determine the Internal Logistic Index of 10 companies in the industrial pole of Manaus. The values of the properties of the component parts of the 10 companies are given in **Table 7**.

The desired Internal Logistics Indexes for the aforementioned companies (supervised training), in order to train the ANN are given in **Table 8**.

In **Figure 4** it is showed the Architecture of the ANN implemented in MATLAB. In order to achieve reliable results, the network was trained five times. In **Figures 5** and **6** the training process is displayed.

4. Result analysis

It was chosen randomly the company three to answer questionnaires regarding the 13 elements or components parts of internal logistics. This company filled the Excel tab, reaching a score in% of each property that was multiplied by the weights assigned in Table to each property. This company reached a general index of 79.17% for Internal Logistics as it is shown in **Table 9**.

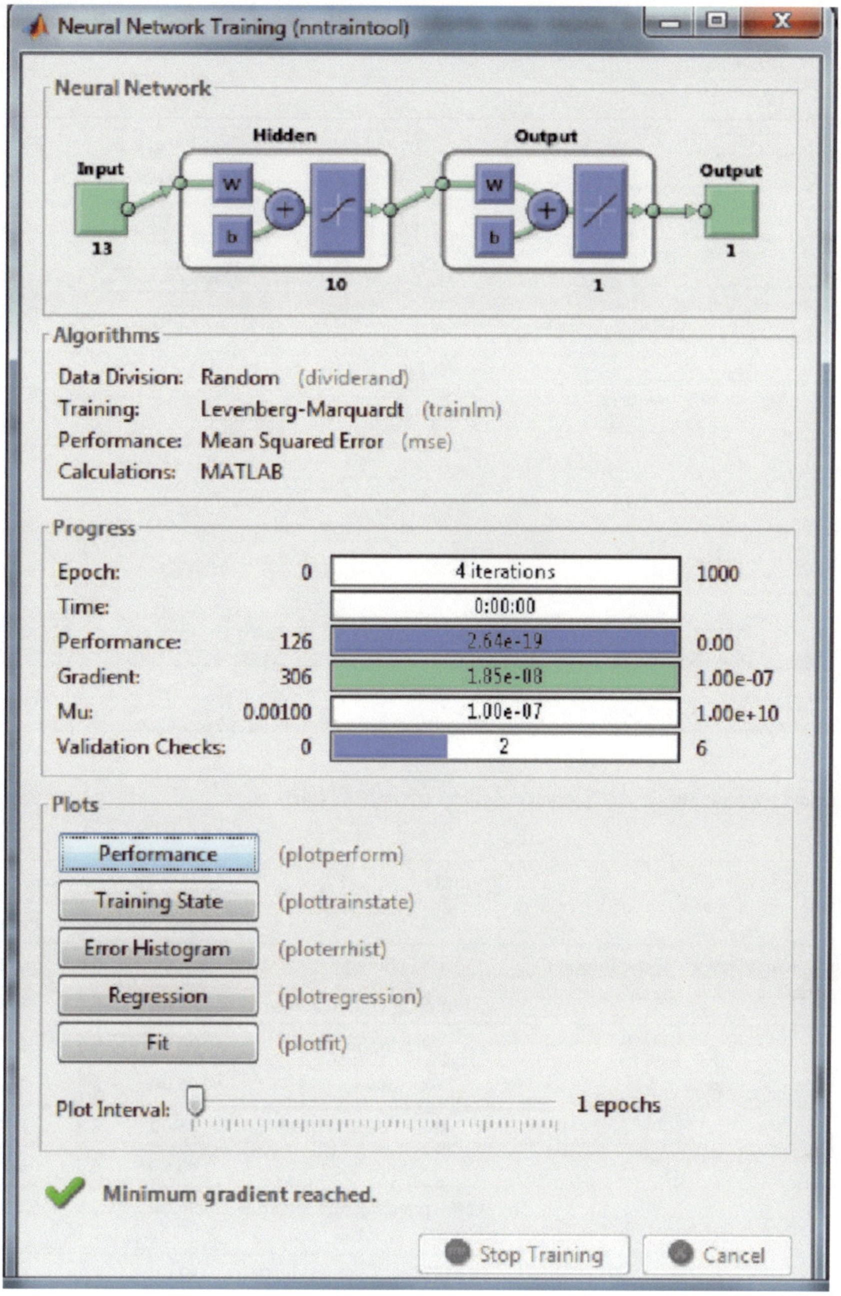

Figure 5.
Neural network training state. Source: Authors (from MATLAB).

It was found that depending on the company and its respective sector, the priorities and the degree of importance may be subject to changes and therefore affect the performance index of internal logistics. The maximum score that each company can get is 65 points, which is the result of the multiplication of the 13 items by the maximum value of each item according to the Likert scale. It is noted for example that the company 1 attributed a very low note for the items: Storage, WIP and internal transport, while companies 2 and 3 attributed notes 5, 5 and 4 respectively for these same items, therefore, it follows which depending on the

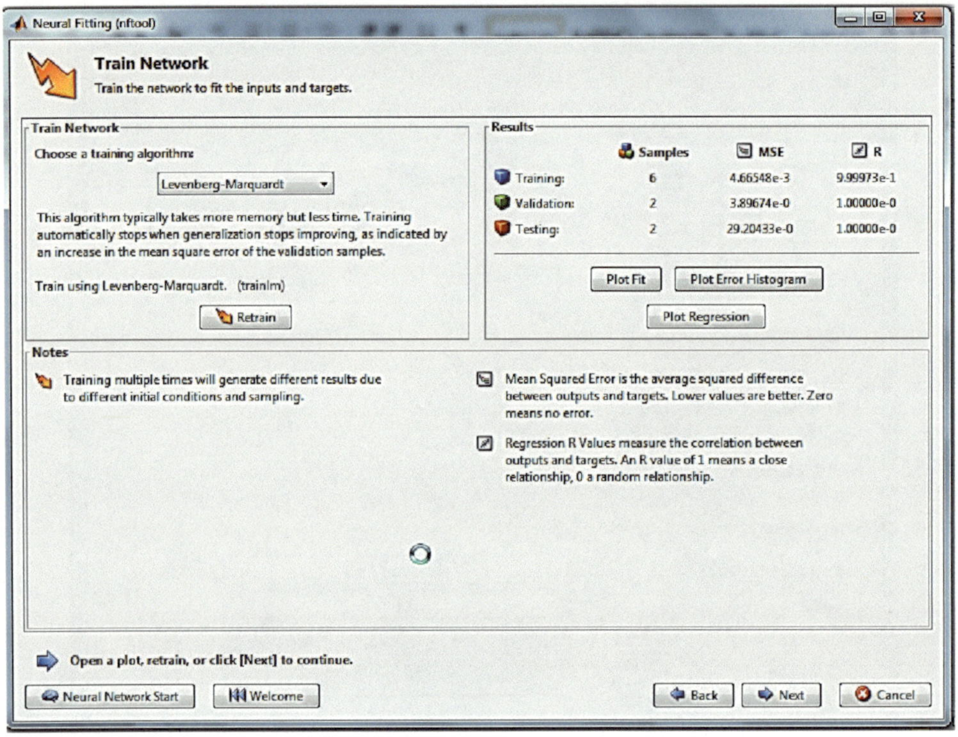

Figure 6.
Training and retraining of the ANN. Source: Authors (from MATLAB).

	Property	Performance		
		Percent	Weight	Points
Component elements of the Internal Logistics	Receipt	96.00	6.8	6.53
	Handling and movement	88.00	5.1	4.49
	Picking/packing	90.00	6.8	6.12
	Storage	86.00	8.5	7.31
	Stocks management	86.00	8.5	7.31
	Supplying	46.00	8.5	3.91
	PMC- planning and material control	94.00	8.5	4.62
	PP - planning and production control	92.00	8.5	4.62
	WIP- working in process	88.00	8.5	7.48
	Order processing	88.00	8.5	7.48
	Internal transports	90.00	6.8	6.12
	Customer support	88.00	8.5	7.48
	I. T. information technology	84.00	6.8	5.71
	General Internal Logistic Index	79.17		

Source: Authors.

Table 9.
General internal logistic index of a company.

sector and the type of production, whether continuous or discrete, the degree of importance may change. An arithmetic mean of the values of the 3 companies was also developed in this tabulation and it was noted that from the maximum possible score of 65 points, the company 1 scored 35 points, followed by 61 points by the company 2, then the company 3 with 59 points, and the arithmetic average was 51.67 points.

4.1 Application of fuzzy logic to determine the internal logistics index of a company in the industrial pole of Manaus. Case study

4.1.1 Internal logistics index company case study

The toolbox of fuzzy logic implemented in MATLAB with the four groups of the diffuse model is shown in **Figure 7**. In **Figure 8** the values of Internal Logistics Index reached according to the input variables are shown. For example, if each group (from A to D) have an average value (5 points), then the Internal Logistics Index reaches a value of 37.4 points. The MATLAB allows vary input values, and consequently modifying the value of Internal Logistics Index.

Another way of demonstrate the results between two groups and the Internal Logistic Index can be analyzed from **Figure 8**, where there are represented the A and B groups with the value of 5 points for each group. The maximal value of the Internal Logistic Index in this case will be of 37.50% as it is shown in **Figure 9**.

It was established a comparative analysis between results of both models: Excel Tab versus Fuzzy Logic. The obtained results of the Internal Logistic Index by using the Excel tab was of 79.17% when assessing the 13 component parts. These

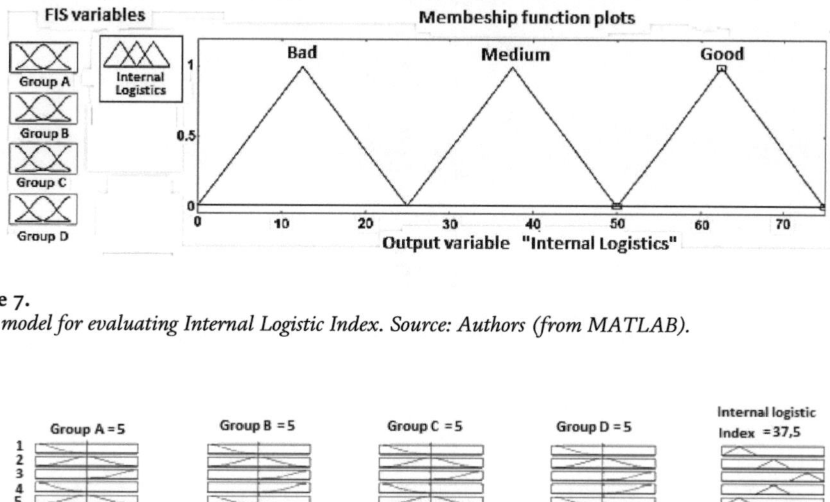

Figure 7.
Fuzzy model for evaluating Internal Logistic Index. Source: Authors (from MATLAB).

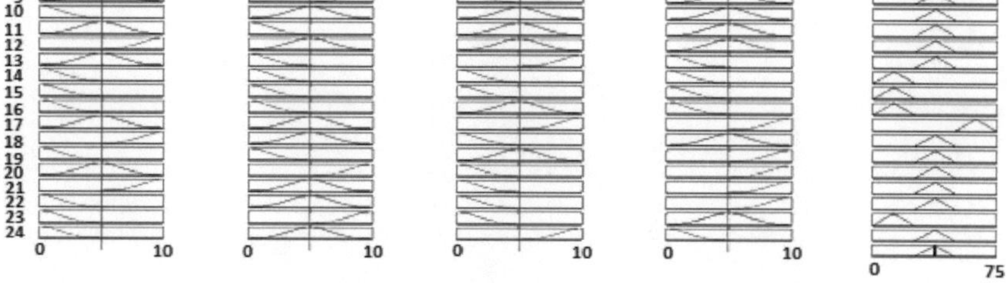

Figure 8.
Internal Logistics Index according to the input variables. Source: Authors (from MATLAB).

component parts of the same company were grouped how was cited before in four groups: A, B, C and D, supported by 24 rules developed and applied in the Fuzzy Logic toolbox from MATLAB. Each input variable can reach a value between 0 and 10. If each input variable reach the average value of 5 points, the Internal Logistic

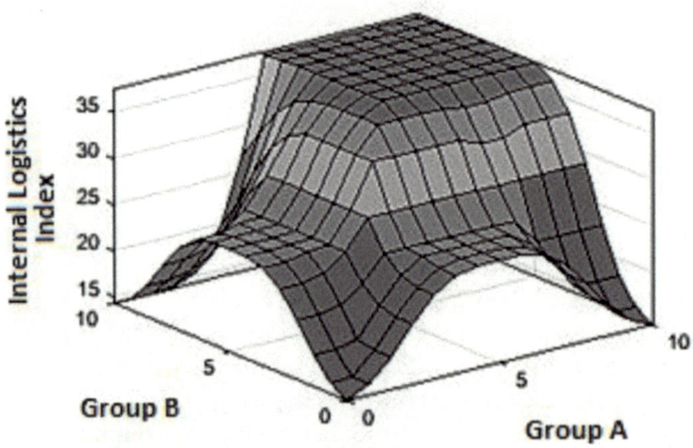

Figure 9.
Surface of values of the internal logistic index according to the input variables. Source: Authors (from MATLAB).

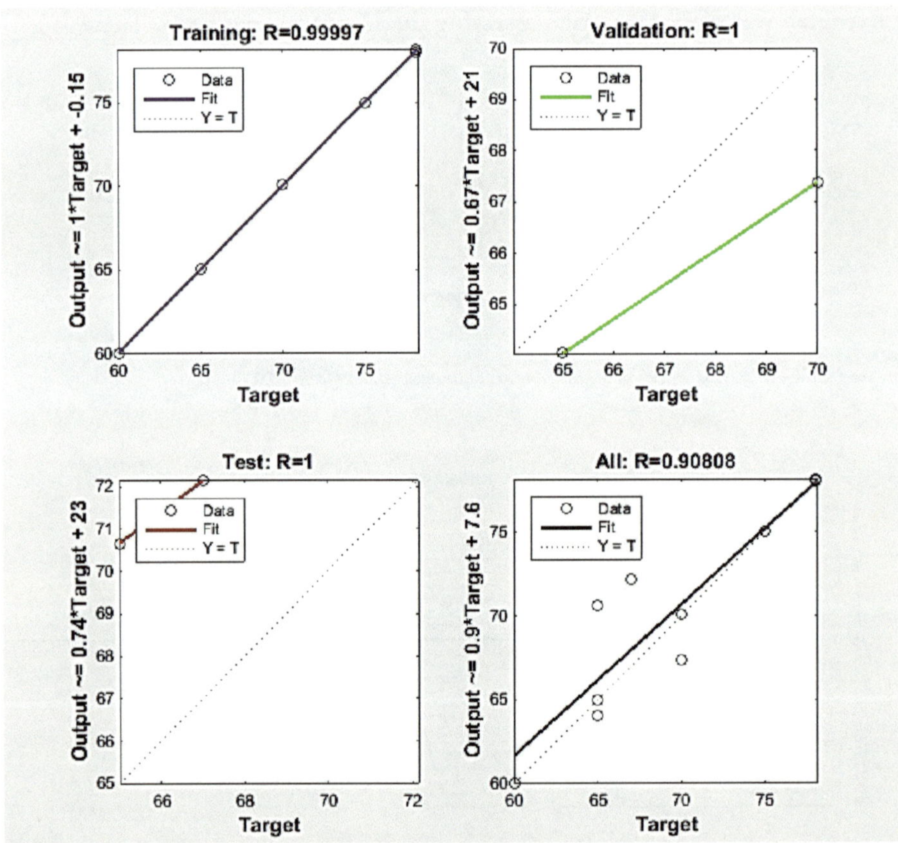

Figure 10.
Validation of the artificial neural network errors. Source: Authors (from MATLAB).

Company	1	2	3	4	5	6	7	8	9	10
ILI	70.65	75.00	67.37	72.13	78.15	60.03	70.05	65.0	78.0	64.04
Error in %	2.65	0.006	2.62	3.13	0.15	0.03	0.059	0.004	0.009	0.95

Source: Authors (from MATLAB).

Table 10.
Obtained values of the internal logistics indexes and errors of these values in the 10 companies studied.

Index will be of 37.50%. Following the same procedure and way of thinking the top possible value of Internal Logistic Index will be of 75%, versus 79.17% obtained by Excel Tab method, demonstrating similarity between the both tools and a precision on the order of 95% of the results.

4.2 Application of neural networks to determine the rate of internal logistics of an industrial company. Case studies

Validation errors of the neural network are shown in **Figure 10**.

4.3 Internal logistic indexes of the studied companies

The ANN enabled in MATLAB with data values of the 13 Internal Logistic Properties from the 10 companies was processed. The values of the indexes of Internal Logistics as well as their possible are given in **Table 10**.

5. Conclusions

In this chapter three approaches and their expressions to assess the internal logistics of a company are established. The first method was based on dividing the internal logistics in 13 properties, having each property 10 indicators that were evaluated between 1 and 5 points. This leads to the maximum value of Internal Logistics Index (ILI) for each company can reach up to 100 points according to the weight stablished for each indicator.

The second approach was based on the fuzzy logic and the third one was based on neural networks.

When assessing the Internal Logistics Index using the Excel tab developed or by the method of Artificial Neural Networks, very similar values consistent with the reality of the companies studied were obtained, demonstrating the validity of both methods.

The methodological approach developed for the definition of the three models contains all the steps and procedures, allowing replication of the research, which is as important as the application of the developed models at the companies.

When assessing this parameter, using the Excel tab or developed by the method of fuzzy logic, similar values were obtained in line with the reality of the company analyzed, indicating the rationality of both methods.

The approach through the Fuzzy logic allows assess the rate of internal logistics for any position of the input variables, which can obtain a value between 0 and 10, depending on the appreciation of the user of this procedure. In the case of the company studied, specialists gave more emphasis to groups A and C.

Acknowledgements

To the "Instituto de Tecnologia Galileo da Amazônia (ITEGAM)," to the "Universidade Federal de Santa Catarina(UFSC)," and the National Council for Scientific and Technological Development (Conselho Nacional de Desenvolvimento Científico e Tecnológico, CNPq-BRAZIL) Productivity of Research Funds Processes 301811/2019-9 for their support to the development of this research.

Conflict of interest

The authors declare no conflict of interest.

A. Appendix

On a scale of 5 (high) to 1 (low), how important are the following issues in your firm's logistics efforts?

A.1 Customer service

1. Is the team geared to provide quality and always prioritize customer?
2. Is there any ERP system that allows integration and exchange of information between companies?
3. Is there an analysis of customer satisfaction?
4. Is there guidance for team procedures?
5. Is there guidance on the importance of customer response speed?
6. Are there a scheduled visit and contacts with current customers and potential and no formal or research and survey on competitors and new niches?
7. Is there a metric to analyze the value perceived by the customer as the supply of goods and services?
8. Are there procedures that allow flexibility for customer service?
9. Is there a division of the teams and the calls of customers by importance – ABC and analysis or particular strategy?
10. All valuable contacts with customers are recorded in minutes or similar?

A.2 PCM – Planning and materials control

1. The PCM is responsible for planning and control of all production inputs of the company?

2. Is there a defined routine priority of PCM activities?

3. Is there an ERP information system that supports the planning materials, e.g. MRP and no analysis of the messages: Action, Exception and Correction?

4. Is there a schedule to run MRP, e.g. weekly, monthly or every 15 days, and metrics to analyze the effectiveness?

5. Is there a time horizon that the PCM plans to purchase inputs; e.g. monthly or every three months or 20 weeks of purchase orders range?

6. Is there a working definition for supply based on MTO - make to order or MTS - make to stock; or customer sales plan?

7. Is there a performance evaluation of the inputs delivery compliments the company (supplier evaluation)?

8. Is there analysis of purchasing volumes for technical ABC analysis or other?

9. Is there a systematic review on the BOM list of theory versus practice?

10. There are policy setting for Horizon Planning, Redesign, expediting and follow-up?

A.3 PCP - Planning and control production

1. Are there management decisions as: quality (what to produce - design and control characteristics); Process (how to produce - facilities, equipment); Capacity (when produce - planning and programming); Inventories (with what to produce and when - requirements of materials and market); Workforce (who produce - qualification, performance, motivation)?

2. Is there a definition as Frequency: Order one or more of an order demand and the demand is constant. Variable demand. Independent demand or demand dependent?

3. The Lead time or supply time are: Lead Team Lead time constant variable and how to support management systems use: Continuous review periodic. Revision or MRP?

4. The PCP is considered along with manufacturing as strategic for the company; and working with: Production Plan. Plan Master Production Scheduling and production?

5. The PCP adopts some criteria sequencing or prioritization techniques in planning the company's production?

6. The PCP has a flow of information and interacts with areas: Production, Capacity, technology, Human Resources, Quality, Engineering, marketing, Maintenance, Logistics and Development?

7. The PCP works with forecasting techniques?

8. The PCP - controls the analysis of the standard cost of open orders and there are closures routines of orders in the month of opening and has OEE indicator?

9. The CFP has the practice of daily or weekly schedule issue and if there is time production monitoring schedule time?

10. The CFP has operations in three hierarchical levels: Strategic, Tactical, and Operational?

A.4 Picking/packing/packaging

1. Is there a practice of picking?

2. Is there the practice of Packing?

3. Is there the practice of obliterating the specifications of inputs to be released to production?

4. The packaging used obey some amount of standardization, weight, color, or other?

5. The separation or preparation of the application or request follows a priority basis the production and or PCP?

6. Is there any guidance on the practice of picking by: zone, batch or discrete?

7. Is there any validity packaging control?

8. The practice of packaging or Packing takes place in the industrial area or shipping?

9. Are there specific areas for these packaging practices?

10. Are there teams trained in these practices? And if there is time control for each operation?

A.5 Order processing

1. Are there set procedures on specific date in the month of receipt of order or forecast customer?

2. Is there a lot of tolerance and or quantity for order fulfillment?

3. The customer demands, both firm orders and forecast are processed by the customer service area and PCP or?

4. Upon receipt of applications and or forecast undergo an analysis sieve critical about the information contained in it?

5. Is there any responsibility of processing the request on the internal expediting to delivery to the customer?

6. Is there any reports on request of the situation and this feedback is sent to the client?

7. Is there the definition of a physical flow and information on the application?

8. The order processing system influences the PCP or logistics performance as a whole?

9. The company uses some technology strategies: Edi- Electronic Data Interchange Barcode RFID- Radio Frequency Identification QR- Quick Response ECR- Efficient Consumer Response CPFR- Collaborative Planning, Forecasting and Replenishment;

10. Is there a setting to work taking into account the order cycle?

A.6 Receiving

1. Are there procedure defining criteria for receiving the aspects: qualitative, quantitative, fiscal and administrative, e.g. variation of tolerance of quantity, quality criteria and time of receipt?

2. All supplies and materials are received only by formal authorization or by order?

3. Is there separation of powers between the fiscal and physical receipt?

4. Is there separation of powers of the incoming teams and warehouse?

5. Is there some defined criteria for quality assurance of production inputs?

6. Is there an area for segregation of inputs nonconforming when detected in receiving?

7. Are there different negotiations for tax problems cases: interstate, local and regional?

8. The flow of receiving inputs and or materials are balanced according to the available warehouse area?

9. Are there appropriate or docks and ramp leveler for the reception of the materials?

10. Are there standard definition pallets and loads unitized arrive?

A.7 Supply

1. The supply of the manufacturing area is performed by the logistics team?.

2. Usually follows a daily schedule: OP, OM or OS?

3. The team that performs the supply meets the inputs?

4. The team practices management view?

5. Is there oriented practice of Housekeeping/5 s?

6. Usually communicate to existing higher losses in the process?

7. The team that performs the supply periodically receives training?

8. Is there control over the supply time to avoid any line stops and or machines?

9. Is there work instructions for each type of supply?

10. Is there a metric to determine the contents of delays and incorrect or supplies?

A.8 Warehouse

1. Is there a procedure that meets the ISO 9000 or 14,000 or similar standards?

2. All warehouse operations are computerized with ERP system?

3. Is there any software that facilitates addressing allocation of material type: WMS or equivalent?

4. Is there a process to expedite the receipt of materials e.g. bar code or RFID?

5. The layout takes into account the minimization of the distance between the area of the warehouse and efficient supply and provides flexibility?

6. The warehouse is suitable for safety standards? And it takes into account at the time of storage: the density, selectivity, frequency output/consumption and costs?

7. The warehouse is structured with metal structures or equivalent within the technical standards and provides earning capacity allowing vertical storage?

8. There flexibility both the fixed storage as random to allow optimizing the use of available facilities?

9. Professionals working in the warehouse are trained, qualified and trained to operate equipment such as forklifts, monorails, and other equipment?

10. The warehouse professionals are aware about the importance of: store, locate, protect and preserve the materials purchased or developed?

A.9 Inventory management

1. Is there inventory policy defined?

2. Is there harmony between: man versus machines versus materials and includes the integration of the materials and information flows?

3. It is known to capacity position and if there is flow control between inputs versus outputs of materials?

4. Studies aimed at harmony between the receipt flow and manufacturing supply?

5. The factory supplies: exit of warehouse supplies takes into account: the criteria: FIFO, LIFO, SHELF LIFE or other?

6. The low of the stock of inputs occur through: PICKING, GOOD, back flushing, or other?

7. Is there an ERP system that enables obtaining daily movement of input and output information?

8. Are there analyzes of: Giro, Coverage, Break, obsolescence, accuracy or automatic inventory replenishment?

9. Is it defined an inventory policy: Rotary, Cyclical, or Monthly and annual.

10. Is there specific team of executors?

A.10 Movement

1. Is there specific training for staff working with handling and movement of inputs?

2. Is there a checklist or guide to guide staff as to the appropriate local drive; attention to inputs and or equipment?

3. The movement of staff is the same answer: the receipt, allocation and industrial supply?

4. In case of handling trucks, the team has CNH or course forklift operator?

5. The staff is trained to obey some as NR or orientation CIPA?

6. Are there training program/periodic retraining the team?

7. All movements of stocks are properly recorded?

8. The movement of staff is responsible for the low inputs of stocks?

9. Is there any training on the observation of the symbols of packaging as stacking, storage and handling?

10. Is there a control or measurement standard time for handling between warehouse receipt versus plant?

A.11 Information technology

1. The Company uses a corporate ERP?

2. The company has systematically to collect, check and update information for decision making and performance?

3. The company's information system is aligned with the Strategic Planning?

4. The information system is available for the entire company and all activities are developed in the ERP environment, avoiding work in parallel spreadsheets?

5. Are there indicators available and understood by employees involved?

6. There ERP system integration with their customers through: EDI, CRM, VMI and or SAP?

7. The MRP and MRP II module is available for business?

8. The factory operating costs are closed monthly and compared with the initially foreseen (e.g. production order closure: standard estimated cost vs. actual cost occurred)?

9. Is there concern in updating and security of ERP (examples: Backup and version upgrades)?

10. Does the ERP provide: integration, speed, availability of information for decision making and the company has access management with logins and responsibilities levels?

A.12 Internal transport

1. Is there a specific area to take care of this activity or if it is subject to internal logistics activities?

2. Are there monitoring the preventive maintenance of equipment?

3. In the case of trucks, repairs and repairs are carried out by specialized companies?

4. Is there an equipment replacement plan when they are with their completed depreciation?

5. The amount of internal transport equipment are adequate demand?

6. People who handle or operate and are trained and qualified?

7. The company has in its internal operations: manual operation of equipment and motorized equipment such as: strollers, manual and motorized lifts, forklifts, and others?

8. The company works with: monorail, monorails, overhead crane, conveyor belts, cranes, roller conveyors, tractors, dynamic roller conveyors?

9. Is there dangerous or flammable cargo transport, if so, follow some NR?

10. The people of this area have some equipment types: EPI and the company provides guidance on its use, compliance and aspects ergonomics?

A.13 WIP - work in process

1. Is there a policy that defines the inventory level in the process: be semi-finished or finished products or raw materials?

2. There's definition to return the inputs to the warehouse when a production order and or service is interrupted before its completion in month?

3. Is there concern at the closure of production orders and or services within the pre-set period?

4. There criterion set when spare or missing inputs for conclusion of production or service order?

5. Are there practical for use of certain order input to meet other unspecified; (In the case of common items)?

6. Are there practices of use of inputs in the process, other than those specified in document production? And if so, are changed in the engineering structure?

7. Are there different control when the packages are not within specifications?

8. Is there control and segregation in the plant for not conforming items?

9. Is there concern in accountability with senior management and or sector responsible for the required inputs versus consumed/met?

10. Is there control and analyze the evolution of inputs processes in monetary values to each inventory?

Author details

Orlem Pinheiro de Lima[1], Sandro Breval Santiago[2],
Carlos Manuel Rodríguez Taboada[3], Jorge Laureano Moya Rodríguez[4*],
Maida Bárbara Reyes Rodríguez[4], Márcia Ribeiro Maduro[1],
Paulo César Diniz de Araújo[1], Nilson José de Oliveira Junior[1],
André Luiz Nunes Zogahib[1] and José Carlos da Silva Lima[1]

1 Amazonas State University, Manaus, Brasil

2 Federal University of Amazonas, Manaus, Brasil

3 Federal University of Santa Catarina, Florianópolis, Brasil

4 Federal University of Bahia, Salvador de Bahia, Brasil

*Address all correspondence to: jorgemoyar@gmail.com

References

[1] Grant, D., Gestão de logística e cadeia de suprimentos. 2017: Saraiva Educação SA.

[2] Tien, N.H., D.B.H. Anh, and T.D. Thuc, Global supply chain and logistics management. 2019, Academic Publications, Dehli.

[3] Campos, G.H.F., et al. AS ATIVIDADES LOGÍSTICAS COMO REDUÇÃO DE CUSTOS E MELHORIA DE SERVIÇOS: ANÁLISES DE TRÊS CASOS. in VIII JORNACITEC-Jornada Científica e Tecnológica. 2019.

[4] Bowersox, D.J., et al., Gestão logística da cadeia de suprimentos. 2013: AMGH Editora.

[5] Ballou, R.H., Business logistics/ supply chain management: planning, organizing, and controlling the supply chain. 2007: Pearson Education India.

[6] Bowersox, D.J., D.J. Closs, and M.B. Cooper, Supply chain logistics management. Vol. 2. 2002: McGraw-Hill New York, NY.

[7] Azadian, F., A. Murat, and R.B. Chinnam, Integrated production and logistics planning: Contract manufacturing and choice of air/surface transportation. European Journal of Operational Research, 2015. 247(1): p. 113–123.

[8] Alkhatib, S.F., et al., A novel technique for evaluating and selecting logistics service providers based on the logistics resource view. Expert Systems with Applications, 2015. 42(20): p. 6976–6989.

[9] Adarme-Jaimes, W., M.D. Arango-Serna, and J.M. Cogollo-Flórez, Medición del desempeño para cadenas de abastecimiento en ambientes de imprecisión usando lógica difusa. Ingeniería y Universidad, 2012. 16(1): p. 95–115.

[10] Dotoli, M., et al., An integrated approach for warehouse analysis and optimization: A case study. Computers in Industry, 2015. 70: p. 56–69.

[11] Simchi-Levi, D., X. Chen, and J. Bramel, The logic of logistics: theory, algorithms, and applications for logistics management. 2013: Springer Science & Business Media.

[12] Bachár, M. and H. Makyšová, Evaluation of the impact of intelligent logistics elements on the efficiency of functioning internal logistics processes. Acta Tecnología, 2019. 5(3): p. 55–58.

[13] Zhang, H., et al., Measuring Logistics Efficiency in China Considering Technology Heterogeneity and Carbon Emission through a Meta-Frontier Model. Sustainability, 2020. 12 (19): p. 8157.

[14] Kisperska-Moron, D. and J. De Haan, Improving supply chain performance to satisfy final customers: "Leagile" experiences of a polish distributor. International Journal of Production Economics, 2011. 133(1): p. 127–134.

[15] Jonsson, P., Logistics and supply chain management. New York, 2008.

[16] Patel, P.C., A. Azadegan, and L.M. Ellram, The Effects of Strategic and Structural Supply Chain Orientation on Operational and Customer-Focused Performance. Decision Sciences, 2013. 44(4): p. 713–753.

[17] Gligor, D.M. and M.C. Holcomb, Antecedents and consequences of supply chain agility: establishing the link to firm performance. Journal of Business Logistics, 2012. 33(4): p. 295–308.

[18] Harrison, A. and R.I. Van Hoek, Logistics management and strategy:

competing through the supply chain. 2008: Pearson Education.

[19] Tan, K.C., A framework of supply chain management literature. European Journal of Purchasing & Supply Management, 2001. 7(1): p. 39–48.

[20] Shepherd, C. and H. Günter, Measuring supply chain performance: current research and future directions, in Behavioral Operations in Planning and Scheduling. 2010, Springer. p. 105–121.

[21] Teixeira, R., X. Koufteros, and X.D. Peng, Organizational structure, integration, and manufacturing performance: A conceptual model and propositions. Journal of Operations and Supply Chain Management, 2012. 5(1): p. 70–81.

[22] Granlund, A. and M. Wiktorsson, Automation in healthcare internal logistics: a case study on practice and potential. International Journal of Innovation and Technology Management, 2013. 10(03): p. 1340012.

[23] Childerhouse, P., et al., Supply chain integration: an international comparison of maturity. Asia Pacific Journal of Marketing and Logistics, 2011. 23(4): p. 531–552.

[24] Winkelhaus, S. and E.H. Grosse, Logistics 4.0: a systematic review towards a new logistics system. International Journal of Production Research, 2020. 58(1): p. 18–43.

[25] Rouwenhorst, B., et al., Warehouse design and control: Framework and literature review. European Journal of Operational Research, 2000. 122(3): p. 515–533.

[26] Öjmertz, B. and M.I. Johansson. Materials handling analysis in supply chain reengineering. in Proceedings of the 3rd International Symposium on Logistics~ Enhancing Competitiveness through Logistics Capabilities.(Eds. M. Muffatto and Pawar, KS), Padua. 1997.

[27] Christopher, M. and D. Towill, An integrated model for the design of agile supply chains. International Journal of Physical Distribution & Logistics Management, 2001. 31(4): p. 235–246.

[28] Chow, G., T.D. Heaver, and L.E. Henriksson, Strategy, structure and performance: a framework for logistics research. Logistics and Transportation Review, 1995. 31(4): p. 285.

[29] Maltz, A.B. and L.M. Ellram, Total cost of relationship: an analytical framework for the logistics outsourcing decision. Journal of Business Logistics, 1997. 18(1): p. 45.

[30] Krajnc, J., K. Logožar, and B. Korošec, Activity-based management of logistic costs in a manufacturing company: a case of increased visibility of logistics costs in a Slovenian paper manufacturing company. PROMET-Traffic&Transportation, 2012. 24(1): p. 15–24.

[31] Dong, Z., The Technical Conditions of Modern Logistics. Open Journal of Social Sciences, 2013. 1(05): p. 19.

[32] Boysen, N., et al., Part logistics in the automotive industry: Decision problems, literature review and research agenda. European Journal of Operational Research, 2015. 242(1): p. 107–120.

[33] Gattuso, D., et al., A freight urban distribution center design with micro-simulation support for city logistics. WIT Transactions on the Built Environment, 2015. 146: p. 303–312.

[34] Caputo, M. and V. Mininno, Internal, vertical and horizontal logistics integration in Italian grocery distribution. International Journal of Physical Distribution & Logistics Management, 1996. 26(9): p. 64–90.

[35] Dapiran, P., et al., Third party logistics services usage by large Australian firms. International Journal of Physical Distribution & Logistics Management, 1996. 26(10): p. 36–45.

[36] Gjerdrum, J., N. Shah, and L.G. Papageorgiou, A combined optimization and agent-based approach to supply chain modelling and performance assessment. Production Planning & Control, 2001. 12(1): p. 81–88.

[37] Gagliardi, J.P., J. Renaud, and A. Ruiz. A simulation model to improve warehouse operations. in Proceedings of the 39th conference on Winter simulation: 40 years! The best is yet to come. 2007. IEEE Press.

[38] Henderson, S., et al. A simulation model to improve warehouse operations. in Proceedings, Winter Simulation Conference. 2012. Citeseer.

[39] Speranza, M.G. and P. Stähly, New trends in distribution logistics. Vol. 480. 2012: Springer Science & Business Media.

[40] Stock, G.N., N.P. Greis, and J.D. Kasarda, Logistics, strategy and structure: a conceptual framework. International Journal of Operations & Production Management, 1998. 18(1): p. 37–52.

[41] Barratt, M., Understanding the meaning of collaboration in the supply chain. Supply Chain Management: an international journal, 2004. 9(1): p. 30–42.

[42] Gunasekaran, A. and B. Kobu, Performance measures and metrics in logistics and supply chain management: a review of recent literature (1995–2004) for research and applications. International Journal of Production Research, 2007. 45(12): p. 2819–2840.

[43] Lummus, R.R., D.W. Krumwiede, and R.J. Vokurka, The relationship of logistics to supply chain management: developing a common industry definition. Industrial Management & Data Systems, 2001. 101(8): p. 426–432.

[44] Rizzi, A. and R. Zamboni, Efficiency improvement in manual warehouses through ERP systems implementation and redesign of the logistics processes. Logistics Information Management, 1999. 12(5): p. 367–377.

[45] Garcia, F.A., et al., A framework for measuring logistics performance in the wine industry. International Journal of Production Economics, 2012. 135(1): p. 284–298.

[46] Govindan, K., et al., Analysis of third party reverse logistics provider using interpretive structural modeling. International Journal of Production Economics, 2012. 140(1): p. 204–211.

[47] Waller, M.A. and S.E. Fawcett, Click here to print a maker movement supply chain: how invention and entrepreneurship will disrupt supply chain design. Journal of Business Logistics, 2014. 35(2): p. 99–102.

[48] Novack, R.A., L.M. Rinehart, and C. J. Langley Jr, An internal assessment of logistics value. Journal of Business Logistics, 1994. 15(1): p. 113.

[49] Huang, G.Q., Y. Zhang, and P. Jiang, RFID-based wireless manufacturing for real-time management of job shop WIP inventories. The International Journal of Advanced Manufacturing Technology, 2008. 36(7–8): p. 752–764.

[50] Gåsvaer, D. and J. von Axelson. Kaikaku-Radical improvement in production. in International Conference on Operations and Maintenance, Singapore. 2012. Citeseer.

[51] Shah, J. and N. Singh, Benchmarking internal supply chain

performance: development of a framework. Journal of Supply Chain Management, 2001. 37(4): p. 37–47.

[52] Leite, A.H.d.L.S., Melhorias de logística interna com recurso a comboio logístico. 2013.

[53] Chikhalikar, P. and S. Sharma, PROCESS ANALYSIS FOR IMPROVEMENT IN VARIOUS PARAMETERS BY IMPLEMENTATION OF LEAN MANUFACTURING IN ENGINE MANUFACTURING UNIT. 2015.

[54] Bertelsen, S. and J. Nielsen. Just-in-time logistics in the supply of building materials. in 1st International Conference on Construction Industry Development, Singapore. 1997.

[55] Lai, K.-h., E. Ngai, and T. Cheng, Measures for evaluating supply chain performance in transport logistics. Transportation Research Part E: Logistics and Transportation Review, 2002. 38(6): p. 439–456.

[56] Klug, F., The internal bullwhip effect in car manufacturing. International Journal of Production Research, 2013. 51(1): p. 303–322.

[57] Rajesh, R., et al., Generic balanced scorecard framework for third party logistics service provider. International Journal of Production Economics, 2012. 140(1): p. 269–282.

[58] Iannone, F., The private and social cost efficiency of port hinterland container distribution through a regional logistics system. Transportation Research Part A: Policy and Practice, 2012. 46(9): p. 1424–1448.

[59] Sheffi, Y. and P. Klaus. Logistics at large: jumping the barriers of the logistics function. in Proceedings of the Twenty-sixth Annual Transportation and Logistics Educators Conference. 1997. Citeseer.

[60] Novack, R.A., Logistics control: an approach to quality. 1987, University of Tennessee, Knoxville.

[61] Skjoett-Larsen, T., Third party logistics-from an interorganizational point of view. International journal of physical distribution & logistics management, 2000. 30(2): p. 112–127.

[62] De Koster, R.B., The logistics behind the enter click, in Quantitative approaches to distribution logistics and supply chain management. 2002, Springer. p. 131–148.

[63] Joong-Kun Cho, J., J. Ozment, and H. Sink, Logistics capability, logistics outsourcing and firm performance in an e-commerce market. International Journal of Physical Distribution & Logistics Management, 2008. 38(5): p. 336–359.

[64] Vernuccio, M., A. Cozzolino, and L. Michelini, An exploratory study of marketing, logistics, and ethics in packaging innovation. European Journal of Innovation Management, 2010. 13(3): p. 333–354.

[65] Kuhn, H. and M.G. Sternbeck, Integrative retail logistics: an exploratory study. Operations Management Research, 2013. 6(1–2): p. 2–18.

[66] Closs, D.J., T.J. Goldsby, and S.R. Clinton, Information technology influences on world class logistics capability. International Journal of Physical Distribution & Logistics Management, 1997. 27(1): p. 4–17.

[67] Lewis, I. and A. Talalayevsky, Third-party logistics: Leveraging information technology. Journal of business logistics, 2000. 21(2): p. 173.

[68] Narasimhan, R. and S.W. Kim, Information system utilization strategy for supply chain integration. Journal of business logistics, 2001. 22(2): p. 51–75.

[69] Lai, F., X. Zhao, and Q. Wang, The impact of information technology on the competitive advantage of logistics firms in China. Industrial Management & Data Systems, 2006. 106(9): p. 1249–1271.

[70] Brah, S.A. and H. Ying Lim, The effects of technology and TQM on the performance of logistics companies. International Journal of Physical Distribution & Logistics Management, 2006. 36(3): p. 192–209.

[71] Prajogo, D. and J. Olhager, Supply chain integration and performance: The effects of long-term relationships, information technology and sharing, and logistics integration. International Journal of Production Economics, 2012. 135(1): p. 514–522.

[72] Cavinato, J.L., Evolving procurement organizations: logistics implications. Journal of business logistics, 1992. 13(1): p. 27.

[73] Gustin, C.M., P.J. Daugherty, and T. P. Stank, The effects of information availability on logistics integra. Journal of Business Logistics, 1995. 16(1): p. 1.

[74] Gupta, M. and *A. kohli*, Enterprise resource planning systems and its implications for operations function. Technovation, 2006. 26(5): p. 687–696.

[75] Karjalainen, I., et al., Materials flow improvement in a lean assembly line: a case study. Assembly Automation, 2007. 27(2): p. 141–147.

[76] Kovács, G.L. and P. Paganelli, A planning and management infrastructure for large, complex, distributed projects—beyond ERP and SCM. Computers in Industry, 2003. 51 (2): p. 165–183.

[77] Umble, E.J., R.R. Haft, and M.M. Umble, Enterprise resource planning: Implementation procedures and critical success factors. European journal of operational research, 2003. 146(2): p. 241–257.

[78] Olhager, J., Evolution of operations planning and control: from production to supply chains. International Journal of Production Research, 2013. 51(23–24): p. 6836–6843.

[79] Becker, T., D. Weimer, and J. Pannek, Network structures and decentralized control in logistics: topology, interfaces, and dynamics. International Journal of Advanced Logistics, 2015. 4(1): p. 1–8.

[80] Lambert, D.M. and R. Burduroglu, Measuring and selling the value of logistics. The International Journal of Logistics Management, 2000. 11(1): p. 1–18.

[81] Stank, T.P., S.B. Keller, and P.J. Daugherty, Supply chain collaboration and logistical service performance. Journal of Business logistics, 2001. 22 (1): p. 29–48.

[82] Boyson, S., et al., Managing effective third party logistics relationships: what does it take? Journal of Business Logistics, 1999. 20(1): p. 73.

[83] Bichou, K. and R. Gray, A logistics and supply chain management approach to port performance measurement. Maritime Policy & Management, 2004. 31(1): p. 47–67.

[84] Gimenez, C. and E. Ventura, Logistics-production, logistics-marketing and external integration: their impact on performance. International journal of operations & Production Management, 2005. 25(1): p. 20–38.

[85] Langley Jr, C.J. and M.C. Holcomb, Creating logistics customer value. Journal of business logistics, 1992. 13(2): p. 1.

[86] Bhatnagar, R., A.S. Sohal, and R. Millen, Third party logistics services: a

Singapore perspective. International Journal of Physical Distribution & Logistics Management, 1999. 29(9): p. 569–587.

[87] Lynch, C.F., T.P. Stank, and S. Scott, Logistics outsourcing. Handbook of Global Supply Chain Management, 2006: p. 373.

[88] Mentzer, J.T. and B.P. Konrad, An efficiency/effectiveness approach to logistics performance analysis. Journal of business logistics, 1991. 12(1): p. 33.

[89] Stock, G.N., N.P. Greis, and J.D. Kasarda, Enterprise logistics and supply chain structure: the role of fit. Journal of operations management, 2000. 18(5): p. 531–547.

[90] Mentzer, J.T., et al., Defining supply chain management. Journal of Business logistics, 2001. 22(2): p. 1–25.

[91] Bruzzone, A. and F. Longo, An application methodology for logistics and transportation scenarios analysis and comparison within the retail supply chain. European Journal of Industrial Engineering, 2014. 8(1): p. 112–142.

[92] Taherdoost, H., Determining sample size; how to calculate survey sample size. International Journal of Economics Management Systems, 2017. 2.

[93] Čulková, K., R. Weiss, and E. Weiss. Economical Analysis of Logistics Processes. in Applied Mechanics and Materials. 2015. Trans Tech Publ.

[94] Hurtado, S.M., et al., Aproximación a la medición del capital intelectual organizacional aplicando sistemas de lógica difusa. Cuadernos de Administración, 2010. 23(40): p. 35–68.

[95] Mamdani, E.H. and S. Assilian, An experiment in linguistic synthesis with a fuzzy logic controller. International journal of man-machine studies, 1975. 7 (1): p. 1–13.

[96] Takagi, T. and M. Sugeno, Fuzzy identification of systems and its applications to modeling and control. Systems, Man and Cybernetics, IEEE Transactions on, 1985(1): p. 116–132.

[97] Sivanandam, S., S. Sumathi, and S. Deepa, Introduction to fuzzy logic using MATLAB. Vol. 1. 2007: Springer.

[98] Egaji, O.A., et al., A comparison of Mamdani and Sugeno fuzzy based packet scheduler for MANET with a realistic wireless propagation model. International Journal of Automation and Computing, 2015. 12(1): p. 1–13.

Chapter 8

Business Process Linguistic Modeling: Theory and Practice Part I: BPLM Strategy Creator

Jozef Stašák, Jaroslav Kultan, Peter Schmidt and Mukhammedov Abu Urinbasarovich

Abstract

The business activities provided within any firm or company should be checked and controlled continuously, while two principal approaches should be applied: (a) qualitative monitoring, (b) quantitative evaluations, while KPI indicators play a role of principle importance within business quantitative evaluation in order to make adequate decisions. However, adequate applications form KPI creation and further processing seem to be very significant and important. We have designed a conceptual model of application denoted as BPLM Strategy Creator in form of expert system (ES) operating based on principles closely related to business process linguistic modeling approach, where linguistic sets and PBPL Equation play a role of principle importance. Our contribution contains such application description from qualitative, quantitative and design point of view. The ES qualitative description contains references to appropriate math relations and algorithms postulated within a subsequent section. Both sections are accompanied by the case study, which indicates how the math relations and algorithms might be applied within BPLM Strategy Creator functionality. However, those sections are accompanied by ES structure and functionality description as well, which represent the BPLM Strategy Creator mean or facility.

Keywords: key performance, indicators, business, process, linguistic, modeling

1. Introduction

The triple known as" people, planet, profit [1] has been chosen as a slogan by many modern businesses, trying to win the support of people and governments. The triple directs organizations to focus not only on the economic value, but also on the social and environmental value [1]. The new focus of organizations stimulates the search for the right measures of organizational success or key performance indicators (KPIs). The KPIs are used almost for any domain of our life, including medicine, education, services and green computing [2]. However, this is only one side of the coin. On the other hand, there are many applications, which enable creating and processing of KPI indicators [3–5].

There are many different tools for KPI generation and processing, but the BSC Designer is considered to be a standardized tool applied, when quantifying business strategy aspects ad relations (BSC Designer). However, there are many different KPI generating tools, which enable establishing objectives performance measurement system through KPIs selection and setting up targets for measuring each KPI and creating SMO (Strategic Management Office) within firm or company as well, while that tool is denoted as Virtual Strategy Creator [6]. All the above-mentioned applications denoted as strategy creators are designed and implemented based different approaches and principles, while the most common ARIS Business Strategy Creator is designed and implemented based on methodology established by Prof. Scheer, while that methodology represents standard in strategy creation and processing [7]. However, there are various approaches, which might be applied to business strategy creator design and implementation [8, 9] as well, while one of them is denoted as business process modeling linguistic approach (BPLM approach), where *linguistic sets* and Principle *Businesses Process Linguistic Modeling Equation* (PBPL Equation) [10, 11] represent the categories of principle importance and are being applied in designing of business strategy creator described within that contribution. The above-mentioned approach is based on sematic analysis related to content of supporting documents for business strategy creation and processing, while that analysis is being done in two phases. In the first phase, two categories of analyzed documents are created: (a) the first category of documents is applied for description of business strategy qualitative aspects and (b) the second category of documents is applied for description of business strategy quantitative aspects incl. Creation and processing of KPI indicators. The applications operating based on that principle had been searched, however no similar applications were found and therefore development of our own approach and methodology was getting started.

The contribution main goal is to design a conceptual model of business strategy creator, which should operate based on business process linguistic modeling principles (hereinafter known as BPLM Strategy Creator). In order to achieve, the main goal, three partial aims should be postulated and fulfilled: (a) to define the BPLM Strategy Creator structure and functionality from qualitative point of view (see also Section 4.1) – it seems to be the first partial aim, (b) to define appropriate math relations and algorithms concerned with linguistic modeling aspects applied in quantification of BPLM Strategy Creator functionality (see also Section 4.2) - it seems to be the second partial aim, (c) to define the BPLM Strategy Creator design and implementation via adequate expert system, the knowledge-base of which contains a set of knowledge represented by appropriate semantic networks (SNWs) and reference databases ((RDBs) (see also Section 4.4) - it seems to be the third partial aim. However, an appropriate case study creates an integral part of that contribution, the aim of which is to show how the derived math relations and algorithms should be applied related to BPLM Strategy Creator functionality (see also Section 4.3).

2. State of the art

2.1 KPI indicator creation and processing, methods and tools

Any firm or company is starting business based on its own business mission statement, business objectives and with the use of appropriate business process. All those categories are being transformed to the firm or company business strategy, which usually consists of two principal sections [12, 13]:

- The first section is concerned with qualitative aspects described via text in natural language (TNL text) and the second one, which is described via set business performance indicators (hereinafter known as KPI indicators). However, both of the above-mentioned section is being prepared based on appropriate documents, which contain adequate supporting data as well, while their content has to be undertaken to preliminary semantic analysis, first of all. This type of semantic analysis indicates, which outgoing documents are closely related to the first and to the second section.

- Subsequently both of those document types are being undertaken to deeper semantic analysis and assigned to the first or the second section. However, the second section document semantic analysis results represent various text string and numeric data stored within sets denoted as the linguistic sets and they usually seem to be market research and the firm or company internal resources analysis results, which create basis for generation of so called initial KPI indicators. On the other hand, the above-mentioned linguistic sets represent the principle elements of linguistic business and business process modeling approach as well, while that approach will be applied, when deriving appropriate math relations and algorithms needed for business strategy creator design and implementation conceptual model discussed in Section 4, while the business strategy creator application program should be implemented and operated with use of graph database algorithms and procedures.

- In general, the KPIs are measures that a sector or organization uses to define success and track progress in meeting its strategic goals. This focus on strategic or long-term goals is what distinguishes KPIs from the wider array of "performance indicators" (PIs) that do not necessarily rise to the attention of policymakers or the public, but may be important for public sector managers [14–16], KPIs are not created in a vacuum. KPIs, thus, should not be thought of as standalone measures, but rather as the product of strategic thinking, analysis and negotiation around policy problems and responses. A useful tool to help conceptualize this production process is the "logic model." In strategic planning, logic models are used commonly to describe the logical linkages between problems and their solutions. The model lays out a three-stage process for [7]:

    - Identifying the problem (s), or the community need.

    - Developing policies or measures to address the problem (s) and

    - Articulating the desired goals—the end-state of affairs or vision for the future.

Strategic planning is a high-level exercise, typically conducted by ministry planning departments in consultation with program managers, staff responsible for stakeholders to define or sharpen focus on strategic goals and policy responses. It is at the program or activity level, however, where the budget comes into focus, and where, ultimately, performance indicators, including KPIs, are most commonly established. Other strategyzer offers real time and asynchronous collaboration to keep everyone on the same page, and one centralized place to collaborate on the firm or company strategy the other Strategic Planning software aggregates historical business performance data and helps with creating predictive models of future performance based on specified business objectives and resource allocations [17].

2.2 KPI modeling

2.2.1 KPI modeling issues

The KPI modeling seems to be one of the most important actions closely related to KPI indicator processing and they have to indicate appropriate properties in order to be denoted as the good KPIs.

However, those properties may be seen as the first aspects concerned with KPI validation as well, while they night be postulated as follows [2].

- KPI should be in a quantifiable form.

- KPI needs to be sensitive to changes of the business process state.

- KPI should be linear, (d) a KPI should be semantically reliable,

- KPI should be efficient,

- KPI should be oriented to improvement, not to conformance to plans

However, the above-mentioned KPI indicator properties seem to be only one side concerned with KPI indicator modeling [15, 16].

The second aspect is closely related to KPI attributes postulated as KPI name, Type, Scale, Source, Owner, Threshold, and Hardness.

The third aspect is the performance indicator expression. It is "a mathematical statement over a performance indicator evaluated to a numerical, qualitative or Boolean value for a time point, for the organization, unit or agent. For example, P I27 $\leq$ 48h."

The **fourth aspect** concerned with KPI formalization is the performance indicator expression. It is "a mathematical statement over a performance indicator. The authors suggest specifying the required values of KPIs as constraints coming from goals. The authors claim that they integrate the performance view with the process, organization and agent-oriented views. However, there is no information about the process semantics used for modeling and no evidence about validation of the PI properties. In any case, the authors write about the process views of the real organizations, not about the abstract processes that are proposed [9, 15, 16].

2.2.2 KPI modeling approaches and methods

2.2.2.1 MetricM method

The method MetricM [18] "is built upon and extends an enterprise modelling approach to benefit from he reuse of modelling concepts to provide relevant organizational context, including business objectives, organizational roles and responsibilities." The method can be adapted to any enterprise modeling approach. The modeling language.

MetricML used in MetricM "adds essential concepts to modelling performance indicators and semantics to key modelling concepts." The concept Indicator is used to present a KPI.

The MetricML Indicator metatype is used for modeling its relations to other indicator types, to reference object types representing organizational context and to goal types [2].

2.2.2.2 Attribute approach

An alternative "attribute" approach conceptualizes performance indicator as (meta-) attribute of metatypes (e. g "average throughput time" of a business process type or "average number of employees" of an organizational unit type). Alternative approach for KPI modeling in our method is used. MetricM uses declarative models. The model of underlying processes needed for validation of KPI properties are not used in MetricM. The two approaches, presented above, build upon ideas of many earlier approaches to KPI modeling. The general tendency is to postpone the validation of the KPI properties to the moment when the process model of the organization is ready.

2.2.2.3 Semantics synchronous and asynchronous modeling

However, the KPIs are defined at a different level of abstraction, namely at the tactical and strategic level, i.e. at the level of observable states of the system and the asynchronous modeling does not provide the right level of abstraction [15, 16].

The synchronous modeling semantics is based on the CSP parallel composition operator defined by Hoare [19]. The operator defines that an event from environment is accepted by the model if all processes of this model are able to accept it. Otherwise, the event is refused.

Although there were many applications of the CSP parallel composition operator in the architecture description languages [20] in programming languages [21] only after the extension of this operator for machines with data, made by McNeile [22] the operator became practical for business system modeling. The Protocol Modeling proposed in enables coping with complexity of business modeling. The reason is that the synchronous semantics decreases the data space of models.

2.2.2.4 KPI indicator linguistic modeling approach

This approach is based on existence of linguistic sets, while they represent KPI modeling static aspects. However, there are many relations among those linguistic sets as well, while they are quantified via PBPL Equation [10, 11, 23, 24]. This approach is discussed in Section 4 in more details.

2.3 KPI indicator decomposition

The KPI indicators are designed and closely related to core business processes implemented and operated at strategic management level and have a nature of so called initial and primary KPI indicators, which should be decomposed to secondary and tertiary KPI indicators. The secondary indicators are closely related to main BP management at tactic level and the tertiary KPI indicators are closely related to subordinated and elementary BP management at operational level. This approach to KPI indicator decomposition is discussed in Section 4. However, the KPI decomposition is closely related to business dashboard existence [13, 25, 26].

A dashboard in business is a tool used to manage all the business information from a single point of access. It helps managers and employees to keep track of the company's KPIs and utilizes business intelligence to help companies make data-driven decisions.

There are 4 general subtypes of dashboards: (a) Strategic - focused on long-term strategies and high-level metrics, (b) Operational - shows shorter time frames and operational processes. (c) Analytical - contains vast amounts of data created by analysts and (d) Tactical - used by mid-management to track performance.

A strategic dashboard is a reporting tool for monitoring the long-term company strategy with the help of critical success factors. They're usually complex in their creation, provide an enterprise-wide impact to a business and are mainly used by senior-level management [27–29].

An analytical dashboard is a type of dashboard that contains a vast amount of data created and used. They supply a business with a comprehensive overview of data, with middle management being a crucial part of its usage.

A tactical dashboard is utilized in the analysis and monitoring of processes conducted by mid-level management, emphasizing the analysis.

Then an organization effectively tracks the performance of a company's goal and delivers analytic recommendations for future strategies [30].

3. Research methodology

In order to achieve the pre-defined main goal and appropriate partial aims a set of adequate research methods should be postulated and applied:

- Business process linguistic modeling (BPLM) approach, where the linguistic sets seem tio elements of principle importance and create basis for design and implementation of reference databases (RDBs) and semantic networks (SNWs), which represent the principal facilities for an appropriate knowledge-based (expert) system structure and functionality

- Design and implementation of an appropriate expert system (ES), where the knowledge stored in the ES knowledge base (ES-KB) are represented with the use of the above-mentioned RDBs and SNWs and a completed ES is being implemented via adequate application program.

- In order to manage that application program implementation the principles and elements of graph databases (GraphDb) are being applied for implementation purposes related to linguistic sets, RDBs and SNWs.

- The designed and implemented BPLM SC application should be utilized as a supporting tool, when designing and updating the actual BS strategy quantitative and qualitative aspects

4. Results

The business process linguistic modeling (BPLM) system represents a complex tool applied for BP linguistic modeling, which consists of the following subsystems: (a) BPLM Strategy Creator, (b) BPLM process analysis and design, (c) BPLM process implementation, which should contain tools for creating of *BP configuration model* (information and knowledge-based support and BP execution model, which includes BP operation and controlling. The BPLM Strategy Creator discussed within presented contribution seems to be the first important component related to the above-mentioned BPLM System.

4.1 BPLM strategy creator – structure and functionality BPLM strategy creator – structure and functionality – qualitative view

4.1.1 Strategic management level

Any business is getting started by business mission statement and business objectives and adequate business process establishment. Those three categories create an integral part of any business strategy. However, before we determine a set of business quantitative and qualitative indicators, real possibilities should be known to apply our business results at an appropriate market area and collect initial information. Usually, the information is stored at different media and documents. However, we have to make a preliminary document content semantic analysis in order to gain a required information and this is an initial action, which should be done with the use of the proposed BPLM Strategy Creator. This type of the document semantic analysis enables providing the document categorization and show use which documents should create basis for processing of business strategy qualitative aspects. Furthermore, we are interested in those documents, which contain data closely related to business strategy quantitative aspects, which might be quantified via indicators denoted as key performance indicators (KPI indicators). However, they usually are not in that form and shape as we need. Therefore, we have to provide the second type of document analysis in order to extract required data – usually denoted as the initial data, which should inform us which products related to our business could be accepted by the market, in which quantity and quality and what about financial assets could be gained. This data type could create content of sets $\{[Y^{Totfin}{}_{Assetst}]\}$, $\{[Y^{Totmat}{}_{Assetst}]\}$. Because the data are of a linguist nature those sets are denoted as linguistic sets. The linguistic set $\{[Y^{Totfin}{}_{Assetst}]\}$, contains data closely related to financial assets and the linguistic set contains data closely related to material assets[1]. This is only one side of the coin, while we to know what about investments (financial costs) are needed in order to pro produce the above-mentioned output products and they are stored within $\{[X^{Totfin}{}_{Costs}(0)]\}$, $\{[X^{Totmat}{}_{Costs}(0)]\}$ linguistic sets. The data represent the first BPLM Strategy creator output, which is called the **basic output** as well, while the financial costs play a role of principal importance, but are not sufficient for production getting started. We have to know what about customers will buy our products, what about human resources with required theoretical knowledge and practical skills, and what about production technological devices and tools are needed, as well. This types of data are being stored in further linguistic sets. The linguistic set $\{[SAD(i,j)]\}$ quantifies potential customers denoted as mainframe customers, the linguistic set $\{[HR(i,j)]\}$ quantifies mainframe human resources and the $\{[TECH(i,j)]\}$ linguistic set quantifies mainframe production technological devices and tools. A qualifier "mainframe" indicates that the linguistic set content is not specified in more details. When adding that linguist sets to the above-mentioned basic output we get the **initial BPLM** Strategy creator output, while formula (1) might be postulated

$$\begin{aligned}\{[KPI(0)]\} = \ &\left\{\left[Y^{Totfin}{}_{Assetst}\right]\right\}, \left\{\left[Y^{Totmat}{}_{Assetst}\right], \left[X^{Totfin}{}_{Costs}(0)\right]\right\},\\ &\left\{\left[X^{Totmat}{}_{Costs}(0)\right], [SAD(i,j)], [HR(i,j)], [TECH(i,j)]\right\}\end{aligned} \quad (1)$$

which represent so called the **total initial KPI** indicators.

---

[1] How many pieces of the actual output products could be produced.

This is the first partial output of the proposed BPLM Strategy creator, which indicates the basic possibilities of our business. However, no business within any firm company might be provided without adequate core, main, subordinated and elementary business processes (BP), while each of those BPs is represent by its own static structure, metrics and dynamic functionality (performance) as well, while the data create basis for determination of the firm or company internal resources and the data are stored within technical and economic standards, which has each firm or company and based on the data a set of appropriate correction coefficients might be calculated. When multiplying the data contained in the above-mentioned linguistic sets with those coefficients, real item values concerned production, customers, human resources, financial and production technology devices and tools might be obtained, based on which our business might be functional and efficient. However, an appropriate algorithm development is the aim of future work as well, while this is a basic principle of for calculation real values of partial KPI indicators and the total KPI indicator/All the above-mentioned KPI indicators are denoted as **primary KPI indicators** and formula (2) might be postulated.

$$\mathrm{KPI}_{\mathrm{prim}}\,(0)) = \{[\mathrm{KPI}\,(0)]\} \otimes \{[\mathrm{CBS}\,(0,1)]\} \tag{2}$$

where {[KPI (0)]} is a linguistic set specified via formula (1) and {[CBS (0, 1)]} is a linguistic set, the subsets of which might quantify adequate core business processes.

It means, we can get BPLM Strategy creator (BPLM SC system) output represented by formula (3).

$$\forall \mathrm{CBP}\,\exists\,\mathrm{KPIprim}\,(0) \Rightarrow \mathrm{CBP} \Leftrightarrow \mathrm{KPI}_{\mathrm{prim}}\,(0)) \tag{3}$$

which says that for any core business process (CBP) might be assigned one total KPI indicator and this is the most important results related to the BPLM SC system functionality and create basis for determination of further KPI indicator values valid for strategic management level and their decomposition related to tactic and operational level.

4.1.2 Tactic management level

The KPI indicators postulated for strategic management level are represented by formulas (25, 26, 28, 29) and (30) create basis for their further decomposition related to tactic management level, where we are operating with main business processes subordinated to appropriate core processes implemented and operated at strategic management level. However, at that level should be the BSC perspectives respected as well.

4.1.2.1 Customer's perspective

The Customer's perspective indicates which of production output product classes (PPCs) will be assigned to which customers, while appropriate contract sets are being created and the linguistic sets applied for quantification of individual customers are postulated via formula (18) and formulas (43–48) and (49, 50) indicate which PPCs will be assigned to which customers, while the first decomposed KPI indicator quantified via {[CONTRACTB (i)]} linguistic set is derived (see also formula (2)), while the {[SAD (i, j)]} linguistic set, which quantifies the mainframe customers creates basis for those purposes.

4.1.2.2 Internal BP perspective

A preparation of contract KPIs represents an initial step of KPI decomposition. In the next step pre-defined PPCs should be produced, appropriate internal BPs should be getting started and operated in order to achieve that aim. As a result of that, the {[BP (i, j_6)]} should be added to formula (51), while the next KPI indicator concerned with internal BP perspective denoted as KPI (i, 3) = {[CONTRACTD (i)]} is derived (see also formula (55), while the {[BP (i, j_6)]} linguistic set, which quantifies the mainframe customers creates basis for those purposes.

4.1.2.3 Financial perspective

A production of pre-defined PPCs and appropriate BPs functionality require an adequate material and financial support, and the production generates closely related assets while both of the above-mentioned support types are being quantified via outgoing linguistic sets {[$X^{Totfinmanp}{}_{Costs}$ (0)], [$X^{Totmamanpt}{}_{Costs}$ (0)], [$Y^{Totfinmanp}{}_{Assetst}$], [$Y^{Totmatmanp}{}_{Assetst}$]} (see also formula (9). With respect to those issues the principal financial perspective KPI indicator denoted as {KPImanp (1)} might be derived.

4.1.2.4 Education and growth and technical perspective

Adequate human and technological resources are required, in order to assure a proper and efficient functionality of BP, while further supplementary KPI indicators KPI (i, 4) = {[CONTRACTE (i)]} might be derived based on similar principles/, while the [HR (i, j)], [TECH (i, j)]} linguistic set, which quantifies the mainframe customers creates basis for those purposes. With respect to previous considerations a set of appropriate KPI indicators related to tactic management level (see also **Table 1**).

4.1.3 Operational management level

However, the tactic management level KPI indicators postulated within **Table 1** are closely related to main BP implemented and operated at that level and to BSC perspectives as well, while the operational management level KPI indicators are closely related to selected business process and its external and internal metrics.

BSC perspective	Tactic management evel KPI indicators	Outgoing linguistic sets
Financial perspective	{KPImanp (1)} = [$X^{Totfinmsnp}{}_{Assets}$ (0)]/ [[$X^{Totfinmanp}{}_{Costs}$ (0)]	{KPI (0)}
Customer's perspective	KPI (i, 1) = {[CONTRACTB (i)]}	{[SAD (i, j)]} {[CUST (i, j)]}
Internal BP perspective	KPI (i, 3) = {[CONTRACTD (i)]}	{[PCP (i, j)]}
Education and growth perspective	KPI (i, 4) = {[CONTRACTE (i)]}	{[HR (i, j)]}
Technical perspective	KPI (i, 5) = {[CONTRACTF (i)]}	{[TECH (i, j)]}

Source: The Authors.

Table 1.
KPI indicators related to tactic management level.

In general, any business process (BP) is characterized via its own internal and external metrics, while the BP external metrics deals with BP inputs and outputs and the internal metrics deals with appropriate human resources, production technological devices and tools. However, the BP external metrics KPI indicators include BP material input costs, BP financial input cost, BP production output material assets and production output financial assets as well, while the internal metrics KPI indicators include human resources theoretical knowledge and practical skill data together with adequate financial costs and assets. On the other hand, the internal metrics technological resources include production technological devices and tools (material aspects) and production technological devices and tools (financial costs and asset aspects).

4.2 BPLM strategy creator – structure and functionality – quantitative view

4.2.1 General overview

The BPLM Strategy Creator (BPLM SC) application represent a relatively independent system, which consist of several subsystems, components and modules postulated as follows:

- the first subsystem should provide selecting of adequate documents closely related to BS qualitative and quantitative aspects and their semantic analysis together with the data storage and processing, which has been generated as a result of the above-mentioned semantic analysis while it is denoted as the BPLM SC 01 Data and document preparation subsystem

- This subsystem contains two components, while the BPLM SC 01–01 component should enable involving the data and document segments concerned with BS qualitative aspects to adequate BS qualitative documents and the BPLM SC 01–02 component enable converting the data and document segments concerned with BS quantitative aspects to adequate BS KPI indicators.

- the second subsystem should provide investigation of the core business process, which create an integral part BS creation and generation of initial KPI indicators and a decomposition of them for tactic and operational management level, while it consists of BPLM SC 02–01 component, which should provide investigation of the core business process from functional point of view, where the core business process (CBP) metrics plays a role of principle importance and the BPLM SC 02–02 component should provide the KPI indicator decomposition related to tactic and operation management level

- the third subsystem should provide BS quantitative aspect simulation and optimization and consists of two components BPLM SC 03–01 component, which should provide the BS KPI indicator and core process simulation and BPLM SC 03–02 component, which should provide the BS KPI indicator and core process optimization.

4.3 KPI indicator quantification, generation and decomposition

4.3.1 KPI indicator quantification

In general, the BP performance is being quantified via key performance indicators (hereinafter known as KPI indicators), which might have a very heterogeneous or varied structure, features, and values. However, the KPI Indicator quantification is a process closely related to the firm or company strategic level and result the **initial KPI indicator** items and values as well, while Consideration no.1 enables deriving them.

5. Consideration no.1 Determination of KPI initial indicators

When considering a top core business process (hereinafter known as CB Process), a vertical structure of which is quantified via linguistic set {[CBS (i, j)]}, where.

i = 0,1, 2,3 .... n is a serial number of the actual business process (BP) within BP vertical structure set.

j = 0,1,2,3 ....m_1 is a serial number of business process function (BPF) within selected BP.

Because the core BP is at the top of BP vertical structure index i = 0 and j = 0 the top business process is being quantified via {[CBS (0, 1)]} linguistic set its performance quantified via KPI indicator with respect to formula (4).

$$\{\text{KPI}\ (0)\} = \left\{\left[\text{X}^{\text{Totfin}}{}_{\text{Costs}}\ (0)\right], \left[\text{X}^{\text{Totmat}}{}_{\text{Costs}}\ (0)\right], \left[\text{Y}^{\text{Totfin}}{}_{\text{Assetst}}\right], \left[\text{Y}^{\text{Totmat}}{}_{\text{Assetst}}\right]\right\} \quad (4)$$

Where

$[X^{Totfin}{}_{Costs}\ (0)]$ is a linguistic subset, which contains elements closely related to CB Process functionality input financial costs

$[X^{Totmat}{}_{Costs}\ (0)]$ is a linguistic subset, which contains elements closely related to CB Process functionality entire input material quantity

$[X^{Totmat}{}_{Assets}\ (0)]$ is a linguistic subset, which contains elements closely related to CB Process functionality entire output material quantity

$[X^{Totfin}{}_{Assets}\ (0)]$ is a linguistic subset, which contains elements closely related to CB Process functionality entire output financial assets

However, a content of linguistic subsets, which create an integral part for {KPI (0)} represents data closely related to market research and the firm or company internal resources analysis as well.

The above-mentioned quantification of the core BP is being done at the strategic management level, while a similar quantification of business processes (BPs) should be done at tactic and operational management levels too.

Let us consider the PBPL equation in a general form with respect to formula (3) and let us assign the linguistic sets with respect to formula (3), while formulas (5) up to (8) might be postulated.

$$\{[\text{Petx}\ (\text{i},\text{j})]\} = \{[\left[\text{X}^{\text{Totfin}}{}_{\text{Costs}}\ (0)\right], \left[\text{X}^{\text{Totmat}}{}_{\text{Costs}}\ (0)\right]\} \quad (5)$$

$$\{[\text{Pe}\ (\text{i},\text{j})]\} = \{[\text{CBS}\ (0, 1)]\} \quad (6)$$

$$\{[\text{Tbex}\ (\text{i},\text{j})]\} = \left\{\left[\text{X}^{\text{Totmat}}{}_{\text{Assets}}\ (0)\right], \left[\text{X}^{\text{Totfin}}{}_{\text{Assets}}\ (0)\right]\right\} \quad (7)$$

where

The $[[X^{Totfin}{}_{Costs}\ (0)]$ and $[X^{Totmat}{}_{Costs}\ (0)]\}$ linguistic sets represent initial inputs for core BP quantified via {[CBS (0, 1)]} linguistic set and the $\{[X^{Totmat}{}_{Assets}$ $(0)]$, $[X^{Totfin}{}_{Assets}\ (0)]\}$ linguistic sets represent subsequent outputs from BP quantified via {[CBS (0, 1)]} linguistic set.

The {[Retx (i, j)]} linguistic set represents relations among the above-mentioned linguistic sets with respect to formula (2d).

$$\{[\mathrm{Retx}\ (\mathrm{i},\mathrm{j})]\} = \{[\mathrm{X}^{\mathrm{Totmat}}{}_{\mathrm{Assets}}\ (0)], [\mathrm{X}^{\mathrm{Totfin}}{}_{\mathrm{Assets}}\ (0)], [\mathrm{CBS}\ (0,1)], [[\mathrm{X}^{\mathrm{Totfin}}{}_{\mathrm{Costs}}\ (0)], [\mathrm{X}^{\mathrm{Totmat}}{}_{\mathrm{Costs}}\ (0)]\} \tag{8}$$

With respect to the above-mentioned formulas the **Primary KPI indicators** might be postulated and have a nature of linguistic sets as well (see also formulas 9, 10).

$$\{\mathrm{KPI}\ (1)\} = [\mathrm{X}^{\mathrm{Totfin}}{}_{\mathrm{Assets}}\ (0)]/[[\mathrm{X}^{\mathrm{Totfin}}{}_{\mathrm{Costs}}\ (0)] \tag{9}$$

$$\{\mathrm{KPI}\ (2)\} = [\mathrm{X}^{\mathrm{Totmat}}{}_{\mathrm{Assets}}\ (0)]/[[\mathrm{X}^{\mathrm{Totmat}}{}_{\mathrm{Costs}}\ (0)] \tag{10}$$

However, those KPI indicators create basis for generation of further KPIs and their decomposition as well, while they have a linguistic set nature. As mentioned above the initial KPI indicators are being quantified via linguistic sets with respect to formulas (9) and (10). However, they are generated at strategic management level and are closely related to core BP quantified via linguistic set {[CBS (0, 1)]} as well, while formula (3) is extended about {[CBS (0, 1)]} linguistic set and formula (11) might be postulated.

$$\{\mathrm{KPI}\ (0)\} = \{[\mathrm{X}^{\mathrm{Totfin}}{}_{\mathrm{Costs}}\ (0)], [\mathrm{X}^{\mathrm{Totmat}}{}_{\mathrm{Costs}}\ (0)], [\mathrm{Y}^{\mathrm{Totfin}}{}_{\mathrm{Assetst}}], [\mathrm{Y}^{\mathrm{Totmat}}{}_{\mathrm{Assetst}}], [\mathrm{CBS}\ (0,1)]\} \tag{11}$$

5.1 Determination of KPI primary indicators

5.1.1 Consideration no. 2

With respect the above-mentioned issues two types of KPI initial indicators are being postulated (see also formula):

$$\{\mathrm{KPI}_{\mathrm{manp}}\ (0)\} = \{[\mathrm{X}^{\mathrm{Totfinmanp}}{}_{\mathrm{Costs}}\ (0)], [\mathrm{X}^{\mathrm{Totmamanpt}}{}_{\mathrm{Costs}}\ (0)], [\mathrm{Y}^{\mathrm{Totfinmanp}}{}_{\mathrm{Assetst}}], [\mathrm{Y}^{\mathrm{Totmatmanp}}{}_{\mathrm{Assetst}}]\} \tag{12}$$

$$\{\mathrm{KPI}_{\mathrm{machp}}\ (0)\} = \{[\mathrm{X}^{\mathrm{Totfinmachp}}{}_{\mathrm{Costs}}\ (0)], [\mathrm{X}^{\mathrm{Totmamachp}}{}_{\mathrm{Costs}}\ (0)], [\mathrm{Y}^{\mathrm{Totfinmachp}}{}_{\mathrm{Assetst}}], [\mathrm{Y}^{\mathrm{Totmatmachp}}{}_{\mathrm{Assetst}}]\} \tag{13}$$

where

the manp index is concerned with utility glass manually oriented production and the manchp index is concerned with utility glass machinery oriented production.

We shall discuss the KPI indicator quantification; generation and decomposition problems for glass utility manual production, while the derived formulas and algorithms might be applied for machinery oriented utility glass production as well.

The partial KPI indicators (see also formulas 14 and 15) indicate financial (Totfinmanp) and material (Totmatmanp) assets and costs and they considered to be results of market and the firm or company internal resources and they might represent initial manual inputs.

$$\{\mathrm{KPImanp}\ (1)\} = [\mathrm{X}^{\mathrm{Totfinmsnp}}{}_{\mathrm{Assets}}\ (0)]/[[\mathrm{X}^{\mathrm{Totfinmanp}}{}_{\mathrm{Costs}}\ (0)] \tag{14}$$

$$\{\text{KPImanp (2)}\} = \left[X^{\text{Totmatmanp}}{}_{\text{Assets}}\ (0)\right] / \left[\left[X^{\text{Totmatmanp}}{}_{\text{Costs}}\ (0)\right]\right. \tag{15}$$

When looking at formula (11), we might see that the initial KPI (0) indicator is closely related to the glass utility production process (GUP process), which seems to be the core process and is being quantified via {[GUPC (0, 1)]} linguistic set and formula (11) is converted to formula (16).

$$\{\text{KPI (0)}\} = \{\left[X^{\text{Totfin}}{}_{\text{Costs}}\ (0)\right], \left[X^{\text{Totmat}}{}_{\text{Costs}}\ (0)\right], \left[Y^{\text{Totfin}}{}_{\text{Assetst}}\right], \left[Y^{\text{Totmat}}{}_{\text{Assetst}}\right] \\ [\text{GUPC}\ (0, 1)]\}\} \tag{16}$$

However, the core process denoted as GUP process consists of main process quantified via appropriate linguistic sets postulated as follows as well.

5.1.2 Determination of Production process ***primary KPI indicators***

Now, let us analyze appropriate main processes denoted as **Production** quantified via {[PROD (i, j_1)]} linguistic set in order to determine the partial KPI indicators, while the initial KPI indicator value could create basis for those purposes. It might be done via following steps:

Step 1

Before providing KPI indicator quantification several auxiliary linguistic subset content should be determined. It might be done within Step 1 and Step 2.

$$\{\left[Y^{\text{Totfinmanp}}{}_{\text{Assetst}}\ (0)\right]\} = \{[\text{Mark\_res\_assets\_fin (0)}], [\text{Mark\_res\_costs\_fin(0)}]\} \tag{17}$$

$$\{\left[Y^{\text{Totfmatmanp}}{}_{\text{Assetst}}\ (0)\right]\} = \{[\text{Mark\_res\_assets\_mat (0)}], [\text{Mark\_res\_costs\_mat(0)}]\} \tag{18}$$

A word Mark_res, which creates basis of an appropriate linguistic set name indicates that items and values contained in there are initial input data acquired as a result of market research.

Step 2

Determination of {[PCPFin (i, j)], [PCPMat (i, j)]} linguistic sets, which are closely related to the firm or company real internal material and financial resources. The normalized internal resource values are postulated within the firm technical and economic standards (Int_res_fin_val_act (0), Int_res_fin_mat _act (0), while based on those values a value of Int_res_fin_koef (0) koeficient, which indicates what about a quantity of financial and material assets might be achieved, when applying real firm or company internal resources (see also formulas (19–22).

$$\{[\text{PCP (i, j)}]\} = \{[\text{PCPFin (i, j)}], [\text{PCPMat (i, j)}]\} \tag{19}$$

$$[\text{PCPFin (i, j)}] = [\text{Int\_res\_fin\_val\_req (0), Int\_res\_fin\_val\_act (0), Int\_res\_fin\_koef(0)}] \tag{20}$$

$$[\text{PCPMat (i, j)}] = [\text{Int\_res\_mat\_val\_req (0), Int\_res\_mat\_val\_act(0), Int\_res\_mat\_koef(0)}] \tag{21}$$

$$\{[\text{Pe (i, j2)}]\} = \{[\text{PCP (i, j)}]\} \tag{22}$$

For {[Petx (i, j1)]} linguistic set content see also formula (2a) and for {[Pe (i,j2)]} linguistic set content see also formula (24).

$$\{[\text{Petx}\ (i, j1)]\} = \{[Y^{\text{Totfinmanp}}{}_{\text{Assetst}}\ (0)], [Y^{\text{Totfmatmanp}}{}_{\text{Assetst}}\ (0)]\} \tag{23}$$

When applying the PBPL Equation its general form, (see also formula (24)) formula (25) might be postulated.

$$\{[\text{Petx}\ (i, j1)]\} \otimes \{[\text{Pe}\ (i, j2)]\} = \{[\text{Tbex}\ (i, j)]\} \otimes \{[\text{Ret}\ (i, j)]\} \tag{24}$$

After installment of previous relations into formula (24) formula (25) might be postulated.

$$\{[Y^{\text{Totfinmanp}}{}_{\text{Assetst}}\ (0)], [Y^{\text{Totfmatmanp}}{}_{\text{Assetst}}\ (0)]\} \otimes \{[\text{PCP}\ (i, j)]\} = \{[\text{Tbex}\ (i, j)]\} \otimes \{[\text{Ret}\ (i, j)]\} \tag{25}$$

where

the {[Tbex (i, j)]} linguistic contains data closely related to quantity of financial and material asset generated based on the actual firm or company internal resources.

$$\{[\text{Tbex}\ (i, j)]\} = \{[\text{Mark\_res\_assets\_fin\_ires}(0)], [\text{Mark\_res\_assets\_mat\_ires}(0)],\} \tag{26}$$

$$\begin{aligned}\{[\text{Retx}\ (i, j)]\} = \{&[\text{Mark\_res\_assets\_fin\_ires}(0)], [\text{Mark\_res\_costs\_fin}(0)],\\ &[\text{Mark\_res\_assets\_fin\_ires}(0)], [\text{Mark\_res\_costs\_fin}(0)]\}\end{aligned} \tag{27}$$

the {[Retx (i, j)]} linguistic contains data closely related to quantity of financial and material asset generated based on the actual firm or company internal resources with respect to appropriate financial and material costs.

In general, the above-mentioned algorithm enables determining KPIprod (0, 1)} and KPIprod (0, 2)}, while formulas (28) and (29) might be postulated.

$$\{\text{KPIprod}\ (0, 1)\} = \{[\text{Tbex}\ (i, j1)]\} \tag{28}$$

$$\{\text{KPIprod}\ (0, 2)\} = \{[\text{Retx}\ (i, j2)]\} \tag{29}$$

where

{KPIprod (0, 1)} = {[Tbex (i, j1)]} –the first Primary KPI indicator, which indicates real possibilities of the firm or company business represented by financial and material assets with respect to the firm or company production internal resources.

{KPIprod (0, 2)} = {[Retx (i, j2)]} – the second Primary KPI indicator, which indicates real possibilities of the firm or company business represented by financial and material assets with respect to the firm or company production financial and material cost. With respect to the above-mentioned issues the following clause might be postulated:

At the strategic management level, the Production main process might be quantified via {[PROD (i, j_1)]} linguistic set and two KPI indicators could be postulated, which indicate that process functionality (performance) (see also formulas (28) and (29) and (30).

$$\{[\text{PROD}\ (i, j1)]\} = \{\text{KPIprod}\ (1, 2)\} = \{\text{KPIprod}\ (0, 1)\} \otimes \{\text{KPIprod}\ (0, 2)\} \tag{30}$$

However, the similar sequence of steps (algorithm) might be applied when quantifying and generating KPI indicators for further main processes, sales and distribution, HR, technological and financial management at the strategic management level.

At strategic management level, a set of core BP are implemented and operated and quantified via appropriate linguistic sets, e.g. the Sales and Distribution BP is quantified via {[SAD (i, j)]}, where the *main frame customers* play a role of principle importance and are being quantified via [MFRC (i, j)]. When applying PBPL equation adequate KPI indicators might be derived.

$$\{[\mathrm{PROD}\ (\mathrm{i},\mathrm{j1})]\}\otimes\{[\mathrm{MFRC}\ (\mathrm{i},\mathrm{j})]\}\otimes\{[\mathrm{SAD}\ (\mathrm{i},\mathrm{j})]\}=\{[\mathrm{Tbex}\ (\mathrm{i},\mathrm{j})]\}\otimes\{[\mathrm{Retx}\ (\mathrm{i},\mathrm{j})]\} \tag{31}$$

$$\{[\mathrm{Tbex}\ (\mathrm{i},\mathrm{j})]\}=\{[\mathrm{PROD}\ (\mathrm{i},\mathrm{j1})],[\mathrm{MFRC}\ (\mathrm{i},\mathrm{j})]\} \tag{32}$$

$$\{[\mathrm{Retx}\ (\mathrm{i},\mathrm{j})]]=\{[\mathrm{PROD}\ (\mathrm{i},\mathrm{j1})],[\mathrm{MFRC}\ (\mathrm{i},\mathrm{j})],[\mathrm{SAD}\ (\mathrm{i},\mathrm{j})]\} \tag{33}$$

$$\{\mathrm{KPIsad}(0,1)\}=[\mathrm{Tbex}\ (\mathrm{i},\mathrm{j})]\} \tag{34}$$

$$\{\mathrm{KPIsad}\ (0,2)\}=\{[\mathrm{Retx}\ (\mathrm{i},\mathrm{j})]] \tag{35}$$

$$\{\mathrm{KPIsad}\ (0\}=\{\mathrm{KPIsad}(0,1)\}\otimes\{\mathrm{KPIsad}(0,2)\} \tag{36}$$

However, at strategic management level, a set of core BP are implemented and operated and quantified via appropriate linguistic sets, e.g. Human resources (HR} management is quantified via {[HR (i, j)]} as well where the *main frame human resources* play a role of principle importance and are being quantified via [HRC (i, j)]. When applying PBPL equation adequate KPI indicators might be derived.

$$\{[\mathrm{PROD}\ (\mathrm{i},\mathrm{j1})]\}\otimes\{[\mathrm{HRC}\ (\mathrm{i},\mathrm{j})]\}\otimes\{[\mathrm{HR}\ (\mathrm{i},\mathrm{j})]\}=\{[\mathrm{Tbex}\ (\mathrm{i},\mathrm{j})]\}\otimes\{[\mathrm{Retx}\ (\mathrm{i},\mathrm{j})]\} \tag{37}$$

$$\{[\mathrm{Tbex}\ (\mathrm{i},\mathrm{j})]\}=\{[\mathrm{PROD}\ (\mathrm{i},\mathrm{j1})],[\mathrm{HRC}\ (\mathrm{i},\mathrm{j})]\} \tag{38}$$

$$\{[\mathrm{Retx}\ (\mathrm{i},\mathrm{j})]\}=\{[\mathrm{PROD}\ (\mathrm{i},\mathrm{j1})],[\mathrm{MFRC}\ (\mathrm{i},\mathrm{j})],[\mathrm{HR}\ (\mathrm{i},\mathrm{j})]\} \tag{39}$$

$$\{\mathrm{KPIhr}(0,1)\}=\{[\mathrm{Tbex}\ (\mathrm{i},\mathrm{j})]\} \tag{40}$$

$$\{\mathrm{KPIhr}\ (0,2)\}=\{[\mathrm{Retx}\ (\mathrm{i},\mathrm{j})]\} \tag{41}$$

$$\{\mathrm{KPIhr}\ (0\}=\{\mathrm{KPIsad}(0,1)\}\otimes\{\mathrm{KPIsad}(0,2)\} \tag{42}$$

Finally, at strategic management level, a set of core BP are implemented and operated (see also **Table 1**) and quantified via appropriate linguistic sets, e.g. Technological TECH management is quantified via {[TECHN (i, j)]} as well where the *main frame technological devices and tools* play a role of principle importance and are being quantified via [TECHNC (i, j)]. When applying PBPL equation adequate KPI indicators might be derived.

$$\{[\mathrm{PROD}\ (\mathrm{i},\mathrm{j1})]\}\otimes\{[\mathrm{TECNC}(\mathrm{i},\mathrm{j})]\}\otimes\{[\mathrm{TECH}\ (\mathrm{i},\mathrm{j})]\}=\{[\mathrm{Tbex}\ (\mathrm{i},\mathrm{j})]\}\otimes\{[\mathrm{Retx}\ (\mathrm{i},\mathrm{j})]\} \tag{43}$$

$$\{[\mathrm{Tbex}\ (\mathrm{i},\mathrm{j})]\}=\{[\mathrm{PROD}\ (\mathrm{i},\mathrm{j1})],[\mathrm{TECHNC}\ (\mathrm{i},\mathrm{j})]\} \tag{44}$$

$$\{[\mathrm{Retx}\ (\mathrm{i},\mathrm{j})]\}=\{[\mathrm{PROD}\ (\mathrm{i},\mathrm{j1})],[\mathrm{TECHNC}\ (\mathrm{i},\mathrm{j})],[\mathrm{TECH}\ (\mathrm{i},\mathrm{j})]\} \tag{45}$$

$$\{\mathrm{KPItech}(0,1)\}=[\mathrm{Tbex}\ (\mathrm{i},\mathrm{j})]\} \tag{46}$$

$$\{\mathrm{KPItech}\ (0,2)\}=\{[\mathrm{Retx}\ (\mathrm{i},\mathrm{j})]\} \tag{47}$$

$$\{\mathrm{KPItech}\ (0\}=\{\mathrm{KPItech}(0,1)\}\otimes\{\mathrm{KPItech}(0,2)\} \tag{48}$$

In order to create a complex set of KPI indicators related to business processes implemented and operated at tactic and operational management level, an appropriate decomposition of primary KPI indicators related to performance of those processes to should be done. However, the KPI indicator decomposition for tactic level will be explained based on Consideration 3 and the KPI indicator decomposition for operational level will be explained based on Consideration 4 as well, while the Consideration no. 3 results the **secondary KPI indicators** and the Consideration no. 4 results the **tertiary KPI indicators** and both considerations are described within subsequent sections.

5.2 KPI indicator decomposition

5.2.1 KPI decomposition related to tactic level

5.2.1.1 Determination of Production process ***secondary KPI indicators*** *Consideration no. 3*

The previous section deals with initial KPI indicator generation and determination of primary KPI indicators for strategic management level. In that section, we shall discuss the KPI indicator decomposition for tactic level, which is based on the following consideration. The outgoing linguistic sets and KPI indicators for KPI decomposition related to tactic level are quantified via formulas (26, 27, 28, 29 and 30). With respect to the above-mentioned issues the following clause might be postulated:

At the strategic management level, the Production main process might be quantified via {[PROD (i, j_1)]} linguistic set and two KPI indicators could be postulated, which indicate that process functionality (performance) (see also formulas (28) and (29) and (30).

However, the similar sequence of steps (algorithm) might be applied when quantifying and generating KPI indicators for further main processes, sales and distribution, HR, technological and financial management at the strategic management level.

The KPI indicators postulated for strategic management level are represented by formulas (25, 26, (28, 29) and (30) create basis for their further decomposition related to tactic management level.

Let us select the [Mark_res_assets_mat _ires(0)] subset from {[Tbex (i, j)]} and assign it to market required output products quantified via {[MROP (i, j)]}, while formulas (28 and 29) might be postulated:

$$[\text{Mark\_res\_assets\_mat\_ires}(0)] \subseteq \{[\text{Tbex}\ (i,j)]\} \tag{49}$$

$$[\text{Mark\_res\_assets\_mat\_ires}(0)] = \{[\text{MROP}\ (i,j)]\}, \tag{50}$$

The {[MROP (i, j)]} contains subsets applied for quantification market required output products classes, e.g. utility glass article classes – bowls, bottles, vases, etc.

$$\{[\text{MROP}\ (i,j)]\} = \{[\text{MROP}\ (i,1)], [\text{MROP}\ (i,2)] \ldots .\{[\text{MROP}\ (i,m_1)]\} \tag{51}$$

Where.

Index m_1 is a number of article classes.

Furthermore, let us create a selected linguistic set {[MROP_{sel} (i, j)]}, a content of which is created by selected classes of [MROP_{sel} (i, j)], [MROP (i, 1)], [MROP (i, 2)], [MROP (i,3)], as for instance (see also formula (53)).

$$\{[\text{MROP}_{sel}\ (i,j)]\}, = \{[\text{MROP}\ (i,1)], [\text{MROP}\ (i,2)], [\text{MROP}\ (i,3)]\} \tag{52}$$

and let us postulate the {CUST (i, j_2)]} linguistic set, the content of which create data concerned with the customers.

$$\{[\text{CUST } (i, j_2)]\} = \{[\text{CUST } (i, 1)], [\text{CUST } (i, 2)], \ldots..[\text{CUST } (i, m_{2)]}\} \quad (53)$$

where Index $m_{2\ \text{means}}$ a number of customers.

In the next step, an appropriate {[MROP_{sel} (i, j)]} set for each customer will be assigned, while formula (19) might be postulated.

$$\forall\{[\text{CUST } (i, j_2)]\} \exists \{[\text{MROP}_{\text{sel}} (i, j)]\} \Rightarrow \{[\text{CUST } (i, j_2)]\} \Leftrightarrow \{[\text{MROP}_{\text{sel}} (i, j)]\} \quad (54)$$

In the next step we shall assign to each {[MROP_{sel} (i, j)]} set a {[$\text{MROP}_{\text{selfinass}}$ (i, j_3)]} and {[$\text{MROP}_{\text{selcosts}}$ (i, j_4)]}, where {[$\text{MROP}_{\text{selfinass}}$ (i, j)]} set quantifies the financial assets related to selected class of any market required output products.

{[$\text{MROP}_{\text{selfincosts}}$ (i, j)]} set quantifies the material costs related to selected class of any market required output products.

$$\begin{aligned}&\{[\text{CUST } (i, j_2)]\} \otimes \{[\text{MROP}_{\text{sel}} (i, j3)]\} \otimes \{[\text{MROP}_{\text{selfinass}} (i, j4)]\} \otimes \{[\text{MROP}_{\text{selfincosts}} (i, j5)]\} \\ &= \{[\text{Tbexc } (i, j)]\} \otimes \{[\text{Retxc } (i, j)]\}\end{aligned} \quad (55)$$

$$\{[\text{Tbexc } (i, j)]\} = \{[\text{CUST } (i, j_2)], \{[\text{MROP}_{\text{sel}} (i, j2)]\} \quad (56)$$

$$\{[\text{CONTRACTB } (i)]\} = \Pi \{[\text{CUST } (i, j_2)], \{[\text{MROP}_{\text{sel}} (i, j)]\} \quad (57)$$

$j = 1 \ldots_{m3}$
$j_2 = 1 \ldots_{m2}$

The {[CONTRACTB (i)]} linguistic set quantifies the basic contract, which indicates relations among customers and selected market required output products, incl. Financial costs and financial assets.

$$\{[\text{Retxc } (i, j)]\} = \{[\text{CUST } (i, j_2)], [\text{MROP}_{\text{sel}} (i, j)], [\text{MROP}_{\text{selfinass}} (i, j)], [\text{MROP}_{\text{selfincosts}} (i, j)]\}$$

$$\begin{aligned}\{[\text{CONTRACTC}(i)]\} == \Pi \{&[\text{CUST } (i, j_2)], [\text{MROP}_{\text{sel}} (i, j_3)], [\text{MROP}_{\text{selfinass}} (i, j_4)], \\ &[\text{MROP}_{\text{selfincosts}} (i, j5)]\}\end{aligned} \quad (58)$$

$j = 1 \ldots_{m3}$ $j_3 = 1 \ldots_{m3}$ $j_4 = 1 \ldots_{m4}$ $j_5 = 1 \ldots_{m5}$
$j_2 = 1 \ldots_{m2}$

Before, we make the final step we have to assign an appropriate group of business processes to each group of selected market required output products, while formula (70) might be postulated.

$$\forall\{[\text{MROP}_{\text{sel}} (i, j)]\} \exists \{[\text{BP } (I, j_6)]\} \Rightarrow \{[\text{MROP}_{\text{sel}} (i, j)], \Leftrightarrow \{[[\text{BP } (I, j_6)]\} \quad (59)$$

$$\{[\text{MROP}_{\text{sel}} (i, j)], [\text{BP } (I, j_6)]\} = \{[\text{Tbexbp } (i, j)]\} \otimes \{[\text{Retxbp } (i, j)]\} \quad (60)$$

{[Tbexbp (i, j)]} = {[[BP (I, j_6)]} – list of BP groups needed for production of market required output products quantified via {[MROP_{sel} (i, j)],

$$\{[\text{Retxbp } (i, j)]\} = \{[\text{MROP}_{\text{sel}} (i, j)], \{[[\text{BP } (I, j_6)]\} \quad (61)$$

Formula (24) indicates a list of relations among BP groups and market required output product group.

$$\{[\mathrm{CONTRACTD}\ (i)]\} = \Pi\, \{[\mathrm{CUST}\ (i, j_2)], \{[\mathrm{MROP_{sel}}\ (i, j_3)], [\mathrm{MROP_{selfinass}}\ (i, j_4)], [[\mathrm{BP}\ (I, j_6)]\} \tag{62}$$

$j = 1 \ldots m_3,\quad j_2 = 1 \ldots m_2,\quad j_3 = 1 \ldots m_3,\quad j_4 = 1 \ldots m_4,\quad j_5 = 1 \ldots m_5,\quad j_6 = 1 \ldots m_6,$
Finally, adequate KPI indicators will be defined.

$$\mathrm{KPI}\ (i, 1) = \{[\mathrm{CONTRACTB}\ (i)]\} \tag{63}$$

$$\mathrm{KPI}\ (i, 2) = \{[\mathrm{CONTRACTC}\ (i)]\} \tag{64}$$

$$\mathrm{KPI}\ (i, 3) = \{[\mathrm{CONTRACTD}\ (i)]\} \tag{65}$$

KPI (i, 3) indicator creates basis for decomposition related to operational level.

5.2.2 KPI indicator decomposition related to operational management level - consideration no. 4

5.2.2.1 Determination of Production process ***tertiary KPI indicators***

Let us consider the {[CONTRACTD (i)]} linguistic set (see also formula (57)), which quantifies order submitted to an appropriate firm or company organization unit to produce adequate products quantified via {[$\mathrm{MROP_{sel}}$ (i, j_3)]} and with the use of business processes (BP), which create an integral part of a given BP group. One of those processes will be selected and demonstrated how the KPI (i, 3) indicator should be decomposed in order to describe the selected BP functionality and performance, first of all. In general any BP is represented by its own internal and external metrics, while the external metrics is concerned with BP outputs and inputs and the BP internal metrics is closely related to appropriate production human resources, production devices and production tools and those aspects are quantified via given linguistic sets.However, that decomposition will be done within several steps as well.

5.2.2.2 BP external metrics KPI indicators

Step 3

In that step, a group of selected products should be created, which is an integral part of products quantified via {[$\mathrm{MROP_{sel}}$ (i, j_3)]} linguistic set, while formula (66) might be postulated.

$$\{[\mathrm{MROP_{sel\_bp}}\ (i, j_3)]\} \subseteq \{[\mathrm{MROP_{sel}}\ (i, j_3)]\} \tag{66}$$

Those products should be produced with the use of the selected BP (see also formula (67)).

$$\{[\mathrm{BP}\ (i, j7)\}\}\ \} \in [[\mathrm{BPG}\ (i, j_6)]\} \tag{67}$$

Now, we have to select set input materials needed for production of the above-mentioned products. We shall apply the [$\mathrm{MROP_{selfincosts}}$ (i, j)]} linguistic set for those purposes, the content of consists of two subsets with respect to formula.

$$[\mathrm{MROP_{selfincosts}}\ (i, j)]\} = \{[[\mathrm{MROP1_{selfincosts}}\ (i, j)], [[\mathrm{MROP2_{selmatcosts}}\ (i, j)]\} \tag{68}$$

Where the linguistic [[$MROP1_{selfincosts}$ (i, j)] subset quantifies financial costs and the [$MROP2_{selmatcosts}$ (i, j)] subset quantifies material costs and create basis for preparation that subset which contains material data needed for production of the above-mentioned products

$$\{[MROP1_{sel\_bp}(i, j_3)]\} \subseteq \{[MROP_{sel\_bp}(i, j_3)]\} \tag{69}$$

$$[MROP2_{selmatcosts}(i, j)]\ \{[MROP_{sel\_bp}(i, j_3)]\}\ \{[[MROP1_{selfincosts}(i, j)], [MROP_{selfinass}(i, j)] \tag{70}$$

5.2.2.3 Applying of PBPL equation solutions

Step 4

When applying the PBPL Equation, KPI indicator for the selected BP functionality and performance might be derived, while the modified PBPL Equation is postulated with respect to formula (61).

$$\{[MROP2_{selmatcosts}(i, j)], [MROP1_{selfincosts}(i, j)], [MROP_{selfinass}(i, j)], [MROP_{selmatass}(i, j)]\} \otimes [MROP_{sel\_bp}(i, j_3)] = \{[Tbex(i, j_8)]\} \otimes \{[Retx(i, j_9)]\} \tag{71}$$

5.2.2.4 PBPL equation solution results

Step 5

$$\{[Tbex(i, j_8)]\} = ([MROP_{selmatass}(i, j)], [MROP2_{selmatcosts}(i, j)]\} \tag{72}$$

$$\{[Retx(i, j_8)]\} = \{[MROP2_{selmatcosts}(i, j)], [MROP1_{selfincosts}(i, j)], [MROP_{selfinass}(i, j)], [MROP_{selmatass}(i, j)]\} \tag{73}$$

5.2.2.5 BP external metrics KPI indicators

Step 6

When dealing with BP External metrics KPI Indicators, so called basic and external KPI indicators will be defined.

$$KPIemb(i, 3) = \{[Tbex(i, j_8)]\} = ([MROP_{selmatass}(i, j)], [MROP2_{selmatcosts}(i, j)]\} \tag{74}$$

$$KPIemext(i, 3) = \{[Retx(i, j_8)]\} = \{[MROP2_{selmatcosts}(i, j)], [MROP1_{selfincosts}(i, j)], [MROP_{selfinass}(i, j)][MROP_{selmatass}(i, j)]\} \tag{75}$$

$$KPIem = KPIemb(i, 3) \otimes KPIemb(i, 3) \tag{76}$$

5.2.2.6 BP Internal metrics KPI Indicators

Step 7

The similar algorithm might be applied, when deriving BP Internal metrics KPI Indicators.

5.3 Case study

5.3.1 Determination of production process KPI indicators

The case study aim is to show how the derived math formulas and relations should be applied in a practice, when calculating actual data. The algorithm proposed within previous sections is passing through the following main phases:

- Determination of initial KPI indicators
- Determination of primary KPI indicators
- Determination of secondary KPI indicators
- Determination of tertiary KPI indicators

Furthermore, appropriate phases will be explained in more details.

5.3.2 Determination of initial KPI indicators

Before that phase is being activated a set of adequate data should be prepared, while the data are categorized as the market research results and the firm or company internal resources. The market research results inform us about possibilities how to apply our production in market and give us information how many products is the market able to accept and in which structure and what about financial assets might be generated as a result of that acceptance. However, the information related to adequate costs plays a role of principle importance as well. It means, we are able to answer the question related to the linguistic set content postulated within formulas (77–80)[2, 3, 4, 5].

$$\left[{Y^{Totmat}}_{Assetst}\right]\} = [100000\ \text{pp}], \tag{77}$$

$$\left[{Y^{Totfin}}_{Assetst}\right] = [1000000\ \text{Euros}] \tag{78}$$

$$\left[{X^{Totfin}}_{Costs}\,(0)\right] = [200000\ \text{Euros}] \tag{79}$$

$$\left[{X^{Totmat}}_{Costs}\,(0)\right] = [50000\ \text{Euros}] \tag{80}$$

With respect to those issues, the initial KPI indicator postulated via formula one might be indicated as follows:

$$\begin{aligned}\{KPI\,(0)\} &= \left\{\left[{X^{Totfin}}_{Costs}\,(0)\right], \left[{X^{Totmat}}_{Costs}\,(0)\right], \left[{Y^{Totfin}}_{Assetst}\right], \left[{Y^{Totmat}}_{Assetst}\right]\right\} \\ &= \{[100000\ \text{pp}], [1000000\ \text{Euros}], [200000\ \text{Euros}], [50000\ \text{Euros}]\}\end{aligned} \tag{81}$$

2 Pp – number of products, which might be accepted in the market in pieces

3 Financial assets, which could be achieved based on acceptance of the product piece amount at the market

4 A need of total financial costs needed for production of products, while that number includes material, technological HR and operational costs

5 This value represents costs for recruitment of adequate material inputs

After having applied formulas (77–80), we shall get values for primary KPI indicators, with respect to formulas (82) and (83).

$$\{\text{KPI (1)}\} = \{[1,000,000 \text{ Euros}], [200,000 \text{ Euros}]\} \tag{82}$$

$$\{\text{KPI (2)}\} = \{[100,000 \text{ pp}]\}, [50,000 \text{ Euros}]\} \tag{83}$$

$$\{\text{KPI (0)}\} = \{[1,000,000 \text{ Euros}], [200,000 \text{ Euros}], [100,000 \text{ pp}]\}, [50,000 \text{ Euros}], [\text{PROD}(\text{i}, \text{j}_1)]\} \tag{84}$$

This expression means that the primary KPI indicator s closely related to the production core process quantified via [PROD (i, j_1)]}. However, that KPI indicator does not consider the firm or company real possibilities related to its internal resources.

As a result of that adequate coefficient should be determined. They might be calculated based on data contained within appropriate technical and economic norms with the use of the following consideration.

5.3.3 Determination of primary KPI indicators

However our disposals are 170,000 Euros only and an appropriate coefficient might be calculated based on formula

$$\text{Coef} = 170,000/200000 = 0,85 \tag{85}$$

Subsequently, we shall get s appropriate values related to {KPI (1)} and {KPI (2)} with the use of formula (8a) (8b). It means 1,000,000 x 0, 85, 200,000 x 0,85, 100,000 x 0,85 and 50,000 x0,85 and we get correct values for {KPI (1)} and {KPI (2)} indicators. Those indicators are denoted **as primary indicators** and they are closely related to core production process running at strategic management level and create bases for determination of the secondary KPI indicator values.

5.3.4 Determination of secondary KPI indicators

Let us consider the {KPI (2)} indicator values, which indicates that costs for production of [100,000 pp]} are [50,000 Euros]}. Now, we shall try decomposing the {KPI (2)} related to actual contracts and groups of business processes assigned to those contracts. However, before we start doing that, we have to manage several auxiliary actions. The first one is closely related to market required output products quantified via {[MROP (i, j)]}. We shall do it within following steps:

Step 1 Determination of product number with respect to adequate Coef value.

$$[\text{Mark\_res\_assets\_mat\_ires}(0)] = \{[100,000 \text{ pp}] \text{ x } 0,85 = 85,000 \text{ pp} \tag{86}$$

$$\{[\text{MROP (i, j)}]\} = 85,000 \text{ pp} \tag{87}$$

Step 2 The {[MROP (i, j)]} content will be divided into subordinated product classes.

$$[\text{MROP (i, 1)}] = [\text{bowls} = 30,000 \text{ pp}] \tag{88}$$

$$[\text{MROP (i, 2)}] = [\text{bottles} = 30,000 \text{ pp}] \tag{89}$$

$$[\text{MROP (i, 3)}] = [\text{vases} = 25,000 \text{ pp}] \tag{90}$$

and appropriate customers {[Cust (1)]}, {[Cust (2)]} and {[Cust(3)]} will be determined Subsequently, we shall assign selected product groups to adequate customers and will be generated adequate orders.

$$\{[\text{Cust }(1)]\} = \{[\text{bowls} = 30,000\text{ pp}], [\text{bottles} = 30,000\text{ pp}]\} \tag{91}$$

$$\{[\text{Cust }(2)]\} = \{[\text{bowls} = 30,000\text{ pp}], [\text{vases} = 25,000\text{ pp}]\} \tag{92}$$

$$\{[\text{Cust }(2)]\} = [\text{bottles} = 30,000\text{ pp}], [\text{vases} = 25,000\text{ pp}] \tag{93}$$

In that step we shall extend the orders and add the data concerned with adequate financial costs (see also formulas 94–97).

$$\begin{aligned}\{[\text{MROP}_{\text{selfinass}}\ (\text{i},\text{j})]\} &= \left[\text{X}^{\text{Totmat}}{}_{\text{Costs}}\ (0)\right] \text{ x Coef} = [50,000\text{ Euros}] \text{ x } 0,85 \\ &= [42,500\text{ Euros}]\end{aligned} \tag{94}$$

$$\{[\text{Cust }(1)]\} = \{[\text{bowls} = 30,000\text{ pp}], [\text{bottles} = 30,000\text{ pp}], [14,000\text{ Euros}\} \tag{95}$$

$$\{[\text{Cust }(2)]\} = \{[\text{bowls} = 30,000\text{ pp}], [\text{vases} = 25,000\text{ pp}], [18,000\text{ Euros}]\} \tag{96}$$

$$\{[\text{Cust }(3)]\} = [\text{bottles} = 30,000\text{ pp}], [\text{vases} = 25,000\text{ pp}], [10500]\} \tag{97}$$

and we shall create the basic order (CONTRACTC).

$$\{[\text{CONTRACTC(i)}]\} = \{[\text{Cust }(1)], [\text{Cust }(2)], \{\text{Cust }(3)]\} \tag{98}$$

However, a set of adequate business processes should be applied in order to manage production of products related to pre-defined orders or contracts as well, while formula (85) should be extended about linguistic sets, which quantify those groups of business processes – see also formula (50).

$$\begin{aligned}\{[\text{CONTRACTD (i)}]\} = \{&[\text{Cust }(1)], \left[\text{BPG1 }\left(\text{i}, \text{j}_{61}\right)\right], [\text{Cust }(2)], \left[\text{BPG2 }\left(\text{i}, \text{j}_{62}\right)\right], \\ &\{\text{Cust }(3)], \left[\text{BPG3 }\left(\text{i}, \text{j}_{6}\right)\right]\}\end{aligned} \tag{99}$$

With respect to the above-mentioned issues the KPI (i, 2) content is defined via formula (88) and KPI (i, 2) content is defined via formula (89) and we have derived a set of the **secondary KPI indicators.**

$$\text{KPI }(\text{i}, 2) = \{[\text{Cust }(1)], [\text{Cust }(2)], \{\text{Cust }(3)]\} \tag{100}$$

$$\text{KPI }(\text{i}, 3) = \{[\text{Cust }(1)], \left[\text{BPG1 }\left(\text{i}, \text{j}_{61}\right)\right], [\text{Cust }(2)], \left[\text{BPG2 }\left(\text{i}, \text{j}_{62}\right)\right], \{\text{Cust }(3)], \left[\text{BPG3 }\left(\text{i}, \text{j}_{6}\right)\right]\} \tag{101}$$

Each business process group consists of appropriate business process quantified via adequate linguistic set (see also **Tables 2** and **3**).

5.3.5 Determination of tertiary KPI indicators

Let us consider a group of business processes represented by **Tables 2** and **3** and let us select one of those business processes in order to derive the tertiary KPI indicators, which are closely related to BP external and internal metrics. Formulas (63 and 64) will be applied for those purposes. When installing adequate data in the

Product class value	BPG number	BP Number	BP Description	BP External metrics	
				Output prod assets	Input mat. Costs
[bowls = 30,000 pp]	[BPG1 (i, 1)],	BP (1,1)	BP (1,1) Description	[bowls = 10,000 pp]	1700 Euros
	[BPG1 (i, 2)],	BP (1,2)	BP (1,2) Description	[bowls = 10,000 pp]	1700 Euros
	[BPG1 (i, 3)],	BP (1,3)	BP (1,3) Description	[bowls = 10,000 pp]	1700 Euros

Source: The Authors.

Table 2.
Business processes as the BP group members (external metrics).

Product class value	BPG number	BP Number	BP Description	BP Internal metrics		
				HR Costs	Dev costs	Tool cost
[bowls = 30,000 pp]	[BPG1 (i, 1)],	BP (1,1)	BP (1,1) Description	1100Euros	800	300
	[BPG1 (i, 2)],	BP (1,2)	BP (1,2) Description	1100Euros	800	300
	[BPG1 (i, 3)],	BP (1,3)	BP (1,3) Description	1100Euros	800	300

Source: The Authors.

Table 3.
Business processes as the BP group members (internal metrics).

above-mentioned formulas we shall get partial KPI indicators related the selected BP external and internal metrics in form of adequate linguistic sets (see also formulas (91) and (92).

$$\{[\text{KPIemb}\ (i,3)]\} = \{[\text{BP}\ (1,1)]\ [\text{bowls} = 10,000\ \text{pp}], [1700\ \text{Euros}]\} \tag{102}$$

$$\{[\text{KPIimb}\ (i,3)]\} = \{[\text{BP}\ (1,1)], [\text{HRcosts} = 1100\ \text{Euros}], [\text{DEVcosts} = 800], [\text{Toolcosts} = 300\ \text{Euros}]\} \tag{103}$$

$$\text{KPIter} = \text{KPIemb}\ (i,3) \otimes \text{KPIemb}\ (i,3) \tag{104}$$

A set of derived KPI indicators via formulas (91, 92, 93) is closely related so selected BP implemented and operated at operational management level and is denoted as **tertiary indicator set.**

5.3.6 Case study – summary

The case study previous sections deals with KPI indicator creation and decomposition steps, while those steps are closely related to three management levels: (a) strategic, (b) tactic, and (c) operational management level. In that section an overview summary with supplementary visual components will be done.

5.3.6.1 Strategic management level

In order to determine, the KPI indicators related to strategic management level, an appropriate data concerned with market research and internal resources needed for providing business. In that case, the business is closely related to the company, which deals with utility glass production, while there are two types of production: (a) manually oriented and (b) machinery oriented production. Further KPI explanations are closely related to manually oriented production, which is considered to be the core business process and the initial KPI indicator KPI (0) is assigned to that process. The KPI (0) indicator has a nature of top linguistic set, which consists of subordinated linguistic sets (see also formula (53).

$$\{\text{KPImanp}\ (0)\} = [\text{Xtotfinmanpassets}\ (0) = 1,000,000\ \text{Euros}],\\ [\text{Xtotmatmanpassets}\ (0) = 100,000\ \text{pp}], [\text{Xtotmatmanpcosts}\ (0) = 50,000\ \text{Euros}],\\ [\text{Xtotfinmanpcosts}\ (0) = 200,000\ \text{Euros}]\} \tag{105}$$

The KPImanp (0) indicator seems to be the initial KPI indicator and creates basis for deriving of primary KPI indicators calculated with respect of actual firm or company internal financial resources and represented by coefficient Coef. A set of initial KPI indicators is shown in **Figure 1**, while a set of the main initial KPI indicator KPImanp (0) is shown in **Figure 2** and its components are determined based on initial KPI indicators (see also **Figure 1**) and coefficient Coef (see also **Figure 2**).

5.3.6.2 Tactic management level

The main business processes and appropriate BSC perspectives should be investigated and discussed at tactic management level. Only two perspectives (customer's and internal BP) perspective are within that case study, while adequate

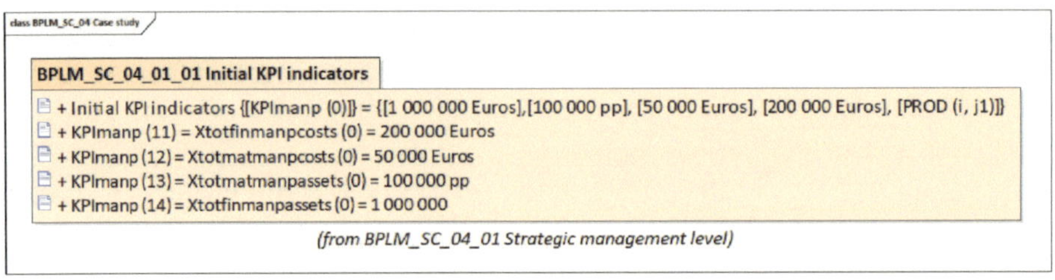

Figure 1.
A set of initial KPI indicators. Source: The authors.

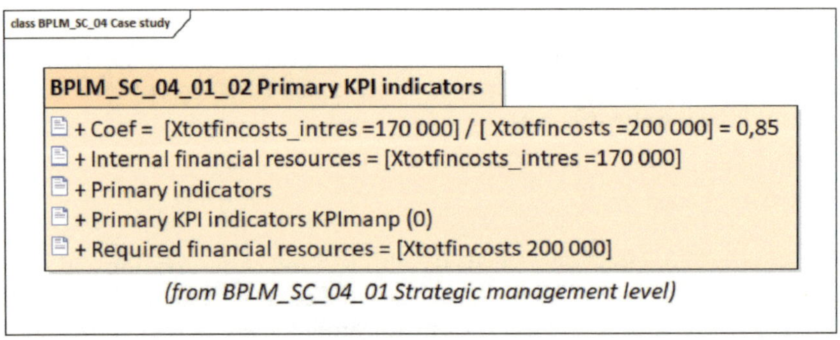

Figure 2.
Inputs for calculating the primary KPI indicators. Source: The authors.

primary KPI indicators represent outgoing inputs for those perspectives as well. The case study is divided in two levels at that management level. The *first level* is closely related to dividing the products to be produced and sold to adequate types and classes (bowls, bottles, vases) and assigned to pre-defined customers Customer 1, Customer 2 and Customer 3, while such assignment enables generating appropriate contract (CONTRACTCS C(I)). On the other hand a set of adequate internal BP is being assigned to assure production of articles, which create an integral part of the (CONTRACTCS C(I)) and creates an extended contract (CONTRACTCS D(I)) business process management point of view (see also **Figure 3**).

5.3.6.3 Operational management level

The subordinated or elementary business processes and their external and metrics should be investigated and discussed at operational management level. In

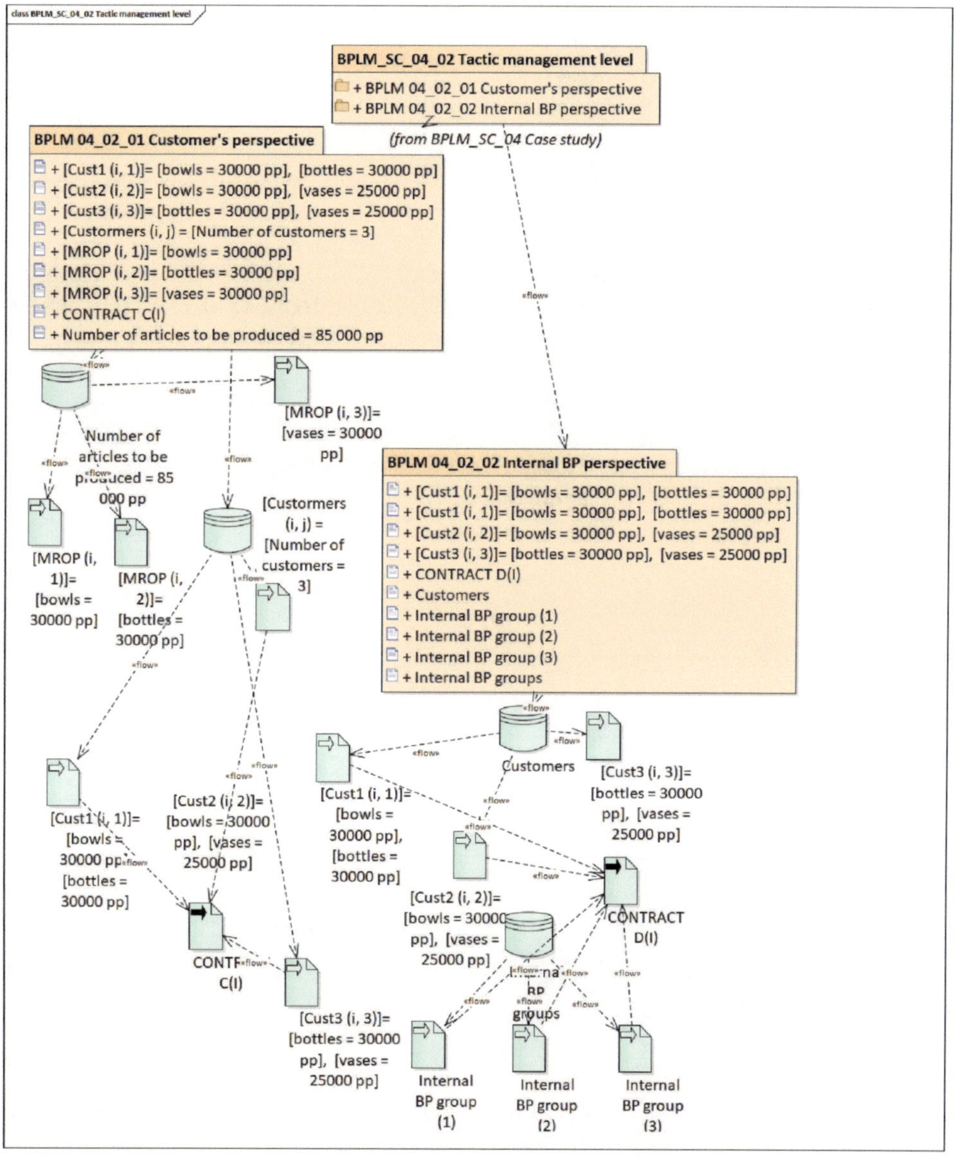

Figure 3.
Example of CONTRACTC (i) and CONTRACTD (i) development. Source: The authors.

general, BP external metrics deals with BP material inputs, while appropriate material costs play a role of principle importance. However, this is only one side of the coin, while the BP output products create an integral part of BP external metrics as well. The KPI indicator closely related to BP external metrics is postulated via formula (106). In order to express adequate numerical values the secondary KPI indicators concerned to output products and input materials should be applied.

$$\text{KPIBP}(1) = \{[\text{KPI\_BPemet}(i,j)]\} = \{[\text{MATfincosts}(i,j)], [\text{MROPselmatassets}(i,j)], [\text{MROPselfinassets}(i,j)]\} \quad (106)$$

However, any BP is represented by its internal metrics as well, while he KPI indicator closely related to BP external metrics is postulated via formula (107). In order to express adequate numerical values the secondary KPI indicators concerned to output products and input materials should be applied.

$$\text{KPIBP}(2) = \{[\text{KPI\_BPimet}(i,j)]\} = \{[(\text{DEV\_costs}(i,j11), (\text{DEV\_assets}(i,j12)], [(\text{TOOL\_costs}(i,j21), (\text{TOOL\_assets}(i,j22)], \{[(\text{HR\_costs}(i,j31), (\text{HR\_assets}(i,j32)]\} \quad (107)$$

The total business process KPI indicator value might be postulated with respect to formula (56). A detailed visualization of the above-mentioned KPI indicator components are shown in **Figure 4**.

However, The sets [(DEV_assets (i,j12)] and [(TOOL_assets(i,j22)] are closely related to depreciation and amortization of devices and tools, as while, [(HR_assets (i,j32)] are closely related to extra contributions generated by employees.

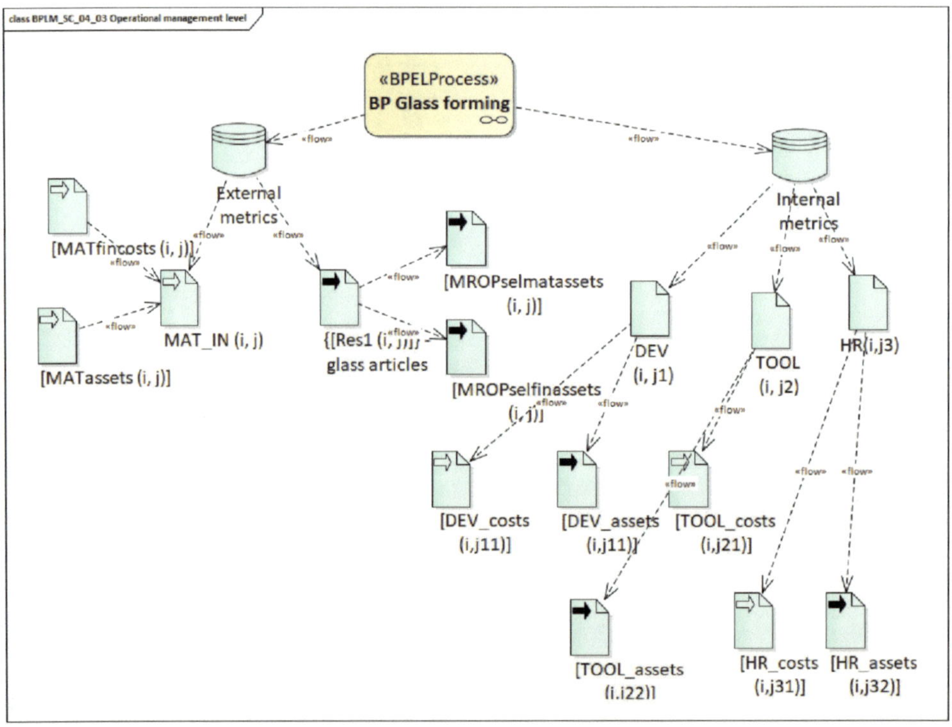

Figure 4.
A detailed visualization of BP KPI indicator components. Source: The authors.

5.4 Design and implementation model

5.4.1 General overview

The BPLM Strategy Creator should be implemented and operated like aim oriented knowledge based or expert system (ES), which consist of an appropriate knowledge base (KB) and inference engine (IE). Both of two subsystems consist of three components closely related to strategic, tactic and operational management levels. However, knowledge contained within KB are being represented via adequate reference databases (RDBs) and semantic networks (SNWs) as well, while IE should enable retrieval and presentation of knowledge contained in ES-KB and generation of new (primary) knowledge based on knowledge actually contained within ES-KB. An interaction between RDBs and SNWs provide transformations rules converted into appropriate transformation functions.

BPLM Strategy Creator is being implemented and operated via adequate knowledge based (expert system), which consists of two subsystems denoted as BPLM_01_06_06_01 ES Knowledge Base and BPLM_01_06_06_02 ES Inference Engine. The Knowledge Base subsystem operates over knowledge base, which contains adequate knowledge, while the Inference Engine subsystem provides retrieval and presentation of knowledge contained within knowledge base and new knowledge discovery based on existed one [25].

When considering the knowledge-based content, we have to talk about knowledge representation. The knowledge representation principle applied within that project is based on existence of reference databases (RDBs), transformation rules (TRrules), transformation tools (TRtools) and semantic networks (SNWs) and is closely related to an appropriate management level (strategic, tactic, operation.

5.4.2 BPLM ES knowledge base

The BPLM ES Knowledge Base functionality is being assured via four modules: (a) Data preparation (b) Reference Database (RDBs) (c) Creation, Semantic Network (SNWs) creation and (d) Import of SNWs to Knowledge Base. The Data preparation component is running within four subordinated steps and modules: (a-1) Data extraction, (a-2) Data transformation, (a-3) RDBs update, (a-4) SNWs creation.

In the first step, an appropriate data is extracted from various documents or they are prepared as a result of the document semantic analysis, while *in the second step* their structure should be transformed to adequate RDBs structure and stored to the RDBs and pre-defined SNWs pointers are being generated. Afterwards, *in the third step*, all the above –mentioned data are stored to linguistic sets and prepared RDBs subsequently. In the fourth step appropriate SNWs are being created and stored to BPLM ES Knowledge base.

5.4.3 BPLM ES inference engine

The BPLM ES Knowledge Base functionality is being assured via four modules: (a) KB content retrieval, (b) Knowledge discovery and (c) Presentation layer. The KB content retrieval operates based on Knowledge general and detailed requirement, which enables selection of appropriate knowledge records, while the selected knowledge record content is visualize via Presentation layer, which consist of the following modules: (c-1) Strategic layer, (c-2) Tactic layer, (c-3) Operational layer and (c-4) Analytical layer

6. Conclusion

We have developed the BPLM Strategy Creator conceptual model, which should operate with several types of KPI indicators and which should provide the KPI creation at strategy management level and decompose them to tactic and operational management levels. With respect to that fact, we define the *initial KPI indicators* created based on research market results, regardless the firm or company internal resources. Afterwards, an appropriate *analysis of the firm or company internal resources* should be provided[6], the results of which could enable generating of so called primary KPI indicators, which reflect real possibilities in providing the firm or company business with respect it actual internal resources. However, the initial and primary KPI indicators should be generated at strategic management level as well, while they create basis for determination of KPI indicators at tactic (*secondary KPI indicators*) and operational management level (*tertiary KPI indicators*). The secondary and tertiary KPI indicators seem to be results of adequate KPI decomposition.

When considering the BPLM Strategy Creator implementation aspects, we have designed structure and functionality of an appropriate expert system (ES), which should be implemented with use of graph databases (GraphDB) based on an adequate logical and physical model[7].

The ES should contain an inference engine, which could contain a presentation layer, which should enable the KPI visualizing in form, which is very closed to text in natural language (TNL text) supplied by adequate graphical information and this is considered to be main significance related to the BPLM Strategy creator utilization in a practice.

Acknowledgements

This work was supported by the Grant Agency of Slovak Republic – VEGA grant no. 1/0339/20 “Hidden Markov Model Utilization in Financial Modeling”.

This work was supported by the Grant Agency of Slovak Republic – KEGA grant no. 019EU-4/2020 Support of distance education through a virtual department.

This work was supported by the Štefan Kassy Foundation and support for science and education.

[6] A development of that type component should be a subject of future work.

[7] A development of ES logical and physical model should be a subject of future work.

Author details

Jozef Stašák[1]*, Jaroslav Kultan[1], Peter Schmidt[1] and Mukhammedov Abu Urinbasarovich[2]

1 Department of Applied Informatics, Faculty of Economy Informatics, University of Economy in Bratislava, Slovakia

2 University of international Business, Almaty City, Kazakstan

*Address all correspondence to: jstasak@atlas.sk

References

[1] Fisk P. People, planet, profit: how to embrace sustainability for innovation and business growth. London: Kogan Page; 2010.

[2] Roubtsova E, Michell V. A method for modeling of KPIs enabling validation of their properties. Proceedings of the 5th ACM SIGCHI Annual International Workshop on Behaviour Modelling - Foundations and Applications - BMFA '13. 2013;

[3] Dimensional Data Modeling - Metrics - Key Performance indicators (KPI) [Internet]. 0AD [cited 2020Nov20]. Available from: https://datacadamia.com/data/type/cube/modeling/metrics

[4] Barr S. A Complete Model of the Types of KPIs [Internet]. Stacey Barr | Performance Measure and KPI Specialist. 2018 [cited 2020Nov20]. Available from: https://www.staceybarr.com/measure-up/complete-model-types-kpis/

[5] McClintock M. Business Process Modeling -KPIs & Benchmarks for Process Improvement [Internet]. ProcessMaker. 2020 [cited 2020Nov20]. Available from: https://www.processmaker.com/blog/business-process-modeling-process-improvement/

[6] Velimirović D, Velimirović M, Stanković R. Role and importance of key performance indicators measurement. Serbian Journal of Management. 2011;6 (1):63–72.

[7] Stašák, J. Modelovanie procesov podnikania s využitím aplikačného programu Aris. Ekonom. 2010

[8] Doug.Z. Key Performance Indicator (KPI) frameworks [Internet]. Douglas Holmes. 2020 [cited 2020Nov20]. Available from: http://douglasholmes.com/key-performance-indicator-kpi-frameworks/

[9] Stoop J. Developing a reference model for KPI and Dashboard reporting in Sales & Marketing . Twente: Universiteit Twente; 2009. https://essay.utwente.nl/60699/1/BSc_Jasper_Stoop.pdf

[10] Stasak J. Business Process Linguistic Modeling – Philosophy & Principles. Computer Science and Information Technology. 2015;3(5):198–213.

[11] Stasak J, Vanickova R, Grell M. Business Process Modeling Linguistic Approach – Problems of Business Strategy Design. Universal Journal of Management. 2015;3(7):271–82.

[12] Kee H, Waters J, Njeru R, Estorilio C. PROPOSE A CHANGE MANAGEMENT PLAN TO REDUCE TIME TO MARKET AT ENTERPRISE X. Bedford : CRANFIELD UNIVERSITY ; 2000.

[13] Staron, M., Meding, W., Niesel, K., Abran, A. A Key Performance Indicator Quality Model and Its Industrial Evaluation [cited 2020Nov20]. Available from: https://gup.ub.gu.se/file/197126

[14] Key Performance Indicator Model [cited 2020Nov20]. Available from: https://seilevel.com/business-analyst-resources/business-requirements-models-templates/key-performance-indicator-model/

[15] Russo M. KPIs in Tabular models for Power BI and Excel [Internet]. SQLBI. 2020 [cited 2020Nov20]. Available from: https://www.sqlbi.com/articles/kpi-in-tabular-models-for-power-bi-and-excel/

[16] Materials Modelling: expectations, benefits and Key Performance Indicators (KPIs) EMMC-CSA – GA N° 723867 [[cited 2020Nov20]. Available from: https://emmc.info/wp-content/uploads/2017/12/Materials-Modelling-KPIs-and-Metrics-update 20180727.pdf

[17] Rozner S. Developing and Using Key Performance Indicators. A Toolkit for Health Sector Managers. Health Finance & Governance Project, Abt Associates Inc. 2013Dec;

[18] Strecker S, Frank U, Heise D, Kattenstroth H. MetricM: a modeling method in support of the reflective design and use of performance measurement systems. Information Systems and e-Business Management. 2011;10(2):241–76.

[19] Communicating Sequential Processes. By C.A.R. Hoare. Prentice-Hall International, London, 1985, viii +256 pages. Science of Computer Programming. 1987;9(1):101–5.

[20] Allen R, Garlan D. Beyond definition/use. Proceedings of the Workshop on Interface Definition Languages - IDL '94. 1994;

[21] Mount S, Hammoudeh M, Wilson S, Newman R. CSP as a Domain-Specific Language Embedded in Python and Jython. Communicating Process Architectures 2009. 2009;67:293–309.

[22] Mcneile A, Simons N. Protocol modelling: A modelling approach that supports reusable behavioural abstractions. Software & Systems Modeling. 2005;5(1):91–107.

[23] Schmidt P, Stašák J. Business process modelling linguistic approach application in public administration (self-governmental). International Journal of Business Performance Management. 2018;19(2):209.

[24] Stašák J, Schmidt P. Semantic technology and linguistic modelling in business strategy design and evaluation. International Journal of Business Information Systems. 2019;31(2):170.

[25] Wannesa A, Ghannouchiab SA. KPI-Based Approach for Business Process Improvement [Internet]. Procedia Computer Science. Elsevier; 2020 [cited 2020Nov20]. Available from: https://reader.elsevier.com/reader/sd/pii/S1877050919322215?token=04D86A99EC0613B93DC883BC20994B51FD67E31EDE377EA62C554B22C 9534411081 81BABCFE0283FC 57E858 FD1D3D9D8

[26] McAuley B, Hore A, West R. Establishing Key Performance Indicators to measure the benefit of introducing the Facilities Manager at an early stage in the Building Information Modelling Process. International Journal of 3-D Information Modeling . 2013;2 (4):38–51.

[27] Zapata-Jaramillo CM, Castro-Rojas LF. Pre-conceptual-schema-based patterns for deriving key ... [Internet]. http://www.scielo.org.co/scielo.php?script=sci_arttext&pid=S0120-56092017000200120. 2017 [cited 2020Nov20]. Available from: http://www.scielo.org.co/pdf/iei/v37n2/0120-5609-iei-37-02-00120.pdf

[28] Perjons E, Johannesson P. A Value and Model Driven Method for Patient Oriented KPI Design in Health Care [Internet]. SpringerLink. Springer, Berlin, Heidelberg; 2010 [cited 2020Nov20]. Available from: https://link.springer.com/chapter/10.1007/978-3-642-18472-7_10

[29] Schulz T, Chelaru A. Model Driven Key Performance Indicators Concepts for Manufacturing Execution Systems [Internet]. Applied Mechanics and Materials. Trans Tech Publications Ltd; 2012 [cited 2020Nov20]. Available from: https://www.scientific.net/AMM.245.173

[30] Dwivedi A, Madaan J. A hybrid approach for modeling the key performance indicators of information facilitated product recovery system [Internet]. Journal of Modelling in Management. Emerald Publishing Limited; 2020 [cited 2020Nov20]. Available from: https://www.emerald.com/insight/content/doi/10.1108/JM2-01-2019-0003/full/html

Chapter 9

Business Process Linguistic Modeling: Theory and Practice Part II: BPLM Business Process Designer

Jozef Stašák, Jaroslav Kultan, Peter Schmidt and Bouafia Khawla

Abstract

The business activities provided within any firm or company should be checked and controlled continuously, while two principal approaches should be applied: (a) qualitative monitoring, (b) quantitative evaluations and getting to know the rules, which regulate structure and functionality of business processes (BPs) implemented and operated there plays a role of principle importance and they are derived based on actual BP models. Therefore we have designed a conceptual model of application denoted as BPLM Process Designer in form of expert system (ES) operating based on principles closely related to business process linguistic modeling approach, where linguistic sets and PBPL Equation play a role of principle importance. Our contribution contains such application description from qualitative, quantitative and design point of view. The ES qualitative description contains references to appropriate math relations and algorithms postulated within subsequent sections. Those sections are accompanied by the case study, which indicates how the math relations and algorithms might be applied within BPLM Process Designer functionality. However, those sections are accompanied by ES structure and functionality description as well, which represent the BPLM Process Designer mean or facility.

Keywords: business, process, linguistic, modeling, designer

1. Introduction

Nowadays, business competition causing companies to optimize existing business processes within the organization. Analysis the business process modeling is a tool to evaluate and make improvements over the business process (BP) there. Through the analysis of business they can decide which one is optimal or not optimal run and give attention to it. Business process modeling is that activity aimed at the representation of all or some elements in order to produce a cohesive model of the behavior required to deliver a service and/or product to a customer or another part of the organization [1]. There are some techniques to model the business process. In practice it is not easy to determine which techniques are suitable and easily understood by stakeholders. However, the similar situation might happen, when considering relations among business process analysts and the people who provide BP implementation and execution as well, while requires research on comparative of business process modeling techniques to overcome the above problems. This research is limited in four business process modeling techniques which often used the comparative analysis phase. Four business process modeling techniques are: (a) Data Flow Diagram (DFD), (b) Business Process Modeling Notation (BPMN), (c) Activity Diagram, or (d) Integration Definition for Function Modeling (IDEF0) and all the above-mentioned techniques together with ARIS methodology [2–4] create an integral part of so called BP modeling standardized approach. On the other hand there are a lot of BP modeling methods and techniques based on semantic and ontology approach or based on analysis of texts in natural language [5–7] (TNL texts), which describe the BP structure and functionality, while BP modeling based on semantic and ontology [5, 8–11] principles play a role of significant importance too. However, there is another one group of BP modeling approaches and methodologies, which are based on so called linguistic set [12, 13] and Principle Businesses Process Linguistic Modeling Equation (PBPL Equation) [14–16], which regulates relations among them. This approach is denoted as business process linguistic modeling (BPLM Modeling) covered by BPLM Process Modeling System, which consists of Business Process Strategy Creator [4, 17–20] and BPLM Designer, the design of its conceptual model seems to be the main goal of that contribution, while adequate BP model views should be respected, it means functional, process, information and knowledge-based support BP model view.

In order to achieve, the main goal four partial and subordinated aims should be postulated and fulfilled. The first partial aim is closely related to BPLM Process Designer qualitative proposal design (see also Section 4.1). The second partial aim is closely related to quantification of BPLM Designer structure and functionality, where adequate linguistic sets and PBPL Equation is applied (see also Section 4.2). The third partial aim is concerned with derivation of BP function (BPF) rules, which regulate the BP and BPF functionality (see also Section 4.3) and the fourth partial aim deals with BPLM Process Designer - implementation and operation (see also Section 4.5). However, an appropriate case study creates an integral part of that contribution, the aim of which is to show how the derived math relations and algorithms should be applied related to BPLM Process Designer functionality as well (see also Section 4.4).

2. State of the art

2.1 Business process modeling -standardized, semantic and ontology approach

2.1.1 Standardized approach

At present, business process management becomes a matter of principal importance all over the world and the development trends indicate leaving for the isolated business process (BP) e. g. sales or purchase and investigation of business process complexity is considered to be more and more important. After WW2, the systems related to managerial quality have been aimed to features of articles in most cases. A need to stabilize a quality of products has flown into management of those business processes, which were closely related to production of products. At present, the firms or companies investigate not only internal processes; however they pay an appropriate attention to external business processes as well [21], while an understanding of BP structure and functionality incl. Creating of appropriate models (modeling) creates an integral part of business process management too. When looking at business process modeling (BPM) modern history, the first steps in these branches are closely related to ARIS methodology created by prof. Scheer [2, 4] who developed an adequate application for those purposes denoted as ARIS System, while that methodology seems to be a standard utilized round the world and this modeling approach is denoted as standardized BP modeling approach [3, 22]. This approach has been applied in seventies of twentieth century, while it was based on the principle that a quality of products is determined by quality of production business process and their check and control and management is considered to be a matter of principal importance. However, there are other approaches based on symbolic, semantic and linguistic methodologies [12] where a text in natural language (English, Slovak, Czech, etc.) – TNL Text, together with methodologies based on BP ontologies play a role of principle importance as well and create basis for establishment of so called BP modeling symbolic, semantic and ontology approach, are very briefly discussed within Section 2.1.2.

2.1.2 Business process models based on symbolic, semantic and ontology approach

In current business process models, the functional perspective (also can be referred to in the literature as business capability, functionality or business function) for each process activity is limited to its label [23, 24], while an appropriate symbol, which creates basis for business process modeling. On the other hand a single label is not enough to describe properly the capability of a particular process element (i.e. activity, fragment or entire process). Using labels only prevents stakeholders from easily and quickly understanding business processes or identifying the differences and commonalities between them in terms of business properties [23–26]. When required, stakeholders need to read the business process documentation in order to find out what a process element does, expressed in terms of business properties. All the above-mentioned create basis for business process modeling symbolic approach. However, in the literature, several languages for BP have been proposed. Such languages can be sketchily gathered in three large groups (a) Descriptive languages, (b) Procedural languages and (c) ontology-based process languages, such as those declared in [6, 7, 24, 26]. This group of languages have a wider scope, aiming at modeling semantically rich processes in an ontological context, and have been conceived not directly connected to the business world [27, 28].

2.2 Business process modeling -linguistic approach: theory

In general, the linguistic modeling approach described and discussed in that contribution is based on the business process, the qualitative and quantitative aspects might be described via standardized TNL text logical sentences, a content of which is quantified via specialized types of sets denoted as linguistic sets, and relations among them are being quantified via PBPL Equation, and any BP to be modeled is represented via specialized linguistic sets closely related to its external and internal metrics [29]. On the other hand, the fact that the modeled BP horizontal structure is created by business process functions, which the BP to be modeled consists of plays a role of principle importance and might be quantified by one linguistic set, which contains three subordinated ones: (a) transformation rules, transformation functions and (c) BPF external and internal metrics. An establishment of transformation rules and transformation functions together with relations among BPF internal and external metrics and BPF transformation functions is discussed within Section 4. However, that section deals with that application implementation and operation as well, while that application is an expert system, where the knowledge stored in an appropriate knowledge base are represented with the use of reference databases (RDBs) and sematic networks (SNWs). This approach is denoted as BP modeling linguistic approach (BPLM Approach) and will be discussed within further sections of that contribution [15].

2.3 Business process modeling -linguistic approach—application programs

Business Process Management (BPM) has been receiving increasing attention in recent years. Many organizations have been adapting their business to a process-centered view since they started noticing its potential to reduce costs, improve productivity and achieve higher levels of quality. However, implementing BPM in organizations requires time, making the automation of process identification and discovery highly desirable. To achieve this expectation, the application of Natural Language Processing (NLP) techniques and tools has emerged to generate process models from unstructured text. However, no BPLM functional application programs were found, which would be similar to system ARIS or Bizagi application programs and which could be tested for practical purposes.

There are techniques applied to the BPM life-cycle phases of process identification, process discovery and process analysis as well as tools to support process discovery. The results of the present study may be valuable to support research in extraction of business process models from natural language text [6, 7, 24, 25, 30].

3. Research methods

In order to achieve, the main goal four partial and subordinated aims a set of adequate research methods should be postulated and applied:

- Business process linguistic modeling (BPLM) approach, where linguistic set seem to be elements of principle importance and the PBPL Equation as well.

- With respect to pre-defined partial aims, functional, process, information and knowledge based support views should be postulated and quantified via linguistic sets and PBPL Equation

- As a result of that, the BPF main linguistic set together with transformation rules and transformation functions should be defined as well as subsets related to internal BPF metrics.

- A set of reference databases (RDBs) and SNWs should be designed in order be possible to generate valid rules, which regulate BP and BPF structure and functionality

- An appropriate expert system - ES (knowledge base and inference engine) should be designed, where the above-mentioned rules could be stored and accessible to authorized users via inference engine components, while the ES should be implemented with the use of Ontotext components.

4. Results and discussion

4.1 BPLM process designer – structure and functionality BPLM process designer – structure and functionality – qualitative view

4.1.1 General overview

A real business is getting started, after an appropriate business strategy creation and implementation. However, any business might be running properly and efficiently without running adequate business process (BP) as well, while their structure and functionality plays a role of principle importance. The BP structure represents BP static aspects, however the BP functionality is closely related to BP dynamics and performance, the result of which are being compared with KPI indicators established within evaluation of the or company business strategy. As a result of that, we have to know the BP structure and functionality, where business process modeling and BP modeling tools seem to be very significant important matter. However, there exist many different approaches BP modeling, incl./ methods and techniques as well, while the ARIS methodology designed by prof. Scheer [2, 4] seems to be the standard applied round the globe. On the other hand, many other approaches and methodologies related o business process modeling exist, while one of the is denoted as business process linguistic modeling (BPLM) based on existence of so called linguistic sets, which create basis for BP static aspects quantifying, while they create basis of Principle Businesses Process Linguistic Modeling Equation (PBPL Equation), which enables quantifying BP functionality aspects. We shall discuss those principles within, next subsections.

4.1.2 Business process model views

However, the ARIS methodology creates basis for BPLM approach as well, while there are defined BP model four views: (a) functional view, (b) process view, (c) data view and (d) organizational view as well, we shall apply and modify them as follows: (a) functional view, (b) process view, (c) information support view, (d) knowledge-based view and (e) organization support view and all those views will be respected, when creating the BPLM Designer, which creates an integral part of the entire business process linguistic modeling system (BLM System). On one hand, the BPLM Designer seems to a subsystem closely related to the BPLM System, while it consist of the following components: (a) BPLM Process Analysis and Design, (b) BPLM Process simulation and (c) BPLM Optimization component, while the BP Architecture model seems to be the main result of BPLM Designer functionality.

4.1.3 Business process model functional view

Business process model functional view represents a functionality of core, main, subordinated, and elementary business processes implemented and operated within actual firm or company and indicates appropriate relations among them, however that view does not indicate any BP outposts and BP inputs. The adequate BPs are being quantified via appropriate linguistic sets and relations among the is quantified via PBPL Equation with respect o Consideration no 1 (see also Section 4.2.1).

4.1.4 Business process model process view

The ARIS methodology describes the BP model process view as a sequence of business process functions (BPFs), which the actual business process consists of and does not consider about BPF structure and functionality. In general, any BPF provides conversion (transformation) of BP inputs to BP pre-defined outputs and consist of: (a) transformation rules, (b) transformation tools and BP internal metrics items, while the BP internal metrics items and values are closely related to BP transformation tools as well and they are quantified via appropriate linguistic subsets, which create an integral part of the actual BPF linguistic set {[BPF (i, j)]}, while formula (1) might be postulated.

$$\{[\mathrm{BPF}\ (\mathrm{i},\mathrm{j})]\} = \{[\mathrm{BPF\_TR}\ (\mathrm{i},\mathrm{j1})], [\mathrm{BPF\_TT}\ (\mathrm{i},\mathrm{j2})], [\mathrm{BPF\_IM}\ (\mathrm{i},\mathrm{j3})]\} \tag{1}$$

where

i = 1, 2 ... n is the index which indicates the BP, which an appropriate BPF is beingassigned to.

[BPF_TR (i, j1)] - Transformation rule linguistic set – the set elements represent math rules and algorithms, which regulate the BPF transformation process

[BPF_TT (i, j2)] - Transformation tool linguistic set – the set elements represent closely related to human resources HRs, production technological device resources PDEV and production technological tool resources PTOOL, while formula (2) might be postulated

$$[\mathrm{BPF\_TT}\ (\mathrm{i},\mathrm{j2})] = [(\mathrm{HRs}\ (\mathrm{i},\mathrm{j21}), \mathrm{PDEV}\ (\mathrm{i},\mathrm{j22}), (\mathrm{PTOOL}\ (\mathrm{i},\mathrm{j23})) \tag{2}$$

where

(HRs (i, j22), – the linguistic set, which contains data closely related to human resources, which participate at transformation operations within actual BPFfunctionality.

(PDEV (i, j21), – the linguistic set, which contains data closely related to production technological device resources, which participate at transformation operations within actual BPF functionality.

(PTOOL (i, j21), – the linguistic set, which contains data closely related to production technological tool resources, which participate at transformation operations within actual BPF functionality.

[BPF_IM (i, j3)]} - BP internal metrics linguistic set, the content of which represent subsets, which contain data closely related to operational and technical parameters of to production technological device resources and theoretical knowledge and practical skills of human resources

However, any BP is represented by external metrics items, which are closely related to actual BP inputs and outputs as well. The actual linguistic sets and algorithms concerned with relations among them are described in Section 4.2.2 via Consideration no. 2.

4.1.5 Business process model information support view

In general, no business process proper and efficient functionality is possible without appropriate information support. At that level, the information support deals with reference database (RDBs) functionality and corresponds with their conceptual, logical and physical model. All linguistic seta related to BPF structure and functionality are stored in those RDBs and are closely related to BPF knowledge based support, while they contain pointers to appropriate semantic networks (SNWs), which create basis of BP knowledge-based support. However, they contain pointers to external data or information support resources (SAP components especially).

The actual linguistic sets and algorithms concerned with relations among them are described in Section 4.2.3 via Consideration no. 3.

4.1.6 Business process model knowledge-based support view

In a previous section, we have postulated that no business process proper and efficient functionality is possible without appropriate information support. However, the same is concerned with the BP knowledge-based support. The BP knowledge-based support provides interconnection between the BP process and information support view at two levels: (a) internal level and (b) external level. On one hand, the internal level is closely related to BPF transformation rules and transformation tools, while the BPF rules regulate the BP tools and the semantic networks are interconnected to RDBs within knowledge representation process. On the other hand, the external level is related to external data sources and transformation of selected from them in order to be possible a generation of new knowledge based on existing one and the above-mentioned data selected and transferred from external data resources. However, at that level adequate linguistic sets and relation among them play a role of principle importance as well, while they are described within Section 4.2.4 and Consideration no. 4 too.

4.2 BPLM process designer – structure and functionality BPLM process designer – structure and functionality – quantitative view

4.2.1 BP functional view consideration no. 1

The functional view deals with the BP vertical structure, which is created by core business processes (CBP), main business processes (MBP), subordinated business processes (SBP) and elementary business processes[1] (EBP).

The view on a process as a structured chain of activities has a direct coupling to coordination as defined by Malone & Crowston. Coordination is simply the management of the dependencies between these activities. This implies that coordination is an activity in itself carried out by some actors. The work object of the coordination activity is coordination manifested as various tangible and intangible elements in the organization.

Now, we shall try quantifying those aspects with the use of PBPL Equation [25, 26].

BP Model Functional view quantification with the use of PBPL Equation.

Let us consider a core business process CBP (0, I) Utility Glass Production represented by the {CBP (I, j)} linguistic set being decomposed into CBP (0, 1) Utility Glass Production Preparation denoted as CBP (i', j') represented by {CBP (i', j')} linguistic set i' = 1… … n' j = 1… ..$m_{1'}$ and CBP (i", j") i' = 1… .n", j = 1… .$m_{1"}$ Utility Glass Production Management represented by {CBP (i", j")} linguistic set, while the {CBP (i', j')} and {CBP (i", j")} are considered to be the linguistic subsets relating to the {CBP (I, j)}, while formula (3) might be postulated.

$$\{\text{CBP (I, j)}\} = \{[\text{CBP (i', j')}\}], [\text{CBP (i", j")}]\} \tag{3}$$

However, the CBP (0, I) business process is represented by its own internal and external metrics as well, while formulas (4) and (5) might be postulated.

$$\{\text{CBP (I, j)}\} \equiv \{\text{CBP}_\text{m}\text{ (I, j)}\} \tag{4}$$

$$\{\text{CBP}_\text{m}\text{ (I, j)}\} = \{[\text{CBP}_\text{mint}\text{ (I, j)}], [\text{CBP}_\text{miext}\text{(I, j)}]\} \tag{5}$$

Where index i' represents a hierarchic level of BP to be investigated and j' index represents a number subordinated processes relating to the BP investigated.

Now, we shall try to investigate how the superior core business process together with its internal and external metrics should be decomposed related to lower levels of management. With respect to this issue, we shall postulate two important questions.

---

[1] The BP, which cannot be decomposed in other subordinated one or its further decomposition is meaningless from practical point of view is denoted as the elementary process.

(A) How the superior business process C (0, I) represented by {[CB (I, j)]} linguistic set should be decomposed to subordinated core business processes related to lower management levels, it means from strategic to tactic and operational management level and how the superior core business functional model should be created.

At first, we shall try to find an answer related to (A) question. In order to achieve that, we have to define the superior core business process in form of adequate linguistic set {CBP (I, j)} and to assign to that set an appropriate linguistic set $\{b_{OI}\}_{=}\{[b_{Ohl}], [b_{nbp}], [b_{inm}])$, denoted as BP Functional View Control Linguistic Set (BP-FWC Linguistic Set), while.

$[b_{Ohl}]$ – is a linguistic subset element, which indicates a hierarchic level of BP to be decomposed.

$[b_{nbp}]$ – is a linguistic subset element which indicates a number of business process stored at subordinated level

$[b_{inm}]$ – is a linguistic subset element which indicates a serial number of that BP at appropriate hierarchic level, which should be decomposed.

Example:

Let us consider a core business process stored at hierarchic level one $[b_{nbp}] = 1$, while a serial number of that BP within appropriate hierarchic level is =1 $[b_{inm}] = 1$ and that BP should be decomposed in 3 subordinated business processes $[b_{nbp}] = 3$. For that case, linguistic set {b0I} elements are represented by formula (6).

$$\{b_{OI}\} = \{b_{OI}\} = \{[b_{Ohl}], [b_{nbp}], [b_{inm}]\} = \{b_{OI}\} = \{[1], [3], [1]) \quad (6)$$

Now, let us consider the superior core CBP (I, j) business process represented by {CBP (I, j) linguistic set, which should be decomposed in two subordinated core processes[2], which operate at strategic management level, while the $\{b_{OI}\}$ linguistic set[3] elements are postulated via formula (7).

$$\{b_{OI}\} = \{b_{OI}\} = \{[b_{Ohl}], [b_{nbp}], [b_{inm}]\} = \{b_{OI}\} = \{[1], [2], [1]) \quad (7)$$

and the $\{b_{OI}\}$ set elements create basis for {[Petx (I, j)]} linguistic set, while formulas (8) and (9) might be postulated.

$$\{[\text{Petx}\ (I, j)]\} = \{b_{OI}\} \quad (8)$$

$$\{[\text{Petx}\ (I, j)]\} = \{[b_{Ohl}], [b_{nbp}], [b_{inm}]\} \quad (9)$$

When applying the PBPL Equation formula (10) might be postulated.

$$\begin{aligned} &\{[\text{Petx}\ (I, j)]\} \otimes \{[\text{Pe}\ (I, j) \\ &= \{[b_{Ohl}], [b_{nbp}], [b_{inm}]\} \otimes \{\text{CBP}\ (I, j) \\ &= \{[\text{CBP}\ (1, i_1, j_1)], [1]\} \otimes \{[\text{CBP}\ (1, i_2, j_2)], [2]\} \end{aligned} \quad (10)$$

This equation corresponds to the first hierarchic level shown in **Figure 1**.

Subsequently, we shall try to decompose the subordinated BP represented by $\{[\text{CBP}\ (1, i_1, j_1)]\}$ into hierarchic level 2, where three subordinated BP should be stored and a number of the BP to be decomposed is 1. The $\{b_{OI}\} = \{b_{OI}\}_{=}\{[b_{Ohl}], [b_{nbp}], [b_{inm}]\}$linguistic set content might be postulated as follows:

$$\{b_{OI}\} = \{b_{OI}\}_{=}\{[b_{Ohl}], [b_{nbp}], [b_{inm}]\} = \{b_{OI}\} = \{[2], [3], [1]\} \quad (11)$$

[2] The terms business process and process are considered to be equivalent from semantic point of view.

[3] The terms linguistic set and set are considered to be equivalent from semantic point of view.

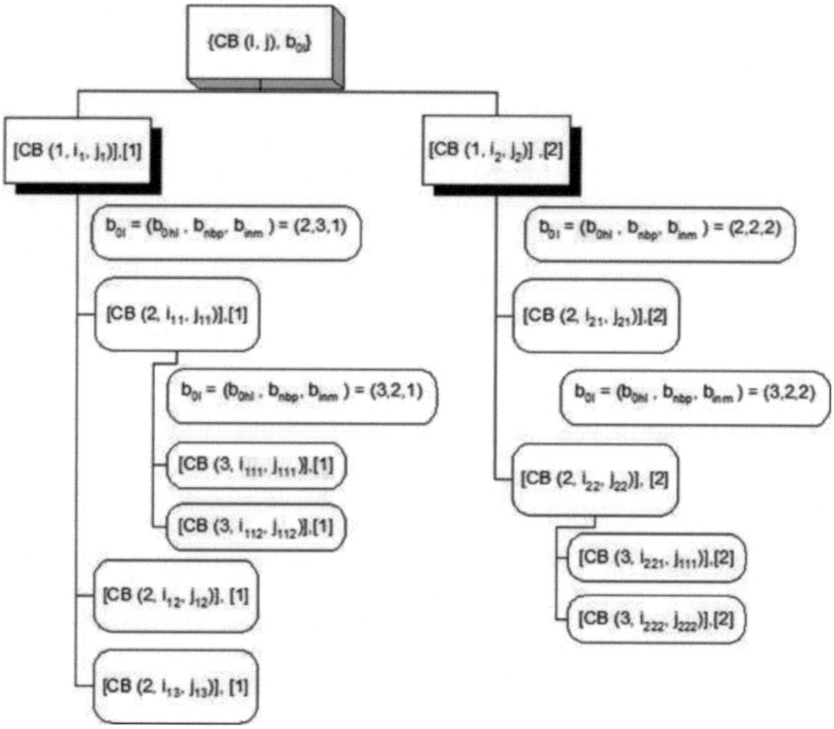

Figure 1.
An example of BP functional model source: The authors.

When applying PBPL Equation, the following result might be generated.

$$\begin{aligned}&\{[2],[3],[1]\}\otimes\{[\text{CB }(1,\mathrm{i}_1,\mathrm{j}_1)],[1]\}\\&=\{[\text{CB }(2,\mathrm{i}_1,\mathrm{j}_1)],[1]\}\otimes\{[\text{CB }(2,\mathrm{i}_2,\mathrm{j}_2)],[2]\}\}\otimes\{[\text{CB }(2,\mathrm{i}_3,\mathrm{j}_3)],[3]\}]\}\end{aligned} \tag{12}$$

Finally, let us consider the business process no.1 located at hierarchic level 2, which should be decomposed into two subordinated business processes located at hierarchic level 3, while two subordinated processes should be stored at that level and the $\{b_{OI}\}$ linguistic set elements are postulated with respect two formula (13).

$$\{b_{OI}\}=\{b_{OI}\}_{=}\{[b_{Ohl]},[b_{nbp}],[b_{inm}]\}=\{b_{OI}\}=\{[3],[2],[1]\} \tag{13}$$

When applying PBPL Equation, the following result might be generated {[3],[2],[1]} ⊗ {[CB (2, i_2, j_2)],[2}]} = {[CB (3, i_2, j_2)],[1}]} ⊗ {[CB (3, i_2, j_2)],[2}]} see also **Figure 1.**

The above-mentioned formulas and relations create basis for BP functional view without BP internal and external metrics linguistic sets.

4.2.2 Process view consideration no. 2

BP Model Process view quantification with the use of PBPL Equation.

However, the BPLM Process View deals with BP horizontal structure as well, an appropriate BP to be investigated and modeled, is selected from set of BP with adequate vertical structure (functional view) and the BP internal and external metrics plays a role of principle importance. Furthermore, a significant role plays BP Input Metrics, which creates an integral part of BP External Metrics {BPEXM (i, j4)} as well (see also formula (14). On the other hand, the BP internal metrics

{BPINM (i, j4)} (see also formula (15) is created by those linguistic sets, which make basis for BP Function (BPF) definition.

$$\begin{aligned} ?F\ (i, j2)\ (j2 = 1 \ldots .m2) &= > [F\ (i, j2)]?[Petxj2\ (i, j1,)]?\{[Petx\ (i, j1,)]\} \\ &= > [Petxj2\ (i, j1,)]?[F\ (i, j2)] \\ &= [Res1j2\ (i, j3,)]?\{[Res1\ (i, j3,)]\} \end{aligned} \tag{14}$$

$$\begin{aligned} ?Pe\ (i, j2)\ (j2 = 1 \ldots .m2) &== > \{Pe\ (i, j2)\}?\{[Petx\ (i, j1,)]\}\&\{[Res1\ (i, j3,)]\} \\ &= > \{[Petx\ (i, j1,)]\}\&\{[Res1\ (i, j3,)]\}?\{BPEXM\ (i, j4)\} \\ &\&\{Pe\ (i, j2)\}?\{BPINM\ (i, j5)\} \end{aligned} \tag{15}$$

4.2.3 BPM information support view

In general, a proper and an efficient functionality of any business process depends on an adequate information support, however the question is: What the term BP information support related to BP functionality does mean? In general, any BP functionality and performance are closely related to BP external and internal metrics. However, the problems of BP external and internal metrics theory are discussed within Section 2 as well, while at that place we shall discuss aspects closely related to so called two stage BP external and internal metrics (see also **Figure 2**). What the term two stage BP external and internal metrics does mean?

In general, the implemented and operated BP is running and generates pre-defined output products (articles) – denoted as the primary products based on appropriated adequate material, information and financial inputs. On the other hand, the investigated BP operates with a set of input and output information generated based on detailed data, e.g. number of good articles $n_{Artgood}$ – a quality of which corresponds to pre-defined requirements, number of repaired articles n_{Artrep} - number of produced articles their quality does not correspond to pre-defined requirements and should be repaired, and number of waste articles $n_{Artwaste}$ - number of produced articles their quality does not correspond to pre-defined requirements – cannot be repaired and should be considered to be a waste. They are considered to be detailed data and have no level of aggregation. The same is concerned with other data closely related to BP external or internal metrics and they are being measured at

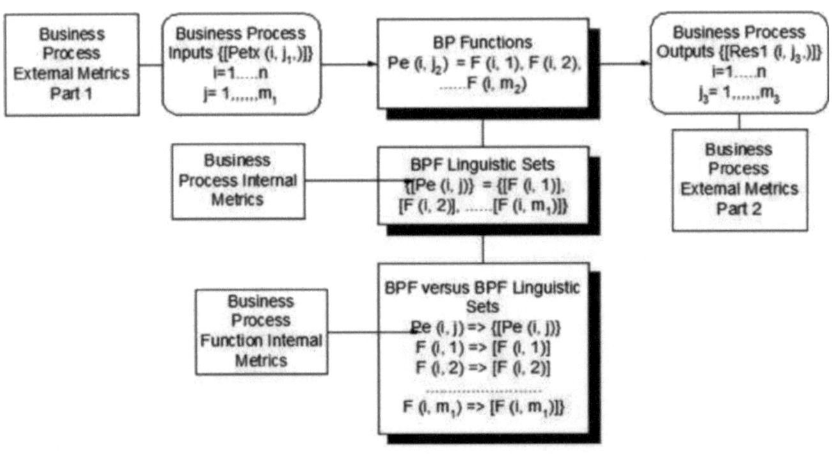

Figure 2.
Business process horizontal structure source: The authors.

pre-defined time points, it means they are time dependent and are called BP external and internal metrics primary data generated at the first stage.

However, that data are undertaken to an appropriate statistic evaluation and analysis as well, while adequate statistic values are being calculated (average and extend of variation) and a predefined time interval should be respected, when calculating those values. Those values are of an aggregated nature and are called BP external and internal metrics secondary data. In the next sections, we shall discuss about that data quantification.

Consideration no. 3a.

BP external metrics primary data – quantification via linguistic sets.

Let us consider the business process, which is of a technological nature[4] and operates with selected material inputs[5] represented by linguistic set {[Petx (i, j)]}, while that set consists of subsets[6] [Petx (i, 1)], [Petx (i, 2)] ... ... [Petx (i, m_1)], which represent one part of BP external metrics.

Where

i. is index closely related to BP serial number with set of processes

j_1- is index, which represents number subsets, the linguistic set {[Petx (i, j_1)]} consists of

However, that BP external metric is represented by adequate outputs as well, while they are quantified via {[Res1 (i, j_3,)]} linguistic set, which contains values concerned with $\mathbf{n}_{\mathbf{Artgood}}$, $\mathbf{n}_{\mathbf{Artrep}}$, and $\mathbf{n}_{\mathbf{Artwaste}}$ items.

With respect to the above-mentioned issues, the PBPL Equation actual version might be postulated.

$$\{[\text{Petx }(\text{i}, \text{j})]\} \otimes \{[\text{Pe }(\text{i}, \text{j}_2,)]\} = \{[\text{Res1 }(\text{i}, \text{j}_3,)]\} \quad (16)$$

A fictive data, which create conted of {[Petx (i, j)]} and {[Res1 (i, j_3,)]} will be discussed within Case study section.

Now, let us have a look at {[Pe (i, j_2,)]} linguistic set, which represents the business process Pe, while that process provides transfer of material input represented by {[Petx (i, j)]} linguistic set into final products (glass articles) represented by {[Res1 (i, j_3,)]} linguistic set.

However, the {[Pe (i, j_2,)]} linguistic set contains subsets, which quantify BPFs, the BP quantified via {[Pe (i, j_2,)]} linguistic set consist of. When selecting one BPF, we can assign to it the linguistic set {[BPF (i, jf)]}, which consists of three subsets, while formula (17) might be postulated.

$$\{[\text{BPF }(\text{i}, \text{jf})]\} = \{[\text{BPF\_TR }(\text{i}, \text{jf}_1)], [\text{BPF\_TT }(\text{i}, \text{jf}_2)], [\text{BPF\_TM }(\text{i}, \text{jf}_3)]\} \quad (17)$$

Where

the subset = [BPF_TR (i, jf_1)] = {[BPF_TR (i, j_1)]} and the content might be quantified via formula (16).

the subset = [BPF_TT (i, jf_1)] = {[BPF_TT1 (i, j_1)]} and the content might be quantified via formula (19)

the subset = [BPF_TM (i, jf_1)] = {[BPFEM (i, j_1)], [BPFIM(i, j_2)]} and the content might be quantified via formulas (14) and (15).

---

[4] Glass Article Primary Production- GAPP

[5] Glass Melt - GM

[6] The terms linguistic set and set are considered to be the same from semantic point of view

$$\{[\text{Pe}\ (\text{i}, \text{j}_2,)]\}' = \Pi\{[\text{BPF}\ (\text{i}, \text{j}_\text{f})]\} \tag{18}$$

i = 1,2 … .n

j_f = 1, 2 … .m1 - number of BPFs, the Pe business process, consists of.

Finally, we shall specify the {[Res1 (i, j_3)]}, the content of which is closely related to number of good articles $\mathbf{n}_{\mathbf{Artgood}}$, number of repaired articles $\mathbf{n}_{\mathbf{Artrep}}$, and number of waste articles $\mathbf{n}_{\mathbf{Artwaste}}$, while formulas (25) and (26) and (28) and (29) might be postulated.

$$\{[\text{Res1}(\text{i}, \text{j}_3)]\} = \{[\text{Pop}(\text{i}, \text{j}_3)], [\text{Article\_good}], [\text{Article\_repair}], [\text{Article\_waste}]\} \tag{19}$$

$$\mathbf{n}_{\mathbf{Artgood}} \in [\text{Article\_good}], \tag{20}$$

$$\mathbf{n}_{\mathbf{Artrep}} \in [\text{Article\_repair}], \tag{21}$$

$$\mathbf{n}_{\mathbf{Artwaste},} \in [\text{Article\_waste}]\} \tag{22}$$

[Pop (i, j_3)] subset contains elements closely related to article type, article class, article name, article, measure unit.

When applying the PBPL Equation in a basic form (see also formula (16), formula (23) might be postulated

$$\{[\text{Petx}\ (\text{i}, 1)], [\text{Petx}\ (\text{i}, 2)] \dots \dots [\text{Petx}\ (\text{i}, \text{m}_1)]\} \otimes \Pi\{[\text{BPF}\ (\text{i}, \text{j}_\text{f})]\} = \text{i} = 1, 2 \dots .\text{n}; \text{j}_\text{f} = 1, 2 \dots .\text{m1}- = \{[\text{Pop}\ (\text{i}, \text{j}_3)], [\text{Article\_good}], [\text{Article\_repair}], [\text{Article\_waste}]\} \tag{23}$$

It should be noted that those linguistic set content is time depended and formula (24) might be postulated.

$$\{[\text{Petx}\ (\text{i}, 1, \text{t})], [\text{Petx}\ (\text{i}, 2, \text{t})] \dots \dots [\text{Petx}\ (\text{i}, \text{m}_1, \text{t})]\} \otimes \Pi\{[\text{BPF}\ (\text{i}, \text{j}_{\text{f},}\ \text{t}]\} = \begin{matrix} \text{i} = 1, 2 \dots .\text{n}; \\ \text{j}_\text{f} = 1, 2 \dots .\text{m1}- \end{matrix} = \{[\text{Pop}\ (\text{i}, \text{j}_3)], [\text{Article\_good}\ (\text{t})], [\text{Article\_repair}\ (\text{t})], [\text{Article\_waste}\ (\text{t})]\} \tag{24}$$

Formula (24) quantifies relation among BP input and output parameters and the actual content of the above-mentioned linguistic set will be discussed within Case study section.

4.2.4 BPM knowledge support view consideration no. 4

The knowledge related to BPM knowledge support view are derived based on appropriate item statistic values mentioned within previous section.

Let us consider the {[Petx (i, j)]}, which contains subsets [Petx (i, 1)], [Petx (i, 2)] … … [Petx (i, m_1)], while any of those subsets contains time depended items and values concerned to actual material input.

$$[\text{Petx}\ (\text{i}, 1\text{a})], = [(\text{mat}\ (1)11\ (\text{t}((\text{k})), \text{mat}\ (1)12\ (\text{t}\ (\text{k}))), (\text{mat}(1)\ 21(\text{t}\ (\text{k})), \text{mat}\ (1)\ 22(\text{t}) \dots .\ \text{mat}(1)\ (\text{m11}\ (\text{t}(\text{k})), \text{mat}(1)\ (\text{k})\ (\text{m12})] \tag{25}$$

$$[\text{Petx}\ (\text{i}, 2\text{a})], = [(\text{mat}\ (2)11\ (\text{t}(\text{k})), \text{mat}\ (2)12\ (\text{t}(\text{k})), (\text{mat}(2)\ 21(\text{t}(\text{k})), \text{mat}\ (2)\ 22(\text{t}) \dots .\text{mat}(2)\ (\text{m11}\ (\text{t}(\text{k})), \text{mat}(2)\ (\text{m12})] \tag{26}$$

$$[\text{Petx (i, m2a)}], = [(\text{mat (m2)11 (t (k)), mat (m2)12 (t(k))), (mat(m2) 21(t), mat (m2) 22(t(k)) ... .mat(m2) (m11 (t(k)), matm2) (m12)}] \tag{27}$$

Now, let us select [Petx (i, 1)] and undertake its content to statistic evaluation[7], while formula (28) and (29) might be postulated[8]

$$[\text{Petx (i, 1b)}] = [(\text{mat (1)11 (mat (1)12Avg, (mat (1)12MMin, (mat(1)12MMax), (mat (1)12Vrp)}] \tag{28}$$

$$[\text{Petx (i, 1ab)}] = [\text{Petx (i, 1a)}], [\text{Petx (i, 1b)}] \tag{29}$$

where

m2a – index, which indicates a serial number of input record within Petx linguistic set.

m12 – index, which indicates a serial number of item and value input record.

Formula (28) indicates statistic values of items assigned to selected input, while formula (29) indicates an extension of [Petx (i, 1a)] linguistic set.

Now let us consider the {[Res1 (i, j_3)]} and let us suppose that the $\mathbf{n}_{\mathbf{Artgood}}$, $\mathbf{n}_{\mathbf{Artrep}}$ $\mathbf{n}_{\mathbf{Artwaste}}$ are time dependent, while formulas (30), (31), and (32) might be postulated.

$$\mathbf{n}_{\mathbf{Artgood}} = \mathbf{n}_{\mathbf{Artgood}}(t) \tag{30}$$

$$\mathbf{n}_{\mathbf{Artrep}} = \mathbf{n}_{\mathbf{Artrep}}(t) \tag{31}$$

$$\mathbf{n}_{\mathbf{Artwaste}} = \mathbf{n}_{\mathbf{Artwaste}}(t) \tag{32}$$

With respect to those issues, appropriate statistic values might be calculated.

$$[\text{Article\_goodst}] = [n_{\text{Artgoodawg}}, n_{\text{Artgoodmin}}, n_{\text{Artgoodmax}}, n_{\text{ArtgoodVrp}}] \tag{33}$$

$$[\text{Article\_repairst}] = [n_{\text{Artrep}} = \quad n_{\text{Artrepawg}}, n_{\text{Artrepmin}}, n_{\text{Artrepmax}}, n_{\text{ArtrepVrp}}] \tag{34}$$

$$[\text{Article\_wastest}] = [n_{\text{Artwasteawg}}, n_{\text{Artwastemin}}, n_{\text{Artwastemax}}, n_{\text{ArtwasteVrp}}] \tag{35}$$

Let us demonstrate previous relations at business process, which deals with forming of glass articles (Ga) from glass melt (Gm), which is represented by three variables: (a) glass melt temperature (Gmtep), glass melt viscosity (Gmvis), and glass melt quantity (Gmquant) and generated glass articles (Gas) represented by three items and values: (a) number of good Gas (n_{Artgood}), number of repaired Gas (n_{Artrep}) and number of waste Gas. The relations among statistic values of selected variables might be defined via: (a) partial rules (see also formulas (36), (37), and (38), (b) complex rule (see also formula (31) and (c) set of complex rules (see also formulas (40), (41), and (42). However, all the above-mentioned rules might be time dependent as well, while they might create pairs (time interval (T(int)), Y (int) and create linguistic subsets, which could quantify a development trend (see also formula (43)).

Partial rules

$$\{[\text{Gmtemp\_awg, GmVrp}]\} = \{[n_{\text{Artgoodawg}}, n_{\text{Artgoodmin}}, n_{\text{Artgoodmax}}, n_{\text{ArtgoodVrp}}], \tag{36}$$

[7] Statistic evaluation = determination of Avg, Min, Max and extent of variation

[8] Vrp – extend of variation

$$\left[n_{Artrepawg}, n_{Artrepmin}, n_{Artrepmax}, n_{ArtrepVrp}\right],$$
$$\left[n_{Artwasteawg}, n_{Artwastemin}, n_{Artwastemax}, n_{ArtwasteVrp}\ \right]\}$$
$$\{[Gmtvis\_awg, GmVrp]\} = \{\left[n_{Artgoodawg}, n_{Artgoodmin}, n_{Artgoodmax}, n_{ArtgoodVrp}\right], \tag{37}$$
$$\left[n_{Artrepawg}, n_{Artrepmin}, n_{Artrepmax}, n_{ArtrepVrp}\right],$$
$$\left[n_{Artwasteawg}, n_{Artwastemin}, n_{Artwastemax}, n_{ArtwasteVrp}\ \right]\}$$
$$\{[Gmtquant\_awg, GmVrp]\} = \{\left[n_{Artgoodawg}, n_{Artgoodmin}, n_{Artgoodmax}, n_{ArtgoodVrp}\right], \tag{38}$$
$$\left[n_{Artrep} = \quad n_{Artrepawg}, n_{Artrepmin}, n_{Artrepmax}, n_{ArtrepVrp}\right],$$
$$\left[n_{Artwasteawg}, n_{Artwastemin}, n_{Artwastemax}, n_{ArtwasteVrp}\ \right]\}$$

Complex rule

$$\{[Gmtemp\_awg,\ GmVrp],\ \{[Gmtvis\_awg,\ GmVrp],\ [Gmtvis\_awg,\ GmVrp]\} = \{\left[n_{Artgoodawg},\ n_{Artgoodmin},\ n_{Artgoodmax},\ n_{ArtgoodVrp}\right], \left[n_{Artrepawg},\ n_{Artrepmin},\ n_{Artrepmax},\ n_{ArtrepVrp}\right], \left[n_{Artwasteawg},\ n_{Artwastemin},\ n_{Artwastemax},\ n_{ArtwasteVrp}\right]\} \tag{39}$$

$$Y(1) = \{[Gmtemp\_awg\ (1),\ GmVrp\ (1)],\ \{[Gmtvis\_awg(1),\ GmVrp\ (1)], [Gmtvis\_awg\ (1),\ GmVrp\ ]\} = \{[n_{Artgoodawg}(1),\ n_{Artgoodmin},\ (1),\ n_{Artgoodmax},\ (1), n_{ArtgoodVrp}(1),],\ [\ n_{Artrepawg},\ (1),\ n_{Artrepmin},\ n_{Artrepmax},\ (1),\ n_{ArtrepVrp}(1),],\ [n_{Artwasteawg}(1),, n_{Artwastemin}(1),\ n_{Artwastemax}(1),\ n_{ArtwasteVrp}(1)\ ]\} \tag{40}$$

$$Y(2) = \{[Gmtemp\_awg\ (2),\ GmVrp\ (2)],\ \{[Gmtvis\_awg(2),\ GmVrp\ (2)], [Gmtvis\_aw\ (2),\ GmVrp\ ]\} = \{[n_{Artgoodawg}(2),\ n_{Artgoodmin},\ (2),\ n_{Artgoodmax},\ (2), n_{ArtgoodVrp}(2),],\ [\ n_{Artrepawg},\ (2),\ n_{Artrepmin},\ n_{Artrepmax},\ (2),\ n_{ArtrepVrp}(2),],[n_{Artwasteawg} (2), n_{Artwastemin}(2), n_{Artwastemax}(2),\ n_{ArtwasteVrp}(2)]\} \tag{41}$$

$$Y(m3) = \{[Gmtemp\_awg\ (m3),\ GmVrp\ (m3)],\ \{[Gmtvis\_awg(m3),\ GmVrp\ (m3)], [Gmtvis\_awg\ (m3),\ GmVrp\ ]\} = \{[n_{Artgoodawg}(m3),\ n_{Artgoodmin},\ (m3),\ n_{Artgoodmax},\ (m3), n_{ArtgoodVrp}(m3),],\ [\ n_{Artrepawg},\ (m3),\ n_{Artrepmin},\ n_{Artrepmax},\ (m3),\ n_{ArtrepVrp}(m3),], \left[n_{Artwasteawg}(m3),\ n_{Artwastemin}(m3), n_{Artwastemax}(2),,\ n_{ArtwasteVrp}(m3)\ \right]\} \tag{42}$$

$$\{[(DevlTrend]\} = \ [(T(1), Y\ (1)], [(T(2), Y\ (2)], \ldots..[(T(m3), Y\ (m3)]\} \tag{43}$$

In general, the knowledge stored with ES knowledge base are represented by semantic networks (SNWs), while partial rules might be compared **with partial SNWs**, complex rules might be compared with **ordinary SNWs** and development trends (DevlTrend) might be compared with **superior SNWs**. This approach will be discussed within Case study in more details and applied when designing and implemented an appropriate knowledge-based or expert system as well.

DOI: http://dx.doi.org/10.5772/intechopen.95350

4.3 Derivation of BPF functionality rules

4.3.1 General overview

In general, a horizontal structure of any business process (BP) is being created via appropriate set of business process functions (BPFs), while the BPF seems to be the principle component of any business process. On the other hand, any BPF might be quantified via multi-layer linguistic set, while at the first layer three significant linguistic subsets might be observed:

- {[BPF_TR (i, j1)]} – a content of which create rules, which regulate a progress related to transformation of BPF inputs into pre-defined outputs
- {[BPF_TT (i, j2)]} – a content of which create transformation functions, which provide transformation of BPF inputs into pre-defined outputs
- {[BPM (i, j3)]} – a content of which create subsets closely related to BPF6external and internal metrics, while BPF external metrics linguistic set and
- [BPFEM (i, j1)]} deals with BPF external metrics and consists of [BPINP (i, j11)] subset the content of which is created by elements closely related to BPF inputs and [[BPOUTP (i, j12), the content of which is created by elements closely related to BPF outputs (see also formula (14) and the {[BPFIM (i, j2)]} deals with BPF internal metrics with respect to formula (15)

However, both the above-mentioned linguistic sets are very closed to {[BPF_TT1 (i, j1)]} the content of which is created by elements closely related to transformation of BPF inputs to predefined BPF outputs as well.

4.3.2 BPF inputs versus BPF outputs

Let us consider the {[Petx (i, j1)]}, which contains a finite number of elements denoted as pt.(i, 1), pt.(i, 2), ... pt.(i, m1), while each of them is created by the element average and element extend of variations value (see also formula (44).

pt.(i, j1), = ((pt(i, j1)avg., (pt(i, j1)vrp)

$$\text{pt.(i, j2)}, = ((\text{pt(i, j2)avg.}, (\text{pt(i, j2)vrp}) \tag{44}$$

pt.(i, jm1), = ((pt(i, jm1)avg., (pt(i, jm1)vrp)

4.4 Case study

Let us consider a statistic file represented by **Table 1** and set of statistic indicators represented by **Table 2**.

Now, let us create a ratio set {Rs} and reference table (**Table 3**), which deals with assignment of words to ratio value intervals (**Table 4**).

$$\begin{aligned}\{\text{Rs}\} &= \{{}^{125}\backslash_{1811,1,}{}^{5,1}\backslash_{4},{}^{10}\backslash_{52,2,}{}^{270}\backslash_{1081,1},{}^{60}\backslash_{49,9}\} \\ &= \{0,069,0,784,0,191,0,249,1202\}\end{aligned}$$

Date	Time	Glm_temp	Glm_vis	Glm_quant	Glar_ident	Glart_good	Glart_repair	Glart_waste
21.11.2020	6:00	1790	6,5	50,3	Vyr1	1000		30
21.11.2020	6:30	1760	6,1	51,3	Vyr1	900		20
21.11.2020	7:00	1780	5,8	50,6	Vyr1	1100		25
21.11.2020	7:30	1800	5,3	51,6	Vyr1	1050		35
21.11.2020	7:30	1820	4,8	49,6	Vyr1	1030		45
21.11.2020	8:00	1810	4,9	52,6	Vyr1	1040		40
21.11.2020	8:30	1815	4,8	51,6	Vyr1	1060		35
21.11.2020	9:00	1830	3,8	53,6	Vyr1	1055		48
21.11.2020	9:30	1835	3,5	54,6	Vyr1	1155		68
21.11.2020	10:00	1885	2,5	56,6	Vyr1	1165		78
21.11.2020	10:30	1865	5,4	57,6	Vyr1	1170		80
21.11.2020	11:00	1765	6,4	47,6	Vyr1	1150		75
21.11.2020	11:30	1795	6,1	48,6	Vyr1	1125		65
21.11.2020	12:00	1805	5,1	49,6	Vyr1	1135		55

Source: The Authors.

Table 1.
Glass article forming -statistic file.

Average	**1811,1**	**5,1**	**52,2**	**1081,1**	**49,9**
Minimum	1760	2,5	47,6	900	20
Maximum	1885	6,5	57,6	1170	80
Extend of variation	125	4	10	270	60
Ratio set {Rs}	0,069	0,784	0,191	0,249	1202

Source: The Authors.

Table 2.
Glass article forming -statistic file values.

Ratio value	**Word**
0–0,3	low
0,4–0,7	middle
0,8 - 1,0	high
1,0 and more	very high

Source: The Authors.

Table 3.
A reference table which deals with assignment of words to ratio value intervals.

IF	**Glass melt**	**Glm_ temp**	**Glm_ vis**	**Glm_ quant**	**Glart ident**	**Glartgood**	**Glartrepair**	**Glart_waste**
		0,069	0,784	0,191	Vyr1	0,249		1202
		low	**high**	**low**		**low**		**very high**

Source: The Authors.

Table 4.
Glass article forming rule table.

4.5 BPLM process designer - implementation and operation

BPLM Process Designer is considered to be the second subsystem related to the BPLM System, while the first one is a subsystem denoted as BPLM Strategy Creator. The BPLM Process Designer consists of three components: (a) BPLM PD_01 Master files, (b) BPLM PD_02 Structure and (c) BPLM_PD_03 Functionality.

4.5.1 BPLM PD_01 master files

However, the BPLM PD_01 Master files component deals with master files needed for BP quantification and modeling as well, while those master files are concerned with production input materials, production output products (articles), production technological devices, production technological tools and production human resources. Each of the above-mentioned master files, which deal with production input materials, production output products, production technological devices and tools are represented by five subsets, which contain adequate types of parameters: (a) general parameters material, product, device or tool identifier, name, text description, drawing or image, (b) technical - height, volume, etc. (c) Operational - temperature, viscosity, quantity, etc. and (d) economic -e.g. different types of prices, and commercial ones, accessibility, vendor, reseller, etc. When considering HR master files, they contain records closely related to personality dispositions, theoretical knowledge and practical skills.

4.5.2 BPLM PD_02 structure

The BPLM PD_02 Structure component deals with modeling of business process static aspects, like business process (BP) transformation rules, BP transformation tools and BP external and internal metrics [30]. Because of that the BPLM Process designer is considered to be the aim oriented knowledge based system (expert system) and the knowledge stored within its knowledge base are represented with the use of appropriate semantic networks and (SNWs) and reference databases (RDBs), the SNWs and RDBs play a role of principle importance within that component. However, the above-mentioned categories create appropriate subordinated modules as well, while they will not be discussed in more details within that contribution. The principal layout of BPL PD_02 component is shown in **Figure 3**.

4.5.3 BPLM PD_03 functionality

However, the BPLM PD_03 Functionality component deals with modeling of business process dynamic aspect modeling as well, while two modules play a role of principle importance: (a) **Static model**, which deals with establishment of new business process or with selection of BP to be modeled from existing business processes (b) Dynamic model, which deals with modeling of primary and secondary external and internal metrics and derivation of transformation rules closely related to selected BP and adequate BPFs. The selected BP and BPFs are considered

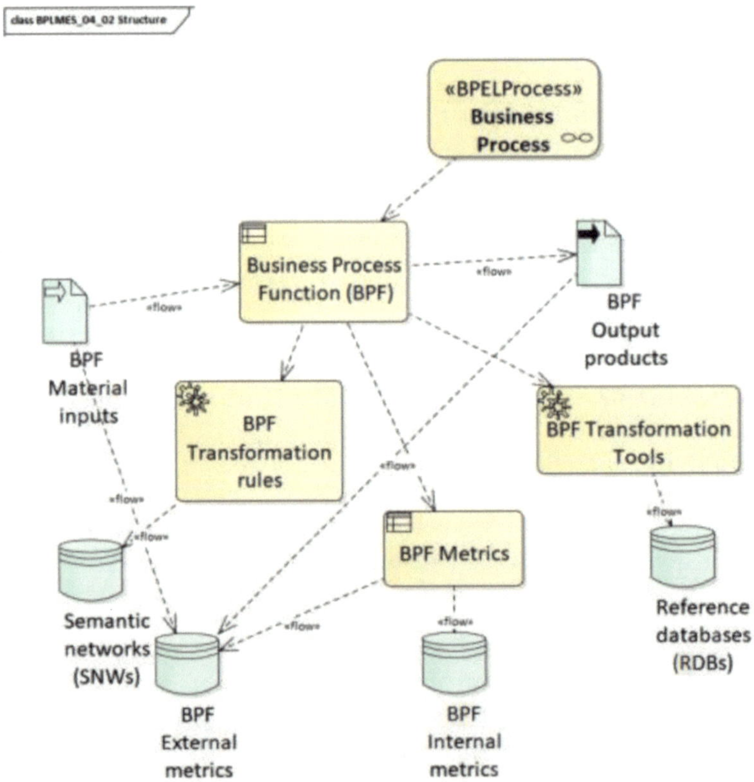

Figure 3.
Principal layout of BPL PD_02 component. Source: The Authors

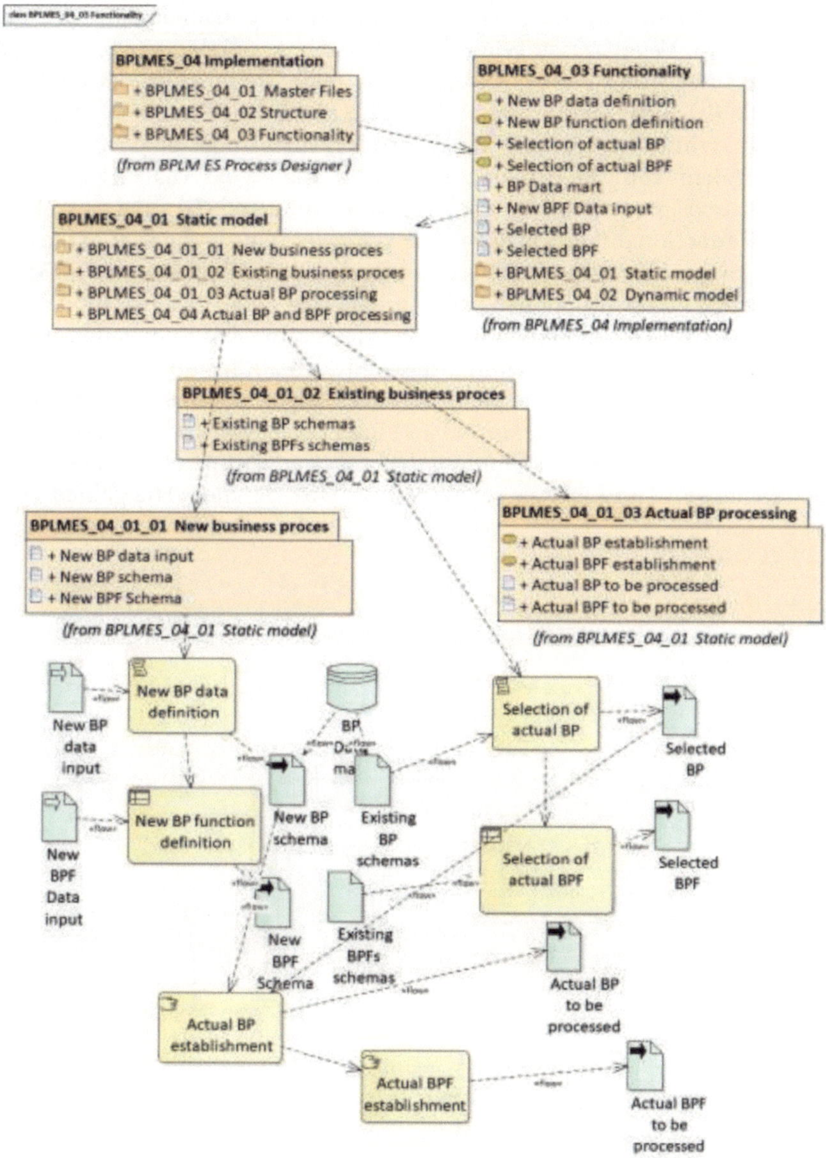

Figure 4.
Principal layout of BPL PD_02 component – Module static model. Source: The authors.

to be outputs from BPLM PD_02 Structure component and module Static model. The principal layout of BPL PD_02 component - module Static model is shown in **Figure 4**.

5. Discussion

We have developed a BPLM Process designer conceptual model, which creates basis for development of tool for analysis and design of business process (BP) models. The conceptual model respects the ARIS methodology, however that methodology is being modified an extended as well, while it operates with business process function (BPF), which creates basis of any BP horizontal structure and seems to be an elements, which provides the BPF input conversion and pre-defined output generation. When quantifying the BPF with use of adequate linguistic set {[BPF (i, j)]}, three subordinated sets (subsets) might be postulated: (a) BPF transformation rule (b) BPF transformation function and (c) BPF external and internal metrics subset, while two types of transformation rules might be postulated: (a) rules overtaken from the firm or company internal or external environment and postulated via text in natural language - *overtaken rules* and (b) rules postulated based on BPF functionality evaluation – *derived rules*. This is the first extension of ARIS methodology. The second one is closely related to BP model views. The ARIS methodology postulates functional, process, data and organizational model view, however the BPLM methodology postulates information and knowledge-based support view (**Figure 5**). When comparing an information support view with standardized data view two principle differences might be observed. *The first difference* is closely related to BPF external and internal metrics, while there is defined so called *primary BPF external and internal metrics* and secondary one, while the primary BPF external and internal metrics deals with detailed data gained within

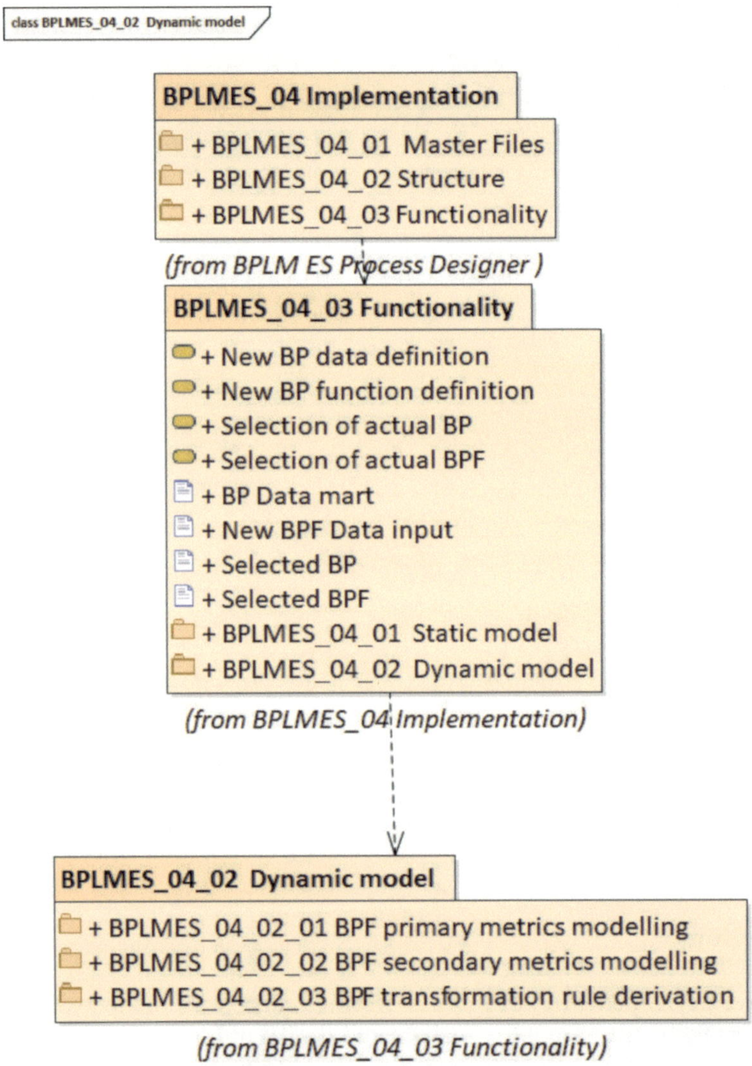

Figure 5.
Principal layout of BPL PD_02 component - module d ynamic model. Source: The authors.

evaluation of BP and BPF functionality and the *secondary BPF external and internal metrics* deals with aggregated data gained as s result of statistic evaluation the above-mentioned detailed data, while that data create basis for derivation of rules within BP and BPF knowledge-based support, which seems to be the next extension of previous BP and model views. However, the second difference is closely related to existence of *reference databases (RDBs)*, which create basis for knowledge representation within BP or BPF knowledge-based support view as well, while the RDBs and semantic networks created based on *secondary BPF external and internal metrics data* might generate knowledge stored in the expert system (ES) knowledge base (KB) and the ES seems to be an application utilized for BP linguistic modeling purposes.

6. Conclusion

The conclusion facts are concerned with modifying and extension of previously developed ARIS methodology and are described within discussion section. We would like to stress the main practical contribution of that system, which deals with a possibility or transformation rule derivation and presentation in form of TNL text, which might read the business analysts and BP managers as well, what generates an easier communication among them too.

Of course, the reader will not find any facts related to BP and BPF simulation and optimization, while those problems are closely related to our research work in the near future. The same is concerned with BP configuration and execution problems being solved within BP implementation and controlling. All the above-mentioned aspects represent objectives of the research work in the near future.

Acknowledgements

This work was supported by the Grant Agency of Slovak Republic – VEGA grant no. 1/0339/20 „ Hidden Markov Model Utilization in Financial Modeling This work was supported by the Grant Agency of Slovak Republic – KEGA grant no. 019EU-4/2020 Support of distance education through a virtual department.This work was supported by the Štefan Kassy Foundation and support for science and education.

Author details

Jozef Stašák[1*], Jaroslav Kultan[2], Peter Schmidt[1] and Bouafia Khawla[2]

1 Department of Applied Informatics, Faculty of Economy Informatics, University of Economy in Bratislava, Slovakia

2 Department of Information System, Eotvos Lorand University, Budapest, Hungary

*Address all correspondence to: jstasak@atlas.sk

References

[1] Aldin, L., & Cesare, S.D. A Comparative Analysis Of Business Process Modelling Techniques. 2009

[2] Sheer A. ARIS-business process modelling. Springer Verlag; 2000.

[3] Stašák, J. (2010). Modelovanie procesov podnikania s využitím aplikačného programu Aris. Ekonom.

[4] Scheer. Die Prozess-Experten. [Internet]. Scheer GmbH. [cited 2020Nov28]. Available from: http://www.ids-scheer.com/

[5] Friedrich F, Mendling J, Puhlmann F. Process Model Generation from Natural Language Text. Notes on Numerical Fluid Mechanics and Multidisciplinary Design Active Flow and Combustion Control 2018. 2011;:482–96.

[6] Bordignon ACDA, Thom LH, Silva TS, Dani VS, Fantinato M, Ferreira RCB. Natural Language Processing in Business Process Identification and Modeling. Proceedings of the XIV Brazilian Symposium on Information Systems - SBSI'18. 2018;

[7] Schlenoff C, Gruninger M, Tissot F, Valois J, Lee J. The Process Specification Language (PSL) overview and version 1.0 specification. 2000;

[8] De Nicola A, Lezoche M, Missikoff: M. 3rd IICAI 2007: Pune, India. Pune, India: dblp; 2007. p. 1794–813.

[9] Engels G, Förster A, Heckel R, Thöne S. Process Modeling using UML. Process-Aware Information Systems. 2005;:83–117.

[10] Fengel J. Semantic technologies for aligning heterogeneous business process models. Business Process Management Journal. 2014;20(4):549–70.

[11] Gruninger Michael, Schlenoff Craig, "Process Specification Language (PSL): results of the first pilot implementation", Proceedings of IMECE: International Mechanical Engineering Congress and Exposition, pp 1–10, 1999.

[12] Stašák, J. Business Process Quantification and Modelling – Linguistic Approach – Several Theoretical Aspects American International Journal of Contemporary Research (AIJCR) 2012,(4):85–100

[13] Stasak J. Business Process Linguistic Modeling – Philosophy & Principles. Computer Science and Information Technology. 2015;3(5):198–213.

[14] Stasak J, Vanickova R, Grell M. Business Process Modeling Linguistic Approach – Problems of Business Strategy Design. Universal Journal of Management. 2015;3(7):271–82.

[15] Schmidt P, Stašák J. Business process modelling linguistic approach application in public administration (self-governmental). International Journal of Business Performance Management. 2018;19(2):209.

[16] Stašák, J., Mazùrek, J. System of Data Transfer from and to Social and Economic Processes via Creative Economy Networks created based on Cultural Heritage Administration Processes andvice versa. IntechOpen 2020

[17] Stašák, J., Schmidt , Kultán, J. Mukhammedova U. Business Process Linguistic Modelling –Theory and Practice Part I: BPLM Strategy Creator IntechOpen 2020, In print

[18] Stoklasa J. Linguistic models for decision support [Internet]. Etusivu. Lappeenranta University of Technology; 2014 [cited 2020Nov28]. Available

from: https://lutpub.lut.fi/handle/10024/102203

[19] Taxén L. Activity Modalities – A Multi-dimensional Perspective on Coordination, Business Processes and Communication Systems. Signs & Actions An

[20] BPM For Your Digital Twin: ADONIS:Community Edition [Internet]. Your free Cloud BPM Tool. 2020 [cited 2020Nov28]. Available from: https://www.adonis-community.com/en/

[21] Kantnerová L, Stašák J, Petrášková V. Procesní řízení a modelování s přihlédnutím k praxi v logistice. České Budejovice, Czech Republik: Jihočeská univerzita v Českých Budějovicích; 2016.

[22] Stašák, J. Princípy modelovania ekonomických objektov (Economic Object Modelling Principles) EKONOM, Bratislava, 2010

[23] Mendling J, Reijers H, Recker J. Activity labeling in process modeling: Empirical insights and recommendations. Information Systems. 2010;35(4):467–82.

[24] Martin D, Paolucci M, Mcilraith S, Burstein M, Mcdermott D, Mcguinness D, et al. Bringing Semantics to Web Services: The OWL-S Approach. Lecture Notes in Computer Science Semantic Web Services and Web Process Composition. 2005;:26–42.

[25] Lautenbacher F, Bauer B, Seitz C. Semantic Business Process Modeling - Benefits and Capability. AAAI Spring Symposium - Technical Report . 2008;:71–6.

[26] Rajasekaran P, Miller J, Verma K, Sheth A. Enhancing Web Services Description and Discovery to Facilitate Composition. Lecture Notes in Computer Science Semantic Web Services and Web Process Composition. 2005;:55–68.

[27] Keller U, Lara R, Lausen H, Fensel D. Semantic Web Service Discovery in the WSMO Framework. Semantic Web Services. :281–316.

[28] Polleres, C. Feier, C. Bussler, D. Fensel, D: Web Service Modeling Ontology, Applied Ontology, 1(1): 77–106, 200

[29] Stašák J. Contribution to Linguistic Modeling of Business Process Structure, Functionality and Performance Metrics. Journal of Scientific Research and Reports. 2017;13(3):1–16.

[30] Leopold H. Natural Language in Business Process Models: Semantic Scholar [Internet]. undefined. 1970 [cited 2020Nov28]. Available from: https://www.semanticscholar.org/paper/Natural-Language-in-Business-Process-Models-Leopold/ec1bb5b68c6a1bc833ce9c089827d877301ebaf8

Chapter 10

Project Management Concepts

Vittal S. Anantatmula

Abstract

This chapter covers fundamentals of project management. It introduces project management concepts and provides a system view of project management plan and processes with which they are implemented. The knowledge areas include scope, time, cost, quality, and risk. The chapter will also emphasize on the interrelated nature of these knowledge areas. In addition to introducing these knowledge areas, the chapter will attempt to develop an understanding of these concepts and the important role of project teams in managing projects successfully. This chapter will also discuss similarities and differences between the plan-driven and change-driven (Agile) project methods.

Keywords: project management, risk management, schedule network, project plan, project teams, agile projects

1. Introduction

It is common that every business faces a situation that compels a change. Some of these changes usually are starting a new office, launching a new product or service, improving an existing work process, installing a new computer system, merging with another company, moving to a new location, entering a new market, meeting a social need and so on. These changes are necessary to meet operational or strategic goals of an organization. And these goals are accomplished using projects. Further, the primary objective of an organization is to create value for its stockholders and it is achieved when the organization creates a healthy profit through the achievement of its strategic objectives [1]. And projects are the instruments by which an enterprise accomplishes its strategic objectives [2].

Organizations design and execute projects that can be classified as internally and externally funded projects based on the nature of their business. In organizations that conduct externally funded projects for a fee (of sorts) on behalf of external clients, efficiency in the conduct of projects is the means by which the amount of real profit is enhanced [2]. If the primary line of business of the organization is service, manufacturing, or research; the projects in the organization are probably internally funded, and the missions of project teams are to create increased operational efficiency, new products, or new markets. Internally funded projects also play an important role in the profit margin, albeit indirectly. In the case of a nonprofit organization, the projects are executed either internally or externally and are conceived to serve its main purpose, which is a social cause and not profit. However, the underlying project management principles of effective and efficient use of resources are still valid and important to expand their services for social cause without increase in their resources [3].

1.1 So, what is a project?

PMBOK® defines project as a temporary endeavor to create a unique product, service, or result [4]. It uses the word, temporary, which sometimes implies negative connotation and low importance. One must remember that "temporary" does not influence the length of the project duration. Also, PMBOK definition does not introduce the concept of value and its contribution to organization's strategic goals. Value can be tangible (e.g., cost savings) or intangible (e.g., increased brand equity). However, value is always associated with benefit. A project must provide value to all its stakeholders; if not, project will not receive adequate support [5].

A project can be seen as a new time-bound effort with several related and/or interdependent tasks to create a unique product or service that adds value. As it is a new effort, often, we do not have complete knowledge or experience about planning and executing the project. Needless to say, projects are characterized with unknown factors and ambiguity, which delay the development of detailed scope and specifications to later stages of project planning. Further, any project requires resources such as materials, tools, equipment, and people to execute it. Considering these additional facets of a project and to extend the definition of a project further, project can be considered as *a complex, non-routine, one-time effort limited by time, budget, resources, and performance specifications designed to meet customer needs and add value to all key stakeholders.*

1.2 Distinction between project and process

It is important understand the distinction between a project and process. Process is repetitive in nature with clearly defined procedures and outcomes. A process will yield the same result be it a product or a service. Project, on the other hand, is unique and new by definition. So, outcome of a project would always be new and unique. However, project is executed through project management processes and project deliverable-oriented processes.

Projects are conceived, created, and managed because an individual or organization other than the project manager or project team identifies a need [6]. As stated earlier, projects are always aimed at fulfilling organizational objectives and/or strategic needs such as market demand, customer request, technological advance, legal requirements, a social need (for non-profit organizations), crisis situation, and obsolete technology or equipment.

1.3 Project characteristics

A project is time-bound, i.e. it has definite beginning and definite ending. However, it does not mean that project is of a short duration, which may vary from 10 days to 10years. However, "temporary" does not apply for the project deliverable. It is interesting to note that project team is certainly temporary (**Table 1**).

- Unique	- Revolutionary improvements
- Defined start and end	- Change
- Transient teams	- Risky
- Little precedence	- Complex
- Unknowns	- Uncertain
	- *Level of difficulty is high*

Table 1.
Project characteristics.

Project outcome can be a product, service, or result.

- *product*: iPhone, power plant
- *service*: online banking, Google search
- *result*: decision to pursue Tokyo airport expansion
- *unique*: something or some aspect that is not repetitive

Progressive elaboration is another characteristic of a project. As requirements are not clear at the initiation stage. Developing the project scope in steps and continuing by increments is a common aspect of a project and it goes through stages such as: scope – scope definition – project plan – detailed WBS. All these project characteristics present a challenge to manage projects successfully.

2. Project management

Managing a project includes identifying requirements, establishing clear and achievable objectives, balancing competing demands of quality, scope, cost, and time, adapting specifications, plans, and approach to meet expectations of all key stakeholders including the client and the end-user. We define project management as *the art and science of using experience, knowledge, skills, tools, and techniques efficiently and effectively to meet stakeholder expectations*.

Current trends in the project suggest that organizations manage several projects simultaneously.

Long-term success in managing projects requires proven and established project management practices and processes and several successful projects to emulate [1]. Therefore, organization-wide resource allocation becomes necessary to succeed, which may not always be possible. Also, required skills and expertise are always not available locally. Consequently, virtual teams are becoming common for project execution. PM tools increasing in complexity and usefulness. Managing project involves:

- identifying requirements
- establishing clear and achievable objectives
- balancing the demands of time, cost, scope, and quality
- adapting to expectations of all stakeholders

Thus, project management is an approach to accomplish project objectives within organizational structural and resource constraints for internal projects. For external projects, political, social, legal and environmental constraints may also have to be considered.

Project is an organization-level effort associated with complexity, uncertainties, and unknowns. Consequently, it requires continual involvement of several functions. Obviously, integration, coordination, and accountability assume greater importance. Project Management tools and techniques will help to accomplish these integration and management functions. Uncertainties and unknowns affect requirements to change dynamically and the project manager has to meet these requirements while meeting the expectations of all the project stakeholders.

2.1 Necessity for project management

Managing project is fast becoming a standard way of executing business strategies. Some of the reasons for deploying project management practices are:

- knowledge economy
- increased competition due to free market philosophy
- constraints of cost, time, and scope (quality)
- client focus
- resource constraints

Nature and frequency of projects determine whether an organization requires an informal or formal project management practices. Some of the following questions will help determine the requirement of a formal project management in organizations.

- Is the effort associated with complexity and uncertainty?
- Are external and technical environments dynamic to compel changes in the organization?
- Is it governed by constraints such as time, budget, scope, and quality?
- Are there multiple functions involved?

Affirmative responses to some or all of these questions will determine the extent to which project management is required to be formalized. However, project Management (PM) needs investment in resources and consequently demands results. Also, organizations are required to train people at all levels. On the positive side, PM facilitates effective and efficient use of resources. PM also provides an opportunity for senior management to realize strengths and weaknesses of an organization.

In many organizations, PM demands changes in policies, practices, and procedures. Further, PM will also influence the function of finance because project-based financial control systems are routinely implemented. PM tools and techniques aid productivity and help us deal with complexity [7].

2.2 Key drivers of project performance

Projects, by definition are new and learning new is associated with change in behavior. Thus, working in teams, integrating, learning, and collaborating are essential characteristics of the project teams. Consequently, effective communication assumes great importance for its success. As a general rule, with the increase in project size, the number of people involved in the project will also increase. Consequently, effective communication becomes complex with the increase in project size. Based on extensive research, the following are identified as the key drivers of project performance and they are also considered success factors [7]. It is the primarily the responsibility of the project manager to:

- define project processes and roles
- implement consistent processes

- communicate expectations and outcomes to all stakeholders
- create clarity in communications
- facilitate organizational support
- establish trust
- manage outcomes be developing performance measures
- PM Maturity and success

Project management maturity is a state determined by fully developed standards and processes, which lead to repeatable success in managing projects. PM maturity ultimately should lead to developing best PM practices. However, you should remember that project success factors are different for different stakeholders. For example, a project, in spite of experiencing time and cost overruns, may still be considered successful by the client.

The distinction between project success and project management success must be understood. Project success is measured against the overall project objectives whereas project management success is a measure of managing project within time, cost, and meeting the scope requirements [8]. Success of a project should be measured by considering three different areas: project meeting its own cost-duration targets, the deliverable meeting enterprise strategic objectives, and the deliverable meeting the enterprise financial objectives.

3. Strategic alignment

Projects are means to accomplish strategic objectives and goals of organizations. Project Management (PM) is an integration of several academic disciplines such as accountancy, decision sciences, economics, finance, and management. It uses systems approach in integrating these disciplines. PM will continue to grow in its importance because it is how business is done, in the present global economy.

Project selection should be based on the corporate strategic plan and investment strategies [1]. Thus, each project must have a business objective that is aligned with the strategic plan; it will serve as the foundation for the project investment proposal. In addition, a project must conform to a broader purpose than a set of internally derived objectives. It must support overarching strategies of their parent company or its clients. To be successful, the project manager and the team should clearly understand the vision and mission of their clients; such understanding is more likely to deliver products and services that meet demanding customer requirements, which eventually results in customer satisfaction.

4. Scope management

Project Scope must include all the work required, and only the work required by the client [6]. Scope management is concerned with defining and controlling the scope. It assumes greater importance than all other knowledge areas of project management. The scope should define the project by what should and should not be planned, budgeted, staffed and executed. Project scope also identifies the boundaries separating excluded activities or resources from those, which must be included for the project execution.

As stated earlier, projects are characterized by unknown factors and ambiguity, which delay the detailed development of scope and specifications to later stage of the project. Consequently, the scope of the project continually changes throughout the project. Sometimes, the original objective of the project may also change as the project progresses and we learn more about what is needed, what is possible, and what the costs are going to be. All these changes will impact on the project scope and its management.

Despite these changes, a clearly defined scope is crucial to developing a project plan, which provides a baseline for managing the schedule, the budget, and the detailed work of the project. Without a clearly defined scope, we have no basis for tracking progress on the project as defined or for altering schedule and cost criteria should the client request changes in the original project plan.

Once a project has been initiated, the project manager and an early-stage project team begin scope management by creating a preliminary scope statement based on the project charter. The preliminary scope statement formalizes between the project and its customer an agreement regarding quantifying project/product objectives and clarifying specific deliverables.

While the preliminary scope statement sets objectives at a higher level of abstraction, these objectives must be defined in as much detail as possible before explicit schedules, budgets, and work plans can be developed during the planning phase of the project management life cycle [9]. At this point, historical information and expert opinion are applied within the constraints and assumptions of the present project to create a body of requirements necessary for guiding project execution. This body of requirements is the foundation of scope definition, which subdivides the major project deliverables into smaller, more manageable components. It is important to translate requirements of the client into specifications, which denote finality and clear definition of project outcomes with specific quality parameters and measures.

This preliminary scope statement document sets various product performance criteria, delineates the initial project organization, boundaries, assumptions and constraints, while providing a high-level WBS and order of magnitude cost estimates [2, 6]. Below is an example of deliverable-oriented WBS (**Figure 1**).

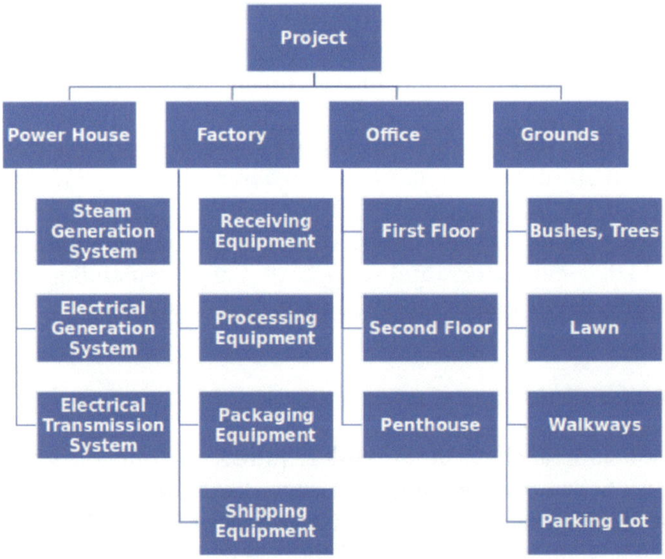

Figure 1.
Deliverable-oriented WBS.

This preliminary will be refined into the formal scope statement as the project moves into its planning phase. This refinement comes in the form of WBS. It provides a cornerstone of the scope management plan, which lays out how objectives and deliverables will be administered and how scope changes will be integrated into the project.

5. Cost management

WBS identifies all project activities to be executed but it does not include estimates of time, resources, or project costs. WBS should be initially developed at a higher level and expanded to include more specifications as you incorporate additional requirements of the project. As a next step, you will develop first two or three levels of the WBS, which essentially contain deliverables. These higher-level work elements are broken down into smaller and smaller work packages so that you can have manageable tasks to complete at any given point in the project [6]. The work package list is both inclusive and exhaustive. Preferably, these tasks should be at level 3 or lower. Remember that a good WBS encourages a systematic planning process, reduces the possibility of omission of key project work elements, and simplifies the project by dividing it into manageable units.

5.1 Cost estimating

To estimate cost for a work element, we need the following information:

- Resources required (people, equipment, tools, materials)
- Duration of usage (time unit)
- Cost rate for time unit

Duration may not be relevant for materials, which are absorbed for completing the work. Number of resources for each resource multiplied by its duration and the rate will provide an estimate of that resource cost. Adding all the resource costs will give you an estimate of the activity cost. We sum up all these activity costs of the project to determine the total estimated project cost. These costs must then be managed and controlled along with the schedule throughout the project execution phase.

The most accurate and most reliable estimate for a project can be developed when all of the elements of the WBS have been identified with a reasonable degree of reliability and when the resource breakdown structure (RBS) has been defined with the desired degree of certainty [2]. This estimate is referred to as the bottom-up estimate and it is derived from detailed information that is contained in the WBS and RBS at the time of the estimate. Resource Breakdown Structure (RBS) is similar to WBS. By resources we mean everything that will cost money to obtain and are necessary for the completion of the project (labor, materials, equipment, licenses, taxes). RBS follows the WBS structure of hierarchy and we can classify resources into major categories such as:

- People
- Materials and installed equipment

- Tools, machinery
- Fees, licenses

To be useful, RBS should be accurate and reliable, so it needs to be updated periodically. An example of RBS is provided below (**Figure 2**).

	Unit of Measure	Cost/Price (Dollars)
R100 Development Staff		
R110 Server Development Personnel		
R111 Systems Analyst	Staff Hour	$70
R112 Application Analyst	Staff Hour	$60
R113 Systems Programmer	Staff Hour	$50
R120 Database Development Personnel		
R121 Data Base Administrator	Staff Hour	$75
R122 Sr. Data Design Specialist	Staff Hour	$65
R123 Data Design Specialist	Staff Hour	$60
R130 Client Development Personnel		
R131 PC Systems Analyst	Staff Hour	$65
R132 PC Systems Programmer	Staff Hour	$55
R140 Network Development Personnel		
R141 Infrastructure Analyst	Staff Hour	$85
R142 Infrastructure Engineer	Staff Hour	$80
R143 Network Engineer	Staff Hour	$70
R200 Product Assurance Staff		
R210 Quality Assurance Personnel		
R211 Sr. QA Specialist	Staff Hour	$65
R212 QA Specialist	Staff Hour	$45
R220 Requirements Management Personnel		
R221 Requirements Manager	Staff Hour	$65
R222 Requirements Specialist	Staff Hour	$45

Figure 2.
RBS sample.

Detailed and accurate estimates require substantial definitive information about the project. Further, they require a relatively large block of time and effort for the estimating task. Therefore, one needs to strike a balance between the time spent on estimating, the accuracy of the results, and the degree of accuracy required by the stakeholders at the point in the project life. Project management methodologies recognize various tools and strategies to support project cost estimating.

Analogous estimating is a top-down approach where managers create estimates based on expert judgment, data from similar projects, industry benchmarks, published or interpersonal research sources. This approach is typically less accurate than bottoms-up estimates, but it may be quicker and less costly to develop.

Parametric estimating uses mathematical models to predict costs. These methods may include specialized computerized estimating tools. Specific metrics may applicable to particular industry sector such as cost per line of code, and square footage construction cost.

Bottoms-up estimating is a method in which the people and other resources assigned to the activity create the cost and schedule estimates. This approach creates optimal commitment from the performing team, and is effective when the team contains appropriate subject matter expertise (domain specific knowledge) and adequate experience to accurately forecast duration estimates. Of all the estimates, this approach tends to be more accurate. An example of bottom-up estimate is shown below (**Figure 3**).

	Unit	Intensity	Duration (Days)	Total (Staff Hours)	Unit Cost	Extension
000 System						**$2,426,760**
100 System Code						**$966,400**
110 Server Code						**$373,160**
111 Server Source Code						**$345,600**
R121 Data Base Administrator	Staff Hour	1	100	800	$75	$60,000
R122 Sr. Data Design Specialist	Staff Hour	1	125	1000	$65	$65,000
R123 Data Design Specialist	Staff Hour	3	125	3000	$60	$180,000
R241 Test Manager	Staff Hour	1	10	80	$70	$5,600
R242 Sr. Test Engineer	Staff Hour	1	25	200	$65	$13,000
R243 Test Engineer	Staff Hour	1	50	400	$55	$22,000
112 Server Object Code						**$27,560**
R121 Data Base Administrator	Staff Hour	1	6	48	$75	$3,600
R122 Sr. Data Design Specialist	Staff Hour	1	12	96	$65	$6,240
R123 Data Design Specialist	Staff Hour	1	24	192	$60	$11,520
R231 Sr. CM Specialist	Staff Hour	1	5	40	$65	$2,600
R232 CM Specialist	Staff Hour	1	10	80	$45	$3,600
120 Client Code						**$139,200**
121 Client Source Code						**$124,600**
R131 PC Systems Analyst	Staff Hour	1	75	600	$65	$39,000
R132 PC Systems Programmer	Staff Hour	2	75	1200	$55	$66,000
R241 Test Manager	Staff Hour	1	5	40	$70	$2,800
R242 Sr. Test Engineer	Staff Hour	1	12	96	$65	$6,240
R243 Test Engineer	Staff Hour	1	24	192	$55	$10,560
122 Client Object Code						**$14,600**
R131 PC Systems Analyst	Staff Hour	1	6	48	$65	$3,120
R132 PC Systems Programmer	Staff Hour	1	12	96	$55	$5,280
R231 Sr. CM Specialist	Staff Hour	1	5	40	$65	$2,600
R232 CM Specialist	Staff Hour	1	10	80	$45	$3,600

Figure 3.
Bottom-up estimate.

Once the costs of all work packages have been estimated, supplementary information documented, and cost accounts assigned, then these costs are rolled up to highest level of the WBS to determine project cost. This is a process of progressively totaling the work package estimates to their higher levels of detail.

6. Time management

Time Management is required to accomplish timely completion of the project and organizes it into six components, five of which occur during the planning phase of project lifecycles:

- *Activity Definition:* identify and document activities to produce the WBS elements
- *Activity Sequencing:* identify and document interactive logical WBS relationships
- *Activity Resource Estimating:* determine what and when resources quantities are needed
- *Activity Duration Estimating:* develop activity durations from scope and resource information
- *Schedule Development:* determine start and finish dates for activities
- *Schedule Control:* manage and control changes to the schedule

In general, scheduling tools include:

- Milestone charts
- Flowcharts

- Bar (Gantt) Charts
- Network Diagrams

Of these, the network diagram shows interdependencies of all tasks and illustrate the workflow of the project and assumes greater importance.

6.1 Network diagram

Network diagrams have evolved as a de facto standard for building project schedules because of their emphasis is on activity dependencies. They depict entire projects, or subprojects, as visual maps that portray activities in their required order of execution. These diagrams can be developed in one of the following two forms:

- Activity-on-Arrow (AOA), also called Activity-on-Line (AOL) or Arrow Diagramming Method (ADM)
- Activity-on-Node (AON) also called Precedence Diagram Method (PDM)

AOA diagramming is almost never seen today in popular software programs due to the fact that additional memory is needed to program the dummy activities.

6.2 Activity on the node (AON) diagramming

AON or precedent diagramming was developed in the late 1960's specifically to "do away" with the dummy activity. In this form of diagramming, the arrow represents "precedence;" it simply identifies what activity must be completed before the next can begin, with the arrowhead representing direction of flow through the network. In precedent diagramming the arrow does not represent work accomplished. Instead, work is represented on the node of the network – and in this case the node is represented by a rectangle with the name of the activity within the rectangle. Other information may also be entered into the rectangular node, including early start, early finish, late start, late finish, duration, cost, etc. The IT industry has adopted precedent diagramming as its predominant form of network diagramming.

6.3 Rules and accepted practices

Network diagramming is a scheduling technique fundamental to the skills of any project manager. There are a few basic rules, or accepted practices, that apply to all network diagrams:

1. An arrow is a straight line and the arrows showing precedence may have right angles built into them. Good practice requires that all such bent arrows be bent at right angles, and only at right angles, so that we can clearly identify that the bend was intended.

2. An arrow is always directional, with the arrowhead demonstrating the direction. There must be an arrowhead on each arrow to indicate its direction.

3. The arrows cross each other frequently – but when good practice is used these arrows will cross each other at right angles. The purpose of this is to help us follow the arrows without getting sidetracked onto another arrow.

4. In well-formatted network diagrams, arrows may proceed from the left to the right, they can proceed up, or they can proceed down on the paper. In the absence of arrows, one can assume that the activities proceed from left to right.

Every project has one, and only one, starting point; and one, and only one, ending point. This is easily accomplished in the AOA format, but it is an accepted practice that is frequently violated in AON networks. The way to correct this error in AON formats is to add two additional nodes, one labeled "start" and the other labeled end. Then, all of your initial activities are initiated from the start node and all of your hanging activities will be tied in to the end node. Of course, the duration of both the start and end nodes will be zero.

6.4 Activity duration estimating

PERT (Program Evaluation Review Technique) has an emphasis on meeting schedules with flexibility on costs and it can be drawn only on an AOA diagram, and therefore, can have dummies.

1. Has 3 estimates per activity: Optimistic (O), Pessimistic (P), and Most likely (M)

2. PERT calculates an estimated time using the following equation: $((P + 4\,M + O)/6)$

3. These estimates can be used to describe a Beta Distribution. You are probably familiar with a Normal Distribution (similar to the shape of a Bell and used in Six Sigma Quality Management). A Beta Distribution has a different shape as shown in **Figure 4**.

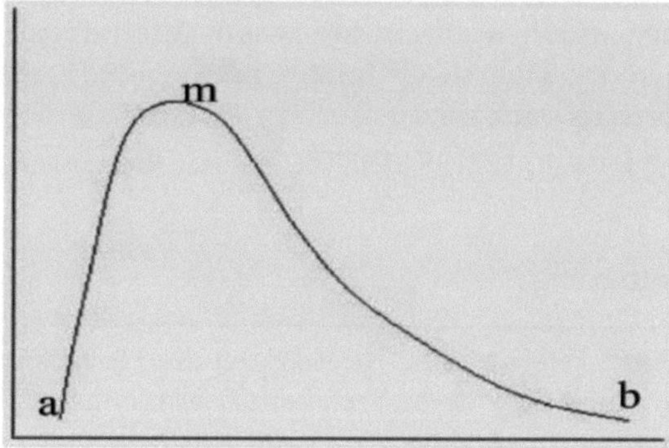

Figure 4.
Estimation.

Notice how there is a long tail to this distribution making different than a normal distribution where the m would be an equal distance between the a and the b and would represent the 50/50 mark of a distribution (50% of sample times falls to the left of m and 50% of sample times falls to the right of m).

6.5 Critical path method (CPM)

Although this technique may use the words critical path, it does not refer to finding the critical path. It refers to estimating based on one time estimate per activity.

- The critical path is the longest path in the network diagram.
- There may be more than one critical path.
- Allows you to identify potential points where project can go "off schedule" and slack or "float time" that allows you to delay some tasks that are not on the critical path.

6.6 Monte Carlo simulation

This method will create a project duration that is closer to reality than CPM or PERT. This method uses computers to simulate the outcome of a project based on PERT estimates and the network diagram, but does not use the PERT formula. Simulation can tell you the probability of completing the project on any specific day, probability of completing the project for any specific amount of cost, probability of any task actually being on the critical path, and overall project risk.

The network diagram shows the sequence of work elements required to reach project completion. Note that the initial stage of the network diagramming does not require time or cost estimates. We simply determine the most logical and technically viable approach to completing the project and use the network format to visualize this scenario. You must remember that initial network diagram logic is dictated by work-related constraints only; you should not introduce resource and time constraints at that time.

Once the network illustrating a logical flow of work elements in the project is developed, you can assign (or assume) resources to each activity and develop initial task time estimates. These estimates are then aggregated across the network to calculate its critical path, and the slack times associated with each activity. These calculations are based on a forward pass to compute the earliest start and finish times for each activity, and then a backward pass to determine the latest start and finish times of the activity. Then the difference between the earliest and latest times for any activity gives us its slack time. All activities with slack times of zero duration constitute the critical path(s) for the project.

7. Quality management

Quality management refers to the processes required to ensure that projects satisfy the needs for which they are undertaken. These processes should make certain that a project does not deviate from agreed-upon specifications. Quality management requires accurate assessment of stakeholder needs and requirements during both project planning and implementation. It is important to remember that quality must be planned in as opposed to inspected in. Without adequate quality management, projects may face the risk of increased costs while compromising the project team morale and customer satisfaction.

Quality planning makes use of some quality management tools and techniques project outcomes (products and tangible deliverables) and project management processes. This is an important distinction to understand.

- *Quality related to project outcomes* (deliverables or services) is determined during the planning phase when requirements and specifications are defined. Quality assurance and control processes are implemented as we develop these project deliverables during the project execution phase. This aspect of quality is primarily the responsibility of the project manager and the project team.

- *Quality related to project management processes* is influenced by the organization's knowledge and maturity in managing projects. However, proper implementation of quality assurance remains the responsibility of the project manager and the project team.

One of the most important tasks a project manager must focus on when planning quality is to define quality and how it will be measured (i.e., What defines Project Success?). When developing these metrics, Project Managers should remember to include both quantitative (benchmark specifications, experiments, and financial/cost metrics) and qualitative (subjective assessment) metrics. *Cost of Quality* includes consideration of conformance costs (both prevention and appraisal) and non-conformance or failure costs (both internal and external). Non-conformance costs tend to be higher.

Quality Assurance is about evaluating overall project performance on a regular basis to provide confidence that the project will satisfy the relevant system of quality standards.

Quality Control is monitoring specific project results to determine their conformance to specific quality metrics and identifying ways to eliminate the causes of unsatisfactory performance. During Quality Control quantitative analysis is applied to specific processes and project protocols. When processes operate beyond predetermined limits, adjustments will have to be swiftly implemented to bring the operations back into conformance with project-defined quality standards. Quality Control tools include:

- *Control charts:* Determines process stability.

- *Cause and Effect Diagrams:* Illustrates how various factors might cause problems or effects.

- *Pareto diagrams:* Identifies quantity of defects assigned by category and cause.

- *Statistical sampling:* Narrows a population of study to a representative sample.

- *Run Charts:* Show the history and pattern of variation.

- *Trend analysis:* Assesses non-conformance over time.

It is important that you understand how these tools operate. Most of these tools are not specific to project environments. They have been developed and utilized to enhance better products and processes in continuous manufacturing environments.

8. Risk management

Risk is an exposure to a situation that is usually associated with unfavorable outcome [10]. Project risk is an uncertain event or conditions that, if occurs, has a positive or a negative effect on at least on one of the project objectives. The possible result of a risk is either loss or gain. If left unaddressed or ignored, specifically, the negative risk, it could potentially interfere with the successful completion of the project and may result in time and cost overruns, or diminished quality of the product or service. Risk is integral to a project as it is a new effort that is associated with unknowns and uncertainties. In general, when you try something new, risk is inevitable.

It is important to make a distinction between a risk and a problem. A problem is identified and known in a current time-frame and it is a certain event. Therefore, it demands an immediate solution. A risk is a probable and a potential event that may occur in future. Even when a risk event takes place, its consequences are uncertain [10]. A risk is also known or unknown event. Risks are problems that have not occurred yet. A risk can become a problem if it is not planned or managed. Occurrence and assessment of a project risk is a combination of an event which is unanticipated or unwanted, likelihood of its occurrence and its impact on project execution. So, identifying, analyzing, and responding are the essential elements of risk planning.

Managing project risk is inevitable and important to project performance. Risk management is the art and science of identifying, analyzing, and responding to risk factors throughout the life of a project and in the best interest of project objectives and metrics for project success. The purpose of risk management is to address risks associated with projects in a systematic manner. If not addressed comprehensively and completely, risks can lead to failure of projects in terms of cost, scope, and project deliverables [11]. The goal of risk management is to maximize the results of positive events and minimize the consequences of negative events. Sources of a risk in a project can be understood from the risk breakdown structure (**Figure 5**).

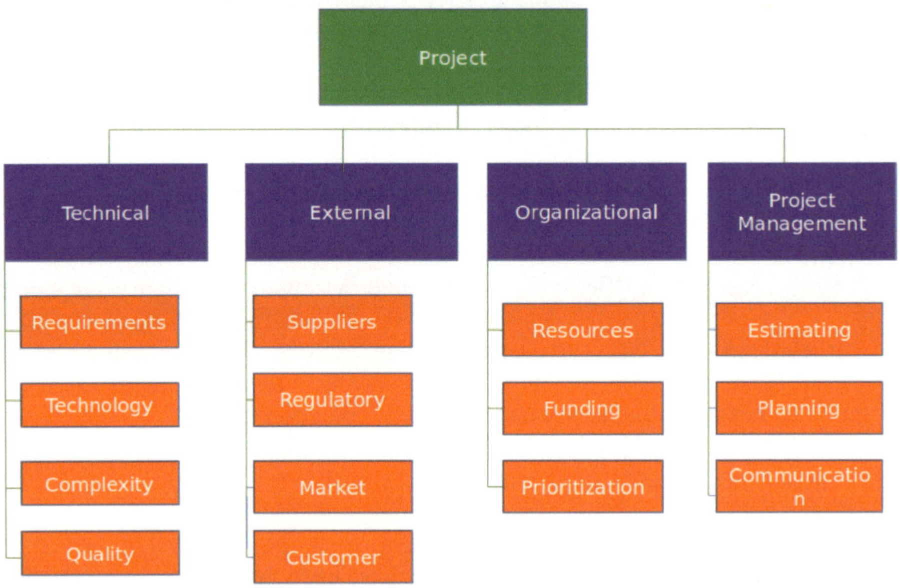

Figure 5.
Risk breakdown structure.

A formal or informal approach to planning and managing risks within a project depends on the impact of risks on project objectives such as financial targets, timely and successful completion of the project. However, it is critical that we understand the formalized approach to risk management. In taking a more formal approach to this process, the PMBOK defines risk management as consisting of (1) Plan risk management (2) Identify risks, (3) Perform qualitative risk analysis, (4) Perform quantitative risk analysis, (5) Plan risk responses, and (6) Control risks. In this learning module, we will focus on the first five of these processes, leaving (6) control risk to be covered later in the program. Various strategies for managing risk are:

- Accept: absorb the potential damages
- Mitigate: develop protective measures
- Transfer: re-route the risk to someone else
- Avoid: use an entirely different approach

The risk management plan is a complex process, influenced significantly by organizational and project context, the cultural and operational styles of the performing organizations, as well as the individual perceptions and risk tolerances of management and of the other project stakeholders. Understanding management and stakeholder risk tolerances is critical: customers, sponsors and senior management are expected to have different perceptions of what constitute acceptable risks. Risk management plan should address each of the above considerations in a manner, which is relevant to the individual project and its unique environment.

9. Importance of project teams

Projects are executed in teams. In general, projects are managed using teams in a work environment that is complex for two reasons: first, each project is unique, and second, conditions for team selection and motivation are often far from ideal [3]. In addition to uniqueness and complexity, unfamiliarity is often described as one of the characteristics of projects and as a result, projects are often associated with change. Consequently, successful project performance requires strong leadership, which provides vision and ability to cope with change.

Projects, by definition, are unique and are often associated with uncertainties and unknowns. It is reasonable to assume that in project management, it is not if the plans will change, it is when, what will change, and by how much. If anything is constant in a project, it is the change. When changes are significant in a project, which is often the case, leadership role assumes greater importance. Leadership has its efforts directed towards convincing people about the need to change, aligning them to a new direction, and motivating people to work together to achieve project objectives under difficult and demanding work environments [12].

Motivating the project team involves getting the team members to do the tasks that need to be done, not because they have to do them but because they understand the value of their work to the overall success of the project and they want to do them. If people want to do what we need to have them do, their commitment is bound to be greater, and the job is likely to be accomplished much better than if they were being forced to do the work.

Project managers must understand the personal aspirations of project team members and support them. The project manager's leadership skills play an important role in motivating and guiding the project team to grow as professionals while accomplishing project goals at the same time. In essence, project managers should ensure that both personal and project goals are accomplished and that the conflict between these two goals is minimized [7].

The project manager's job is to get others to do what the project manager needs them to do, not because they have to but because they want to. Empowerment means providing freedom to people (not control) so that they can successfully do what they want to do. This is very different from making them do what you want them to do. Empowerment makes sense on projects. Projects typically employ a multidisciplinary approach, requiring subject matter experts from different functional areas. Each person brings unique expertise, and experience to the project team. By letting project team members make decisions and set goals pertaining to their roles and functions, the project manager can empower the team.

10. Plan-driven vs. change-driven projects

Plan-driven or traditional project management presents a transformation view of production. It is based on principles of: managing the project as planned, using the plan for execution, monitoring and for control. The plan includes measures for monitoring progress and completion. The plan-driven or traditional approach is suitable when we can define the scope and specify deliverables precisely. In this approach, uncertainties and unknowns may still exist but they may not be challenging or disruptive to the project deliverable [13]. However, the traditional or plan-driven project management approach may be challenged if a project is complex and changes are major.

When a project lacks complete understanding of the project deliverables or outcomes, or when a project is associated with fast-paced technological changes, or needs of the client are unclear, change-driven or Agile method is preferred. These projects are associated with uncertainties and change. The client may reveal functional outcomes incrementally as the project makes progress and results are assessed. The project adapts change-driven approach and project team's immediate focus is on generating value to the client.

The main difference between the plan-driven and change-driven approaches is that instead of freezing specifications early and developing a fixed plan, the Agile approach adapts flexibility to modify and alter project plans to address critical and changing business needs [13].

Author details

Vittal S. Anantatmula
Western Carolina University, USA

*Address all correspondence to: vittal@wcu.edu

References

[1] Rad P, Anantatmula V. Successful Project Management Practices. Bingley, UK: Emerald Group Publishing; 2010

[2] Rad P, Anantatmula V. Integrated Project Planning. Berkeley Heights, NJ: Project Management Excellence; 2009

[3] Anantatmula V. Project Teams: A Structured Development Approach. New York, NY: Business Expert Press; 2016

[4] Project Management Institute. A Guide to the Project Management Body of Knowledge (PMBOK®). 6th ed. Project Management Institute; 2017

[5] Anantatmula V, Rad P. Attributes of Project-Friendly Enterprises. New York, NY: Business Expert Press; 2016

[6] Rad P, Anantatmula V. Project Planning Techniques. Vol. 2005. Vienna, VA: Management Concepts; 2005

[7] Anantatmula V. Role of technology in project manager performance model. Project Management Journal. 2008;**39**(1):34-48

[8] Anantatmula VS, Rad PF. Role of organizational maturity factors on project success. Engineering Management Journal. 2018;**30**(3):165-178

[9] Kloppenborg T, Anantatmula V, Wells K. Contemporary Project Management. 4th ed. Boston, MA: Cengage Learning; 2017

[10] Anantatmula V, Fan Y. Risk management instruments, strategies and their impact on project success. International Journal of Risk and Contingency Management (IJRCM). 2013;**2**(2):27-41

[11] Steffey WR, Anantatmula V. International projects proposal analysis: Risk assessment using radial maps. Project Management Journal. 2011;**42**(3):62-74

[12] Anantatmula V. Project manager leadership role in improving project performance. Engineering Management Journal. 2010a;**22**(1):13-22

[13] Anantatmula V, Kloppenborg T. Being Agile and Then Doing Agile. New York, NY: Business Expert Press; 2020

Chapter 11

A Modern Approach for Maintenance Prioritization of Medical Equipment

Călin Corciovă, Doru Andrițoi and Cătălina Luca

Abstract

Maintenance is a crucial topic in the life cycle management of medical equipment. Evidence-based maintenance consists of continuous monitoring of equipment performance, starting from evidence—the current state from the point of view of fault history—and improving its efficiency through the necessary modifications. This process is very important for optimizing the use and allocation of the resources available by the clinical engineering departments. Maintenance of medical equipment consists of two basic activities: scheduled maintenance and corrective maintenance. The purpose of this chapter is to present document-based methods to evaluate every aspect of the medical equipment maintenance process and to provide a correct, objective and standardized approach that supports clinical engineering activities. Following the analysis, the results show that the combination of the use of the two methods provides an overview, in a periodic manner, of maintenance performance that indicates the use of the most appropriate procedures.

Keywords: medical equipment, maintenance strategies, life cycle, health technology management, prioritization

1. Introduction

Within the large and modern hospitals, an increasingly common problem is the efficient management of the maintenance of the medical equipment, the quality of the assistance and the profitability. If effective management of medical equipment maintenance is to be applied, the management structure should apply appropriate planning, management and implementation processes. This is essential for providing quality health services while saving resources. Medical equipment management includes inspection and preventive and corrective maintenance operations [1].

The efficient management of maintenance and repair work must be planned and implemented using appropriate maintenance strategies to keep the devices safe and functional in accordance with the basic functional specifications. In addition to the high initial investments, medical equipment requires continuous and costly maintenance during its useful life. The issue of maintenance is the main point of discussion of the management of medical devices. Studies have shown that the most frequent cause of stopping of medical equipment is poor maintenance, planning and management. To solve this problem, it is necessary to establish and regulate an adequate system for the proper maintenance and use of medical

IntechOpen

equipment. Perfect maintenance is the equation of performance, risk, resources and costs to achieve this goal [2, 3].

The first maintenance policies developed consist of interventions on equipment, which run until it stops accidentally (breakdown) in place due to wear or because of defects. The intervention is considered satisfactory as long as the equipment/system is operating at a minimum acceptable level (reactive maintenance). The development and increase of the complexity of medical equipment and devices have led to modernizing and updating maintenance techniques and policies. Depending on the costs related to the spare parts and materials, respectively to the losses due to the time spent in repair, several types of maintenance policies have been developed [4].

Due to the way the health services are organized, the technical staff in the health units should not only perform maintenance and repair work but also be actively involved in the acquisition and management of the equipment. For example, they can plan equipment services and manage stocks; they can provide technical consultancy for procurement and can develop technical cost estimates. I can also make budget forecasts regarding the maintenance costs of medical equipment.

2. Maintenance organization: objectives and responsibility for medical equipment

In providing high-quality health services, medical equipment plays an essential role, because when the equipment is not used or properly maintained human damage can occur. In many situations, noncalibration, modification or repair of medical equipment by unqualified personnel can result in injury to the patient or loss of medical record. Preuse testing, preventative maintenance, malfunction reports (and incident reports) and repair procedures are just a few of the necessary actions prior to performing the medical act, to avoid injury caused by the use of medical equipment.

Even if the medical equipment used in the hospital is purchased, rented or borrowed, the commitment to safety is an essential element of any process related to the use of medical equipment. Proper maintenance and proper use of medical equipment ensures maximum efficiency and increased availability of equipment, at optimal costs and under satisfactory conditions of quality, safety and environmental protection [5].

In order to make the process of maintenance of the medical equipment more efficient, it is necessary to consider the use of a maintenance program of the equipment that takes into account its characteristics and the defects that appear to the medical equipment. The application of such a program of maintenance of medical equipment could be effective in applying correct maintenance strategies for the management of the older technological devices and the new high-tech devices, due to their different characteristics.

Maintenance was long considered as a subordinate function, entailing an inevitable waste of money. There was a tendency to lump it together with troubleshooting and repairing machinery that was subject to wear and obsolescence. However, hospitals today are realizing that maintenance is not merely a 'partner' in medical services: it is an indispensable requirement for quality medical services [6]. Its relation with equipment performance is a question of integrated strategy at senior management level. As such, the maintenance function becomes a management responsibility.

The structure that determines the goals and objectives of maintaining medical equipment is very important. If the goals and objectives are progressive, then the maintenance structure is recognized as a contributor to the hospital's foundation line, and thus, variations can be used on some of the more conventional organizational structures.

Objectives of maintenance management: the more specific objectives of maintenance management are as follows [7]:

- To optimize the reliability of equipment and infrastructure
- To ensure that equipment and infrastructure are always in good condition
- To carry out prompt emergency repair of equipment and infrastructure so as to secure the best possible availability for medical use
- To improve operational safety
- To train medical personnel in specific maintenance skills
- To advise on the acquisition, installation and operation of medical devices
- To ensure medical environmental protection

Within the clinical engineering department of any hospital, a crucial aspect of the activities is the activity of maintenance and preventive maintenance of the medical equipment, because it involves significant human and financial resources. Therefore, the optimization of the use of the resources available in the clinical engineering departments is done by evaluating the efficiency of the preventive maintenance programs of the medical equipment [8].

Forms of maintenance:

Maintenance has three major forms:

a. Design-out maintenance

b. Preventive maintenance, which includes systematic (periodic) maintenance and condition-based maintenance

c. Corrective maintenance

These are illustrated in **Figure 1**.

Maintenance can also be divided into planned and unplanned maintenance (or scheduled and unscheduled) (**Figure 2**). The following chart highlights the relation to the previous chart.

Table 1 briefly explains the terms used in the two charts.

The seven forms of maintenance distinguished above are the main types currently used in practice. Although preventive and predictive maintenance strategies differ in many ways, a maintenance program comprising both strategies yielded positive results. The maintenance strategy evaluation demonstrated that strategies based on performance verification and safety testing results and the manufacturers' recommendations led to a significant reduction in equipment failures and a significant increase in corrective maintenance.

An efficient strategy in the correct application of the maintenance of medical devices consists of the use of a maintenance strategy for older devices and another strategy for high-tech devices. We must keep in mind that older medical devices to which only corrective maintenance has been applied cannot be included in

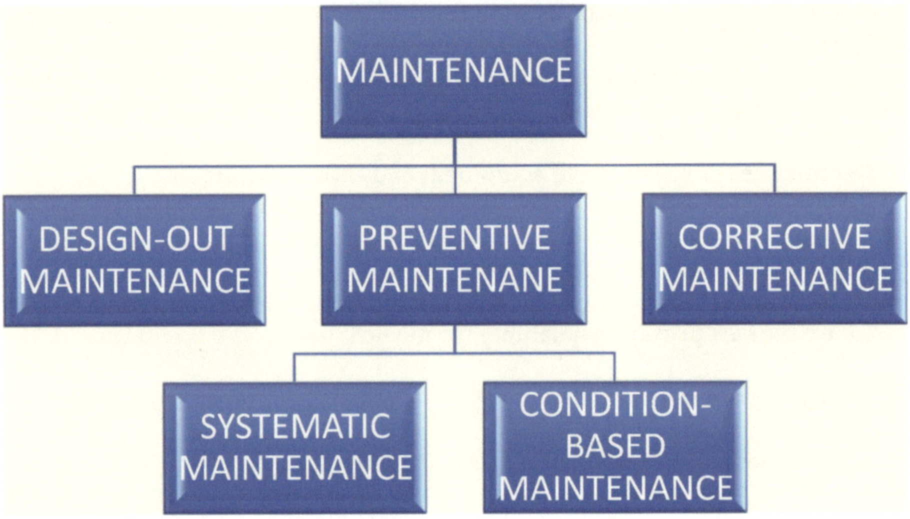

Figure 1.
The forms of maintenance.

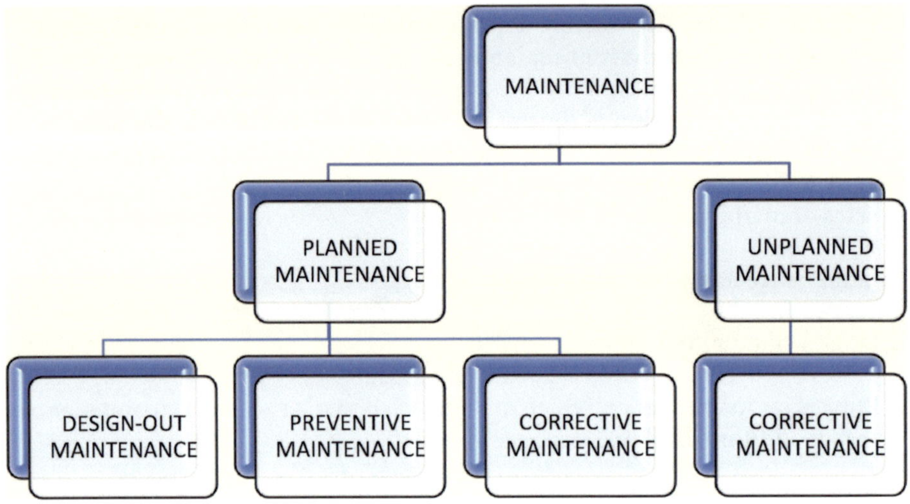

Figure 2.
Planned and unplanned forms of maintenance.

Maintenance	Maintenance is the function whose objective is to ensure the fullest availability of production equipment, utilities and related facilities at optimal cost and under satisfactory conditions of quality, safety and protection of the environment.
Design-out maintenance	This is also known as plant improvement maintenance, and its object is to improve the operation, reliability or capacity of the equipment in place. This sort of work usually involves studies, construction, installation, start-up and tuning.
Preventive maintenance	The principle of preventive maintenance is anticipation. It is put into practice in two forms: systematic (periodic) maintenance and condition-based maintenance.
Corrective maintenance	This is also called breakdown maintenance, palliative or curative maintenance. This form of maintenance consists of the following: • Troubleshooting machines whose poor condition results in stoppage or in operation under intolerable conditions • Repairs

Systematic maintenance	This consists of servicing equipment at regular intervals, either according to a time schedule or on the basis of predetermined units of use (hours of operation or distance traveled). The aim is to detect failure or premature wear and to correct this before a breakdown occurs. The servicing schedule is usually based on manufacturers' forecasts, revised and adjusted according to experience of previous servicing; this information is recorded in the machine's file. This type of maintenance is also called periodic maintenance.
Condition-based maintenance	This type of maintenance of the medical equipment is easy to apply because it does not require the disassembly of the equipment, the same technique based on the inspection by listening to the equipment involved. Predictive maintenance requires continuous observation of equipment to detect possible faults or to monitor its condition.
Planned maintenance	This is maintenance that is known to be necessary sufficiently in advance for normal planning and preparation procedures to be followed.
Unplanned maintenance	This is maintenance that is not carried out regularly as the need for it is not predictable; it is sometimes called unscheduled maintenance.

Table 1.
Short description of the terms of maintenance.

preventive maintenance strategies, such as new high-tech devices. Maintenance costs would increase greatly if we also reported old devices in the maintenance process [9, 10].

The access today is put on performance verification and safety testing in the use of medical equipment, which leads to a change in the maintenance strategies of the devices without necessarily taking into account the manufacturer's recommendations. Also, the decision-making in the management of medical equipment should be based on all the results of medical equipment malfunctions and the existence of a detailed history for each medical device.

3. Performance measurement and maintenance productivity in medical equipment

Performance measurement is a key management tool. In terms of maintenance management, an essential issue is to ensure that the planned and executed maintenance activities have given the expected results. Efficient use of indicators can facilitate this fact. Such an indicator, represented by key performance indicators (Kpi) is able to evaluate important aspects of the maintenance function. To this end, it has been shown that the measurement of maintenance performance is dominated by delay indicators (equipment, maintenance costs and safety performance).

The reduced use of the peak indicators in the maintenance process can also be observed. The obtained results did not show direct correlations between the maintenance objectives pursued and the Kpi used. Subsequent analyzes revealed that only a small part of the companies involved have a high percentage of decisions and changes caused by the use of Kpi and only a few are satisfied with their performance measurement systems. By analyzing the correlation, a strong positive linear relationship was identified between the degree of satisfaction and the changes/decisions of the process that are triggered by the use of Kpi, the people least satisfied with the least decisions and changes triggered by the use of Kpi. These observations indicate some inefficiency of performance measurement systems in improving driving performance [11].

The components of a system, such as pumps, electric or hydraulic motors, transmission systems, etc. as integral parts of it, must operate at optimal parameters to ensure that the overall performance of the device is achieved. Addressing the maintenance problems and establishing the procedures and the maintenance strategy for equipment must therefore take into account both monitoring and diagnosing at the level of each component, but also the influence of the system variables. Most of the time, the cause of a defect is found in the variations of the process parameters, and a nonintegrative approach to monitoring and diagnosing the system can lead to inefficient actions. Thus, in addition to the most popular techniques of monitoring and diagnosis (vibration monitoring, thermography and tribology), other parameters of a system such as flow rates, voltages, currents, temperatures, etc. must be considered.

In systems equipped with computer control or semiautomatic control, most of these parameters are purchased and used in the command and control process. Their type and number vary from system to system, but the algorithm for applying the monitoring and diagnostic procedure is similar. The collection of these parameters, together with the application of the traditional technologies of predictive maintenance, will provide all the necessary data for the analysis of the state and the performances of the system [12].

Since a large part of the equipment used in the medical field belongs to the category of electromechanical systems, the analysis of the maintenance technologies will focus on these, from the simplest (examples: electric motor-pump type drive systems) to complex devices.

It should be kept in mind that, in any system, the maintenance program will focus on its critical components. A critical component is defined as the element directly involved in the proper functioning of the device, on which the entire system depends, its efficiency and, last but not least, the quality of the product.

Some of the technologies for monitoring and diagnosing the state of a system are set out in the following. Vibration analysis is one of the most widely used detection methods to diagnose defects in electromechanical systems. This method measures the vibrations of the system, usually with an accelerometer, and then examines the frequency spectrum generated to identify significant frequencies from the point of view of the state of the equipment. Certain frequencies are typical of the system in normal operation. Changing the amplitude of certain harmonics, for example, can mean the presence of a defect. The data can be collected periodically, using a portable system, or continuously, by installing a continuous monitoring system. A major advantage is that the measurements are fast and noninvasive, and the functioning of the tested system is not disturbed [13].

Another key parameter that can provide information about one's status of equipment/system is temperature. This is an important indicator of the mechanical, electrical or load conditions applied to a component. Thermography is a predictive maintenance technique that uses instruments that can monitor infrared energy emission to determine operating conditions.

Infrared scanning is recommended as a regular maintenance procedure in many situations, extracting solid results as quickly as possible and without interrupting process flow, a key benefit to the industry, regardless of the age of the equipment. As an advantage of scanning a large area in a very short time, the ease with which data can be stored and processed for further analysis of images, the high mobility of the thermography camera that can be positioned at any time and place, the thermographic evaluation that is done uninterrupted and equipment inspection staff who are out of danger are emphasized.

Lubrication fluid analysis can be used to determine mechanical wear, lubrication or fluid condition. The presence of metallic particles in the lubricating fluid suggests the existence of a wear, their analysis providing information on the part subjected to wear. For fluid analysis, it uses complex equipment, which is why this method is not so often used in practice.

This strategy prioritizes the training of technicians to maintain an optimal number of actions, very important for essential medical equipment frequently used in medical institutions.

4. Classification and prioritization of medical equipment for maintenance activities

Prioritization of medical equipment maintenance should be performed for each new type of device during the inspection received when the device is added to the inventory. The device will then be assigned a test frequency. Subsequently, the maintenance history of the device will be monitored to evaluate the effectiveness of the maintenance program.

The end point of providing an organizational tool to the biomedical or clinical engineer would ensure the safe and efficient performance of medical equipment. The system must be evaluated on criteria such as:

- Data management for medical devices, manufacturers and suppliers
- Acquisition conditions
- Implementation and management of quality and safety protocols and procedures, including necessary documentation and data
- Carrying out corrective maintenance activities
- Routine procedure planning, such as acceptance testing, preventive maintenance, quality and safety inspections
- Management and monitoring of training provided by manufacturers or technical staff including biomedical engineer or clinical engineer [14]

The risk assessment was divided into four main areas: clinical function, failure avoidance probability, history of incidents and regulatory or manufacturer requirements. Devices would be evaluated on the aforementioned criteria and be assigned a score. The values would be added and a cumulative score is given for each device type. The total score would act as a quantifiable indicator for the maintenance policy. A total score of 12 or more would indicate a semiannual testing, a score between 9 and 11 would require annual testing, whereas a score of 8 or less would suggest a lesser necessity for annual testing, either biannual or no schedule, depending on clinical use. The end result would be an increase in the cost-effectiveness of the test program, less equipment downtime leading to improved patient care and a higher financial return to direct patient care activities.

To illustrate the applicability of risk assessment criteria, we evaluated two types of devices extensively used in healthcare: the defibrillator and the enteral feeding pump. Defibrillators are devices that correct or prevent arrhythmias (e.g., ventricular fibrillation and ventricular tachycardia) by sending an electrical impulse to the heart. External defibrillators, in particular, send high electrical impulses through the thoracic wall, stopping the independent action of the individual myofibers, so that the intrinsic pacemaker can take over. A set charge, between 0 and 360 J, is generated and delivered through paddles or disposable electrodes through the chest wall to the heart, determining a global contraction. Most defibrillators include an electrocardiograph to monitor the patient's rhythm, while others even include the pacer function. The clinical use is typically for emergency heart pacing such as severe bradycardia, asystole, pacemaker failure or ventricular fibrillation.

For this particular type of device, the assessment should include electrical safety evaluation—ground wire resistance, chassis and lead leakage—and inspection of parameters' performance, which includes measuring the energy output of the defibrillator throughout its range. This would include determining the value output at the lowest, midlevel and highest settings. The range of error should be with 15% of the set energy level (for 360 J, the output should be ranging from 206 to 414 J). Other performance tests would be determining the output levels at maximum setting for 10 charge cycles. The final output should still be within 15% of the recommended setting and charge time should not exceed 15 seconds. The appraisal for functional assessment frequency would be twice a year (**Table 2**) [15].

Enteral feeding pumps are used in patients who have gastrointestinal complications and who cannot consume adequate nutrients for certain reasons. The feeding solutions are transmitted to the patient through temporary feeding tubes or surgically implanted. These pumps can precisely control the flow of liquid supply solutions that are administered entirely through the digestive tract. These pumps are based on a pump mechanism such as a rotary peristaltic pump, a linear peristaltic pump or a volumetric pump. Most pumps record the dose frequency, dose settings and volume infused into memory. Audible and visual alarms alert the user to flow changes or malfunctions.

The quantity of volume delivered must be within 10% of the established volume. Thus, for a set volume of 10 ml, the measured volume must be between

Criteria	Risk	Score
Clinical function		
No patient contact	1	
Device may make contact with the patient who is noncritical	2	
Device is used for patient diagnosis or direct monitoring	3	
Device is used to deliver direct treatment to the patient	4	
Device is used for a life support	5	5
Problem avoidance probability		
Maintenance would not impact reliability of the device	1	
Common device failure modes are unpredictable	2	
Common device failure is predictable and can be avoided by preventive maintenance	3	
Specific regulatory requirements dictate preventive maintenance or testing	4	4
Incident history		
No history	1	
A significant history of incidents exists	2	2
Manufacturers/regulatory requirements for specific schedules		
No requirements	1	
There are requirements for testing	2	2
Total		**13**
Times per year tested	**2 (hight level)**	

Table 2.
Sample risk assessment for defibrillator.

Criteria	Risk	Score
Clinical function		
No patient contact	1	
Device is in contact with the patient who is not critical	2	
Device is used for patient diagnosis or direct monitoring	3	
Device is used to deliver direct treatment to the patient	4	**4**
Device is used for a life support	5	
Problem avoidance probability		
Maintenance would not impact reliability of the device	1	
Common device failure modes are unpredictable	2	**2**
Common device failure is predictable and can be avoided by preventive maintenance	3	
Specific regulatory requirements dictate preventive maintenance or testing	4	
Incident history		
No history	1	**1**
A significant history of incidents exists	2	
Manufacturers/regulatory requirements for specific schedules		
No requirements	1	**1**
There are requirements for testing	2	
Total		**8**
Times per year tested	**1 (normal)**	

Table 3.
Sample risk assessment for enteral feeding pump.

9 and 11 ml. The measured occlusion pressure must be within 1 psi of the pump occlusion pressure. For an occlusion pressure of 20 psi, the measured pressure must be between 19 and 21 psi. The recommended frequency of the functional test is annually (**Table 3**).

Before returning the equipment to medical personnel, it must be ensured that it has been adjusted to the original specific settings. Make sure that the volume of the audible alarms is loud enough to be heard under normal operating conditions [15].

5. Maintenance optimization models

Maintenance costs represent a large part of total cost functioning of health systems. Depending on the specifics of each device, the costs of maintenance can represent from 15 to 60% of the value of the expenses. For the situation in which the equipment works in safe conditions until a certain level of wear or a defect in the initial state has been established, we discuss about preventive and predictive maintenance. In such cases, the equipment will be stopped at an early date, and the repair will only be done where needed. This type of maintenance allows the early detection, localization and identification of the defect or the worn part, as well as the calculation of the operating life in safe conditions of the device. The activity of preventive and predictive type makes possible the planning of the stop, the preparation of the intervention team, the provision of the necessary spare parts and respectively the minimization of the parking time for repair [16].

Predictive maintenance represents a superior qualitative leap in a modern maintenance system, regardless of the domain or the specific production, because it offers all the information needed for the following:

- Early detection of the defects
- Location
- Diagnosis of defects
- Calculation of the operating life in safe conditions of the medical equipment

The common premise from which the predictive maintenance starts is that the periodic or continuous monitoring of the mechanical, electrical or other indicators of the functioning of the systems or processes can provide the data necessary to ensure the maximum interval between the repair and maintenance works, respectively, to minimize the cost of interruptions of maintenance. Unplanned maintenance can be the cause of possible failures, sometimes major. However, predictive maintenance is more than that. It is in fact the means of improving and increasing the productivity, product quality and overall efficiency of the systems in question. Predictive maintenance is actually a philosophy or attitude that, based on operating conditions, allows the optimization of the entire medical system. A comprehensive management of predictive maintenance uses the best methods to obtain the operating parameters of the component subsystems of a medical system, on the basis of which it will schedule maintenance and repair activities. Including predictive maintenance in the general maintenance program optimizes the availability of devices and equipment and greatly reduces maintenance costs. By using the records of the entire care of historical repair components and maintained maintenance, we can make a mathematical prediction model for the entire world.

Classifications of different types of failures and the establishment of policies for analysis involve three different levels: system level, failure peak and component level. Results analyzed can be set for a model for optimizing maintenance/inspection.

5.1 Rejects detection model

It is considered a continuous process so that it can be put into operation or rejected (scrap). The way of monitoring the functionality is as follows: first, check each product; continue checking until the consecutive k linear products are reached (full inspection). From this point, the inspection of the equipment is no longer deterministic, "piece by piece"; they will be chosen randomly, independently of the other, with probability α. Continue random monitoring (partially verified) until a defect is discovered, and then revert to previous monitoring and so on. Suppose the probability of a product being defective is q. It is understood that if a problem is found, the item is removed temporarily or permanently.

$$\frac{\left(\frac{1}{p}\right)^k - 1}{q}, \text{ where } p = 1 - q \tag{1}$$

Average of all relevant product cycle is equal to:

$$P_1 = \frac{\left(\frac{1}{p}\right)^k - 1}{q} + \frac{1}{\alpha q} \tag{2}$$

It can be shown that the proportion of undetected defective items is given by:

$$P_2 = \frac{q(1 - P_1)}{(1 - qP_1)} \tag{3}$$

Another model that offers good results when used in this field is known as "replacing a durable good." This is based on the assumption, for example, that the service life of the equipment is represented by a continuous random variable with the distribution function H and the density h and that a policy to replace the good says that it will happen if it has a major failure or if it is still in operation, it is acceptable to reach a certain "age," say the T years. We assume that the price of similar new equipment is C1, and when the equipment fails, we seriously consider a C2 amount, corresponding to the provision of the equipment.

The average length of a life cycle of equipment can be expressed as:

$$\int_0^T xh(x)dx + T[1 - H(T)] \tag{4}$$

Depending on the distribution of H, which is usually uniform $(0, T_0)$, where T_0 is a standard period, depending on the case, for example 10 years and costs C_1 and C_2, one can estimate the value of T, which will reduce to a minimum the cost of having an older, optimizing device. As in the field of health care, failure prevention is more effective than focusing on remedying them. Repairs are almost always expensive, requiring overspecialized personnel and often expensive parts. However, corrective maintenance is a permanent component of medical technology management.

Corrective maintenance allows a device to maintain its full performance of functions, through effective interventions at the time of a problem. However, this action must be well planned, because it acts not only on the level of symptoms, but also on the level of finding and solving the cause of the defect itself.

6. Life cycle of medical equipment

Users and technical staff have the obligation to maintain medical equipment at a level of safety as high as possible, compared to other types of usual equipment. Most complex medical equipment works, for example, in the intensive care unit. They have an electrical connection that in certain situations of first defect can create injuries or even death of the patient by electric shock. Patients connected to such medical equipment are not able to respond to dangerous conditions or pain. Other types of medical equipment work to support life, and a problem, sometimes even minor in some respects, can lead to the death of the patient when the equipment is used incorrectly or is poorly maintained. The life cycle of medical equipment, from the point of view of media technology management, comprises 4 stages and 9 themes according to current standards (**Figure 3**) [17–20].

An important stage in the life of medical equipment is that of maintenance and repairs that involve certain assumptions and challenges.

Some assumptions are as follows:

- Maintenance culture exists and is respected by the technicians, users and other staff.

- Technical staff are present, trained and know how to maintain and repair the equipment.

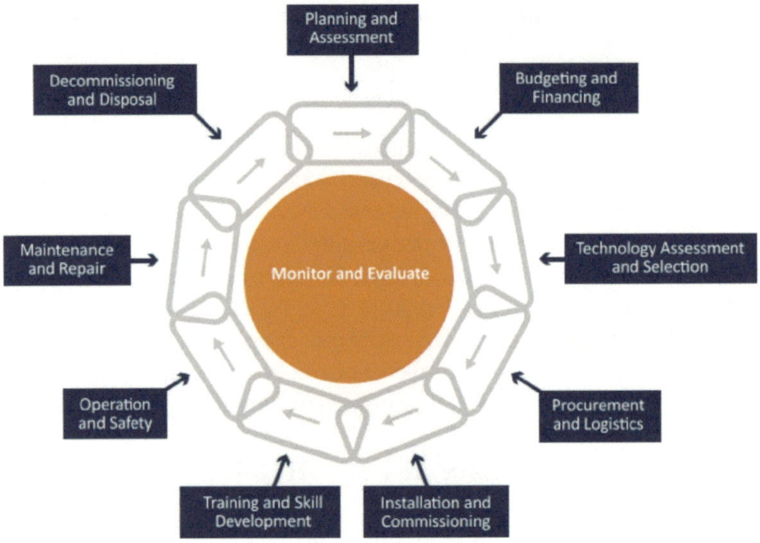

Figure 3.
Life cycle of the medical equipment.

- Preventive maintenance schedules exist and they are performed regularly.
- Technicians have access to spare parts, on stock in the hospital or ordered in and spare parts are delivered within 24 hours if necessary.
- Technicians have access to and know how to use test equipment to calibrate and test medical equipment.

7. Maintenance control system

A maintenance system of medical equipment should be considered as a simple system with inputs/outputs. Inputs to the system are data of defective equipment, materials and spare parts, consumables, data and information on its use, local and global policies and procedures. The result is reliable and well-configured medical equipment that can be achieved only by efficient planning of maintenance and service. The system to be functional has a set of rules that must be implemented. These activities include planning, scheduling, executing and controlling.

The control is performed having as objective the organization and functioning of the maintenance system. The objectives coincide with the organization's objectives and include equipment availability, costs and quality. An important role is played by the feedback that is used to improve the performance of the medical system/equipment [21].

The existence of an effective maintenance control system improves the reliability of the equipment and increases its service life without having unscheduled shutdowns. Maintenance control contains a set of activities, tools and procedures used to coordinate and allocate maintenance resources, including those for specialized personnel, to achieve the objectives of the system, including the following:

1. Work control

2. Quality control and processes

3. Cost control

8. An efficient reporting and feedback system

An essential element of maintenance control is the work order system used for planning, executing and controlling maintenance work. The work order system consists of the necessary documents and the well-defined workflow process. The documents provide means for planning and collecting the information needed to monitor and report maintenance work.

Currently, the process of controlling the maintenance of medical equipment involves four stages:

1. Concrete and coherent setting of objectives and standards: the control process begins with planning; the objectives and performance standards to be pursued are established. Performance objectives must be clear results that must be achieved.

2. Methods of measuring effective performance: the purpose is to accurately determine the results of performance (output standards) and/or performance efforts (input standards). Quantification must be accurate to identify significant differences between what was actually achieved and what was originally planned at the beginning of the process [22, 23].

3. An important role is played by the comparison of the results obtained following the measurements with imposed objectives and standards. This stage is expressed by the control equation: Need for action = Desired performance – Actual performance. Sometimes, a comparison with data from the history of equipment use, data collected from the medical device file, can be taken into account for an evaluation of current performance. Or you can use a relative comparison that tracks the performance of other equipment in the same model, meeting the same standard, used by people with similar training. In comparison, maintenance standards are scientifically established by methods such as time and motion studies. Preventive maintenance routines, for example, are measured in terms of expected time in each routine performed, depending on operating hours or time intervals.

4. Carrying out corrective actions: the last step in the control process is to take all necessary measures to correct problems, nonconformities or improvements. Effective management is one that pays attention to situations that show the greatest need for correction. It saves time, energy and other valuable resources, focusing on critical and priority areas. Maintenance managers must pay special attention to two types of situations: a problematic situation in which the real performance is below the imposed standard and a second situation, of opportunity, in which the real performance is above the standard.

9. Maintenance strategies

The oldest and most common maintenance and repair strategy is "fix it when it breaks." The appeal of this approach is that no analysis or planning is required. The problems with this approach include the occurrence of unscheduled downtime at times that may be inconvenient, perhaps preventing accomplishment of committed production schedules. These problems provide motivation to perform maintenance and repair before the problem arises. The simplest approach is to perform maintenance and repair at preestablished intervals, defined in terms of elapsed or operating hours. This strategy can provide relatively high equipment reliability, but it tends to do so at excessive cost (higher scheduled downtimes) [24]. A further problem with time-based approaches is that failures are assumed to occur at specific intervals. The only way to minimize both maintenance and repair costs and probability of failure is to perform ongoing assessment of machine health and ongoing prediction of future failures based on current health and operating and maintenance history [25–27].

This is the motivation for prognostics: minimize repair and maintenance costs and associated operational disruptions, while also minimizing risk of unscheduled downtime. Preventive maintenance is the strategy organized to perform maintenance at predetermined intervals to reduce the probability of failure or performance degradation. It can be classified into constant interval, age-based or imperfect maintenance:

1. Constant interval maintenance: as the name suggests, it is done at fixed intervals (in addition to any maintenance prompted by failure that is performed when it manifests). Intervals are selected to balance high risk of failure with long intervals and high preventive maintenance costs with short intervals.

2. Age-based maintenance: in this strategy, preventive maintenance at fixed intervals is carried out only after the system has reached a specific age.

3. Imperfect maintenance: in the above to be restored to its original condition after a preventive maintenance. However, it may be the case that the condition of the system is in between good (original) and bad (failure). This is the premise of imperfect maintenance strategies, which take into consideration the uncertainty of the current state of the equipment while scheduling future activities [28–29].

10. Conclusions

Providing quality medical services involves correct and efficient resource management and planning. An important element in achieving this is a balance between costs involved in the investment of new equipment and its maintenance. Proper use and proper maintenance of medical equipment must be supported by a clear policy in the field, technical guidance and practical tools for maintaining the functional parameters of media equipment. By using functional medical equipment, it will be possible to significantly improve the quality of the medical act and the efficiency of such a service. Consistent management practices in this area will help increase efficiency in the field of health.

An analysis of the maintenance of medical equipment is made to assess the lifespan of that equipment, which can be extended or shortened depending on the actions taken. Equipment maintenance is crucial for its lifespan. If maintenance periods are not met, on time and on a regular basis, medical equipment will be damaged to the point where it will cost more to repair than to replace. If no decisions are made at all in the maintenance of medical equipment, it will degrade irreparably. The importance of maintenance activities consists in the efficient management of the equipment; this task requires extensive information about the medical device. Thus, it is necessary to know the history of the equipment, how it has been exploited in the past, to say if the situation is improving and to learn from previous situations.

Finally, records provide staff with valuable technical information and evidence that they can use when they need arguments or need help or additional resources. The maintenance of the database system helps to keep track of repair services and other actions for optimal operation of medical equipment.

Conflict of interest

The authors declare no conflict of interest.

Author details

Călin Corciovă*, Doru Andrițoi and Cătălina Luca
Medical Bioengineering Faculty, University of Medicine and Pharmacy "Grigore T. Popa", Iasi, Romania

*Address all correspondence to: calin.corciova@bioinginerie.ro

References

[1] Miniati R, Dori F, Gentili G. Design of a decision support system for preventive maintenance planning in health structures. Technology and Health Care. 2012;**20**:205-214. DOI: 10.3233/THC-2012-0670

[2] Wang B, Rui T, Balar S. An estimate of patient incidents caused by medical equipment maintenance omissions. Biomedical Instrumentation & Technology. 2013;**47**(1):84-91. DOI: 10.2345/0899-8205-47.1.84

[3] Ni J, Jin X. Design support systems for effective maintenance. CIRP Annuals—Manufacturing Technology. 2012;**61**:411-414

[4] Sezdi M. Two different maintenance strategies in the hospital environment: Preventive maintenance for older technology devices and predictive maintenance for newer high-tech devices. Hindawi Publishing Corporation—Journal of Healthcare Engineering, Research Article | Open Access. 2016;**7267983**:16. DOI: 10.1155/2016/7267983

[5] Justin J, Madhukumar S. A novel approach to data driven preventive maintenance scheduling of medical instruments. In: International Conference on Systems in Medicine and Biology. Kharagpur, India: IEEE Engineering in Medicine and Biology Society; 2010. pp. 193-197

[6] Wireman T. Maintenance Work Management Processes: Maintenance Strategy Series, Volume 3, Chapter 1. Copyright Industrial Press Inc; 2008

[7] Duffuaa S, Al Ghamdi A, Amer A. Quality function deployment in maintenance work planning process. In: 6th Saudi Conference. Vol. 4. Dhahran, Kindom of Saudi Arabia: KFUPM; 2002. pp. 503-5012

[8] Shen X, Tan K, Xie M. An integrated approach to innovative product development using kano's model and QFD. European Journal of Innovation Management. 2000;**3**(2):91-99

[9] Wang B, Furst E, Cohen T. Medical equipment management strategies. Biomedical Instrumentation & Technology. 2006;**40**(3):233-237. DOI: 10.2345/i0899-8205-40-3-233.1

[10] Deglado D, Aspinwall E, et al. Quality function deployment in construction. Construction Management and Economics. 2007;**25**:597-609

[11] Adnan A et al. Building medical devices maintenance system through quality function deployment. Jordan Journal of Mechanical and Industrial Engineering. 2012;**V6**(N1):25-36

[12] Nataly Y, Hyman W. A medical device complexity model: A new approach to medical equipment management. Journal of Clinical Engineering. 2009;**34**(2):94-98. DOI: 10.1097/JCE.0b013e31819fd711

[13] Jamshidi A, Abbasgholizadeh RS, Ait-kadi D, Ruiz A. Medical devices inspection and maintenance: A literature review. In: Proceedings of the 2014 Industrial and Systems Engineering Research Conference. Montreal, Canada; 2014

[14] Khalaf AB. Maintenance model for minimizing risk and optimizing cost-effectiveness of medical equipment. Journal of Clinical Engineering. 2004;**14**:3649-3653

[15] Corciova C, Andritoi D, Ciorap R. Elements of risk assessment in medical equipment. In: The 8th International Symposium on Advanced Topics in Electrical Engineering. Bucharest: ATEE; 2013

[16] Zheng LY, McMahon CA, Li L, Ding L, Jamshidi J. Characteristics management in product lifecycle management: A survey of methodologies. Proceedings of the IMechE Part B: Journal of Engineering Manufacture. 2016;**222**(8):989-1008

[17] Bronzino JD. The Biomedical Engineering Handbook. 3rd ed. Connecticut, USA: CRC Press in Taylor and Francis Group; 2006

[18] Chien C, Huang Y, Chong F. A framework of medical devices management system for in-house clinical engineering department. In: 32nd Annual International Conference of the IEEE EMBS. Buenos Aires, Argentina; 2010. pp. 6054-6057. DOI 10.1109/IEMBS.2010.5627617

[19] Bronzino JD. Management of Medical Technology: A Primer for Clinical Engineer. USA: Butterworth-Heinemann; 1992

[20] Dyro JF, Cheng M. Good management practice for medical equipment. In: Dyro JF, editor. Clinical Engineering Handbook. San Diego: Elsevier Academic Press; 2004. pp. 108-110

[21] Iadanza E, Dori F, Biffi Gentili G. The role of bioengineer in hospital upkeep and development. IFMBE Proceedings. 2007;**14**:3641-3644

[22] Barringer H, Weber D. Life cycle cost tutorial. In: 5th International Conference. Westside, Houston, Texas, USA; 1995. DOI: 10.1.1.546.7966

[23] Yeh TH, Deng S. Application of machine learning methods to cost estimation of product life cycle. International Journal of Computer Integrated Manufacturing. 2012;**25**:340-352

[24] Luschi A, Marzi L, Miniati R, Iadanza E. A custom decision support information system for structural and technological analysis in healthcare. IFMBE Proceedings. 2014;**41**:1350-1353

[25] Baretich MF. How to use financial benchmarks. Biomedical Instrumentation & Technology. 2011;**45**(5):405-407

[26] Biomedical Engineering Advisory Group (BEAG). Life Span of Biomedical Devices. USA: BEAG; 2004

[27] Cohen T. AAMI's benchmarking solutions: Analysis of cost of service ratio and other metrics. Biomedical Instrumentation & Technology. 2010;**44**(4):346-349

[28] Miniati R, Dori F, Iadanza E, Fregonara MM, Biffi Gentili G. Health technology management: A database analysis as support of technology managers in hospitals. Technology and Health Care. 2011;**19**(6):445-454

[29] Guide 5: How to organize the maintenance of your healthcare technology. In Townsend S, editor. 'How to Manage' Series for Healthcare Technology. UK: Ziken International; 2005

Chapter 12

Operations Knowledge Management in Health Care

Ann Svensson and Eva Hedman

Abstract

The aging population of the western world poses a medical challenge for the society of today and the future. The pressure on health care and its organization is increasing as the demand for health care is growing at the same time as the costs are continuously rising. There seems to be consensus regarding bottlenecks in health-care production systems, and that knowledge is needed in order to increase insights about operational decisions. This chapter is based on a qualitative case study conducted at a hospital in western Sweden. Eleven CEOs together with their production controllers were interviewed. The chapter aims to analyze knowledge management mechanisms in the infrastructure of a health-care organization. The analysis shows how mechanisms have significant impacts on the knowledge management practice in the operations management. A learning and knowledge creation culture, together with an organizational architecture for adaptive and exaptive capacity, and a business model for knowledge capitalization could support the production of smooth and effective health care in society, which is of high quality.

Keywords: health care, knowledge management, operations management, capacity planning, health-care sector, hospital

1. Introduction

Management of health care has received attention over some years [1], and the interest in this topic has increased during the last years, both in practice and in the literature. The COVID-19 pandemic, especially, has brought immediate focus to the capacity of resources in health care and its processes that are required to provide efficient and effective health care. The aging population of the western world also poses a medical challenge for the society of today and the future [2, 3]. The pressure on health care and its organization is increasing as the demand for health care is growing at the same time as the costs are continuously rising [4]. In hospitals, which is the focus of this study, health-care professionals aim to provide health care of high quality using the limited resources available, even though community-based services could be involved in the provision of health care for the inhabitants in a society. As such, a hospital can be described as a "network of service units with finite capacity through which patients are flowing" [5].

Operations management (OM) refers to the planning and control of processes that transform inputs into outputs. The term also applies to management of health care where patients who have requested health care are diagnosed, referred to a further service, or cured [6]. Resources at the hospital have to be managed to transform inputs to outputs. It is challenging to manage the health-care processes in order to ensure an effective health care that uses the available resources efficiently [7]. Moreover, it is challenging to balance resources with fluctuating and uncertain demand as in health care [8]. There seems to be a consensus regarding bottlenecks in health-care production systems, and that knowledge is needed in order to increase insights about operational decisions [8]. There are, however, flaws in the understanding and management of the variations in demand and capacity [9]. Some studies claim that problems exist due to poor demand and capacity management practices [10, 11]. Even though the planning and management of health care has gained more interest during the last 20 years, both within research and practice, very few concrete steps have been take in order to improve the OM in health-care practice [1]. Appropriate knowledge about accessibility, demand, and capacity variations could improve patient processes [12] as well as planning and modeling processes [13]. Knowledge management (KM) in the health-care sector is intended to support the operation of planning, performance, and control of health care. Health-care systems need to be designed to achieve smooth flow of patients so that timely and appropriate care can be provided [14].

Massaro et al. have identified a research gap regarding KM in the public sector [15], even though organizational and managerial knowledge processes in health care have been increasingly studied during the past 20 years [16, 17]. Fugate et al. [18] claim that effective communication and KM play key roles in OM and the improvement thereof. However, most of this literature has not affected the development and organizational goals in the health-care sector, even though health-care organizations have adopted KM strategies to a certain extent [16]. The main success factors that influence the implementation of KM in the health-care sector are considered to be the infrastructure capability (such as information technology, organizational culture, and organizational structure) as well as the performance evaluation and measurement [16]. Developing an appropriate KM infrastructure is considered crucial for the improvement of quality in the health-care sector [19]. Few studies have, to this date, explicitly addressed the specific mechanisms in the KM practice systems in the health-care sector that are related to the infrastructure at the organizational level. Therefore, this chapter contributes with a qualitative analysis of the KM mechanisms in the infrastructure of a health-care organization. The analysis shows how mechanisms have significant impacts on the KM practice in the OM.

The remainder of this chapter is structured as follows: Section 2 provides a literature review of the OM and KM in the health-care sector. Section 3 provides the empirical setting and describes the methods used for data collection and data analysis. Section 4 presents the results of the study, which are discussed in Section 5. Section 6 concludes the chapter, together with the suggestions of a few topics for future research.

2. Literature review

2.1 Operations management in health-care organizations

OM is an interdisciplinary field that often uses mathematical modeling, analytic methods, and statistics to create production and capacity plans for controlling and improving an organization's ability to reinforce rational and meaningful management decisions at all levels [7]. Historically, these models and methods have mainly been used in the manufacturing industry. Recently, however, such models and methods have begun to be widely applied in the public sector as well, not least in the health-care sector. Production and capacity planning processes involve communication and coordination of information between hierarchical and horizontal levels. Planning and control involve deciding in advance which activities to do, how to do the activities, who should do what, when activities should be done, and what capacities are needed. A variety of decisions related to the hospital health care have to be made based on medical resource and financial aspects [20]. Decision-making in production and capacity planning requires coordination and management of information and knowledge at different hierarchical levels and in different time frames. Thus, management of information and knowledge is also needed between various health-care services within a hospital, as well as between other health-care providers, at a horizontal level. Capability of KM adds value to operations performance and in effective use of knowledge through acquiring, sharing, and applying knowledge across various health-care services [21].

Within a health-care organization, there is need to balance the available capacity of various resources. The resources also need to be coordinated to manage different types of medical activities and treatments. Departments are also sharing resources, and the demand and availability is fluctuating and uncertain [8]. The resources and the available capacities need to be managed in order to match the demands. The goal is to deliver health care of high quality, using the limited resources available. Designing and organizing health-care processes implicates planning and controlling activities. The process of designing and organizing in turn also implies setting goals for the activities and planning and controlling the operations. Planning requires information and knowledge at different organizational levels. Planning and management in health care can, for example, refer to the planning of operating rooms, the need for nurses and scheduling patients [22]. Dexter [23], among others, has for example studied planning and scheduling of operating rooms. The management and planning of health care thus comprise dimensioning, planning, scheduling, monitoring, and controlling resources [7]. However, there is an apparent gap between the demand and the available capacities [8, 10, 24].

The efficiency of health-care processes is a result of planning and management at different managerial levels. Developing effective plans for using resources and capacities requires understanding and knowledge of the dynamics in the hospital system and the flow of patients through it [8]. The quality of decision-making in each department depends on the information available for planning and decision-making, also, in relation to other departments in the hospital system. Managers and professionals who are planning and making decisions need to have knowledge about many different aspects within the organization, even beyond departmental borders. Sub-optimization is also a threat against effective planning and management, as decisions are made within different departments. Hulshof et al. [7] found that higher efficiency can be achieved if decisions are made from a more integrated perspective. The integrated planning and management of health care also faces challenges as different departments can have different goals, and conflicts can exist between different goals. Complex relationships between decisions within different departments exist, and this complexity has been identified as the most significant hindrance to effective OM. Hulshof et al. [7] claim that models are still missing for the management of health-care processes, for example flows of patient between different departments. Therefore, models for OM and KM need to be created [22, 25], also for the infrastructure that should be the basis for the processes [16].

2.2 Knowledge management in health-care organizations

Knowledge-intensive and professional organizations as hospitals are complex in their nature and require attention to aspects relating to their complex dynamics [26, 27]. Traditional management and leadership models are no longer entirely suitable for such knowledge-intensive organizations as health-care organizations [28]. In order to effectively create sustainable operations performance in health care, developments in knowledge capability is important [18, 21]. KM adds value to operations performance in organizations concerning cost, quality, flexibility, and delivery. Thus, knowledge is a strategically significant resource. KM can be viewed from two perspectives: process and infrastructure [21, 29, 30]. The process perspective reflects the capability in an organization, but the process is based and dependent on the infrastructure. The infrastructure perspective defines the technology, structure, culture, and mechanisms that enable the configuration of resources and operational routines in the organizational processes [21].

Technology has made it easier to transmit information in organizations, and is an important part of the infrastructure [18]. Many sophisticated information systems are used within health care, and they have a tremendous impact on the complex organizational context. Information systems also include different structures of their own. Information systems thus create structures for how people perform their work, and how they interact with the systems [20, 31]. By understanding the complex relationships between the health-care organization and the information systems that are used, it is possible to get a better understanding of how information systems can support OM and KM [32]. However, effective communication and a shared interpretation of the knowledge are critical for the performance of a health-care organization.

Three mechanisms for KM practice systems have been identified by Loon [33]; (1) learning and knowledge creation culture; (2) organizational architecture for adaptive and exaptive capacity; and (3) "business model" for knowledge capitalization and value capture. *Learning and knowledge creation culture* is based on culture theories, in which learning is seen as a set of values among a group of professionals, which underpin their behavior in creating knowledge. Those values are shaped by organizational structures. This mechanism influences the importance placed on formal and informal learning in the organization, and includes reward schemes, coaching programs and other formalized KM-specific roles and operations that strengthen the learning and knowledge creation culture. The *organizational knowledge architecture for adaptive and exaptive capacity* consists of the design of organizational systems, technologies, practices, skills, and behaviors, that for example facilitate sharing of knowledge. This mechanism shapes the orientation of technology use, to primarily codify knowledge or to use technologies to connect people to exchange tacit knowledge. Appropriate and relevant structures, technologies, and processes have to be developed to allow knowledge to be stored, transformed, and exapted to facilitate the performance of the organization. The *business model for knowledge capitalization and value capture* describes how an organization benefits from its KM practice system. This mechanism is directing how new knowledge is embedded in the organization's value proposition, as the organization has to be aware of how newly created knowledge will be characterized as useful and appropriate for its outcomes and defined goals.

3. Research method

3.1 Empirical setting

This study was conducted at a hospital in the western part of Sweden. The health care produced at this hospital consists of both acute and planned health-care services, divided in three different medical areas, each of which is further subdivided into smaller units; 17 in total. Acute and planned health care respectively, call for different planning methods. The patients' paths through the different health-care departments of the hospital is also planned and controlled, regardless of whether the health care is of an acute or planned character. The possibility for booking each patient is set by the detailed planning in each health-care department. To achieve a patient path that is as smooth and effective as possible, coordination of plans between different departments is often needed.

The mission, orders, and economical frames of Swedish hospitals are decided by the political governance. A document called "the health-care agreement" prescribes the overall assignments for the hospitals. This framework has to be transformed into terms that are useful for the planning and control processes within the hospitals, and communicated to the hospitals [20]. The admission of patients is also an important aspect that determines the need for resources and capacity. Moreover, there are databases of waiting lists with different patient groups, and different economical efforts at regional and national levels that affect the production of health care at the hospitals. Other regional health-care organizations, as well as the home health care conducted in the municipalities, also affect the production of health care at the hospitals.

At the studied hospital, the planning and control process generates forecasts, production plans, capacity plans, etc. The forecasts and plans aim to balance the needs for care within the frames given by the political governance. The different plans are then broken down through the organization, into more details regarding time horizons, care services, and resources at each department and service unit [20].

The processes of creating production plans differ to some extent between different levels in the organization. The collection of data and the elaboration of plans are performed by staff members with different positions in the organization. At the operations level, nurses often collect the data, create the plans, and book the patients. At the clinical department level, the manager and the production controller create the production plans. The plans cover one financial year and are adopted at the clinics board meetings once per year. At the hospital level, the production controllers from the different hospital departments together with the hospital's chief economic controller are responsible for creating the plans. The plans are adopted at the hospital board meeting once per year. The follow-up is normally done routinely each month, both at the levels of departments and hospital. Different actions are then taken when there are deviations from the defined production plans.

3.2 Data collection

A qualitative approach was used in this case study [34, 35]. The qualitative case study approach is a strategy often used in studies of the use of information systems and KM within organizations [36, 37]. The case study produces context-dependent knowledge and experience without any attempt to make generalizations based on the collected data [38]. The study focuses on the planning and control of the production and capacity within the hospital health care as the case. Since the study focuses on a specific work practice, it is of great importance to use a methodological strategy that is adapted for research of the applied nature of the empirical setting [39]. The approach was to interpret the social phenomena in order to show what is socially constructed by the health professionals in their natural work practice.

Data were collected through qualitative interviews that focused on views and experiences expressed by people [37]. What types of managerial actions were performed within planning and control of hospital health care and how the work was conducted were of interest in this study. The data collection focused on acquiring a deep insight into the planning and control of production and capacity of the health care provided at the hospital. A semi-structured interview guide was constructed based on a literature review of OM as well as the knowledge and experience that one of the authors has gained from her ordinary work practice of OM at the hospital.

Eleven interviews were held with 11 chief executive officers (CEOs) and their respective production controllers. The interviews lasted for about 2 hours each. Each CEO is responsible for one health-care service unit at the hospital, and they all collaborate with a production controller. Each production controller works at about four to five different health-care service units, and each health-care service unit consists of several departments. The 11 CEOs were chosen from units that need to cooperate with other units in some way. In total, six more CEOs work at the hospital, but their activities are less extensive, and a few have no direct contact with patients. By collecting data through interviews, it was possible to maximize the exploration of different perspectives and activities within the planning and control of production and capacity. The interviews were used as the primary data source. In addition, internal documents used in the planning and control of production and capacity were requested and used as a secondary data source. Both authors conducted all the interviews, which were tape-recorded and afterwards transcribed verbatim.

3.3 Data analysis

Thematic analysis was used for the qualitative analysis of the collected data [40]. This method was used to interpret the various aspects of KM mechanisms in the infrastructure of the health-care organization and its OM. By using a deductive thematic analysis, different codes were found in the empirically collected data. Then, a detailed analysis of the specific data that mapped to the aim of the study was conducted. Since the analysis focused on the specific theoretical aspects of the collected data, there is a clear connection between the analysis method and the theoretical perspective [39]. The aim was to analyze the collected data in depth as the data contain meanings that the respondents expressed as their experiences of the work practice rather than measure data quantitatively.

The coding process aimed to fit the theoretical preconception of the study. As such, the coding was theory driven, as we searched in the transcripts for codes within the predefined theory-based themes. The analysis resulted in codes at both a semantic and a latent level. Codes were identified both in the explicit meanings of the data and in the underlying ideas and assumptions. The latent codes are thus based on interpretations of the data during the analysis [40]. The analysis was inspired by a constructionist perspective since it was assumed that meanings and experiences are socially constructed within the context.

Based on the aim of the study, three different KM mechanisms were analyzed in the collected data: (1) learning and knowledge creation culture, (2) organizational architecture for adaptive and exaptive capacity, and (3) "business model" for knowledge capitalization and value capture.

4. Results

The health care provided at the hospital is based on needs, forecasts, and the division of resources between the departments. The operations at each of the three areas at the hospital affect each other. Different activities, together with the culture and infrastructure, affect the outcomes of the OM, and knowledge is thus managed

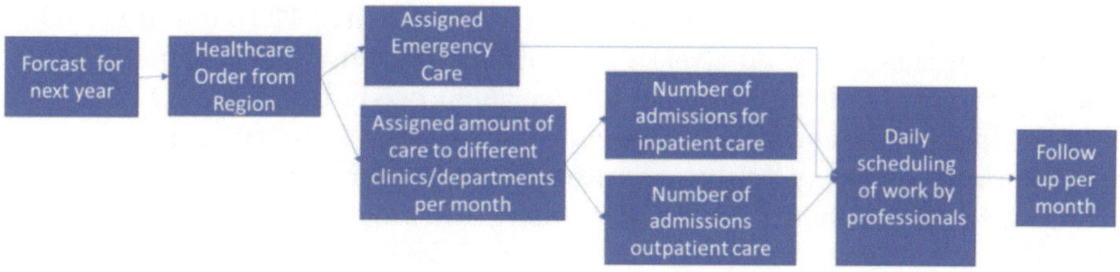

Figure 1.
Schematic view of OM for health-care planning at the hospital.

in different ways. **Figure 1** shows the OM planning process where knowledge mechanisms have to be in place for a smooth work.

4.1 Learning and knowledge creation culture

Middle managers would like to have data about the performance in real time, in order to be able to take appropriate actions if the situation starts to be worse, instead of saying the following day that: "oh well, now this was damn crazy yesterday". It should be possible to follow up and document lead times at the triage work, if it is going up and down. This is now just recognized based on experience of the staff. Thus, there is a prevalent culture where the staff adapts to changes in the admission of patients. Moreover, knowledge of how to match the schedule to the needs is important and it makes the planning much smoother. Knowledge and experience of where there are shortcomings and what needs to be changed in the planning saves a lot of resources. However, professionals need a lot of education to acquire such knowledge. The departments also have whiteboards for planning and following up on the daily management. The resource utilization can also be reassessed if anyone anticipates changes in the workload. Moreover, oral analyses are made at the departments on a daily basis. These analyses focus on what worked and what did not work, deviations and so on, and aim at learning and creating knowledge for the future operations and capacity planning at the department. These oral analyses are however not documented and deeper analyses are sometimes needed. Moreover, the results from the previous day are sometimes more or less obvious. There is a desire at the hospital to work with, and follow up, based on real-time data, instead of using data from the day before, in order to learn and create more appropriate knowledge, and be able to put in efforts a little earlier. A suggestion is to have a visual and real-time based monitor that shows the inflow, the operations, and outflow of the hospital as support for the daily planning.

Since the operations at the emergency department cannot be planned in advance, resources are allocated based on historical numbers of patients. It is somewhat unclear how many patients the emergency department can handle, as there is no defined assignment. The professionals work toward a goal of shortening the waiting times at the emergency department based on requirements from the regional level. Forecasts are dependent on whether someone discovers changed patterns in an ad hoc basis. It is difficult to compare outcomes since there have been so many changes at the hospital during the last years.

A certain number of operating rooms with staff and other resources must be available at the hospital. The planning of resources is not directly related to which operating activities will be conducted in these operating rooms. Instead, the operating rooms with staff and other resources are prepared for immediate use. The number of prepared operating rooms is measured in retrospect, based on how much the operating rooms were actually used at a specific time, or a specific day. The managers have discussed whether they could do as much in 10 rooms as they do in 12 rooms, so it is not only a matter of the number of rooms. The dimensioning of capacity is thus not only based on knowledge about needs: traditions are also crucial. Different doctors or departments are, for example, used to be in one specific operating room, and may not want another doctor or department to use it. There are huge discussions about how the scheduling of the operating rooms should be arranged, and how the availability and the use of the operating rooms can become more effective. There are some ideas about how to solve this situation. As doctors often have other things planned at the same time, and the time is not clearly set aside for the operation, the schedules for doctors could be more strict. Learning how to manage the required resources has to improve.

Even though collaboration and knowledge sharing with other health-care organizations is highly prioritized, as it contributes to decreasing the emergent inflow of patients, the extent of collaboration and knowledge sharing within and between departments largely depends on staff availability. A coordinator at each department is responsible for having daily meetings with other coordinators, in order to gather knowledge from the doctors and distribute knowledge to the departments about what is going on at the moment, and what inflow and outflow of patients to expect during the day.

There is a test package for each group of disease, such as X-ray, blood tests and other screening activities. These tests need to be taken from each patient and analyzed. However, there is no follow-up if those standardized test packages are needed every time. Thus, the knowledge about the effectiveness of such test packages is not acquired at the hospital. This is one example of an area for improvements in the learning and knowledge creation culture, where the health-care OM could be more appropriate.

The learning and knowledge creation culture does not seem so formally structured in the organizational culture at the hospital. Planned activities that support learning and knowledge creation, especially activities that could contribute to improving the performance of the organization, do not seem to exist. KM practices do not seem to be an implicit assumption that guides how the professionals act. The interviewees did not discuss any specific or formalized KM-specific roles or operations which suggest that any formal KM initiatives exist. Yet, there are different informal learning and knowledge creation activities going on. Therefore, the learning and knowledge creation culture is not at all missing at this hospital.

4.2 Organizational architecture for adaptive and exaptive capacity

Three different IT systems are used for forecasting, planning, and follow-ups, however, they are used in different ways at different departments. The emergency department uses one more IT system that makes it possible to compare numbers in all the regional emergency departments, also at other hospitals. One IT system together with Excel sheets are used for scheduling the staff. Since the IT system alone does not give a clear picture of the schedule in order to plan, Excel sheets are also used. Information about activities conducted in relation to the patients is registered in another IT system. Some information is also registered in the patients' medical record system. These IT systems are not integrated and do not have all the functions that are actually needed. Since the staff must always be reminded of manual routines, the statistics cannot always be trusted. The data quality thus partly depends on how well the professionals remind each other to register information. For example, doctors have to manually register when meeting a patient in a planned visit and for how long time. This system also makes it difficult to detect deviations in the reported information and what has happened in reality.

The production of health care at the hospital depends on how the professionals are scheduled to work, as patients have planned visits to a specific department or visit the emergency department. This is a way of production, capacity planning, and dimensioning that is based on old traditions. Different amount of time is devoted to scheduling the staff, depending on how experienced the planning staff is in identifying shortcomings. The staff should have knowledge to schedule their own work, but sometimes a manager has to decide. Based on experience, the number of patients, and how their medical needs have to be followed up, the situation is monitored based on what is working and what is not working at each department. Managers want to follow up the daily work in order to analyze the performance, but this is only done occasionally based on knowledge about a normal situation. Since there are no routines for documenting the work performance, analyses and follow-ups are not documented.

Forecasts are reviewed once per year. The health-care agreement specifies a number of admissions; a number which is independent of the number of average admission days. The number of admissions is monitored and communicated to the decision makers. The use of resources cannot be estimates based on the specified number of admissions, since admissions can last from 1 day to a few months. This situation puts pressure on physicians to discharge patients before they are fully investigated. Activities related to patients with planned visits to specific departments need to be considered and planned to ensure a smooth workflow with appropriate capacity each day.

Readmissions do not take into account on what medical basis the patient was previously admitted, and there are no regulations of how readmissions are defined at or between hospitals. The health-care professionals have to use their own knowledge and experience in order to share knowledge about the medical status of the patient, and if the patient seems to be readmitted, or admitted for the first time. There are also daily meetings about the current situation among all the departments. Each discharge of a patient needs to be planned right from the time of the admission. Since many patients are frail elderly who need interventions from the municipality health care and the primary health care at home, other health-care providers need to be involved in planning the discharge.

The departments within each area have one daily meeting where the managers and coordinators meet for their common planning. It is important to exchange information about the discharge of patients, in order to reach the goals for the outflow. Both the common inflow and the common outflow are important to monitor within each area. Especially, the flow of patients at a medical elderly health-care department is important. At the same time, as a patient is admitted, the staff has to plan the discharge of the patient, and therefore it is important to quickly estimate the number of admission days for the patient. The coordinators also have contact with the physicians in order to allocate the patients between the different departments depending on their medical needs. Whether the patient is still admitted at a department related to the medical needs is also followed up after 1 day. Knowledge about the patients, their medical status, and number of admission days, is continuously shared. Knowledge about the status of the patients is also shared with nurses and other professionals in the municipalities where the patients live in order to plan for supporting them in their homes after discharge.

Improvement work, for example, regarding more effective working methods, is primarily initiated by department managers. Ideas often come from the staff. The ideas that come directly from the professionals are often the best ideas. A lot of brainstorming takes place at department meetings, and there are well-established routines and professional skills for how to transform the organization and achieve knowledge diffusion.

4.3 Business model for knowledge capitalization

The middle managers normally follow up on the statistics on a monthly basis. This could, however, in some situations be changed. Especially, if changes have been made, follow-ups could be done more often. The departments follow up on both a daily and a weekly basis, especially the number of discharges. Three different IT systems are used to follow up. In the case of warnings, or if the staff recognizes bottlenecks, analyses of what is happening and follow-ups will be conducted. Of course, all follow-ups result in some learning, but the knowledge is not diffused and implemented in the organization in a predefined way. However, the analyses further guide how the capacity is dimensioned. People at all levels, from the political and management levels, to the staff level at the hospital, who all naturally want to contribute to developing and improving the health care, get support from the data that the IT systems provide. Especially, the internal work is much in focus, to avoid queuing of patients.

The regional administration orders the amount of health care that should be produced at the hospital. The provision of health care is measured by the number of admission sessions rather than the number of admission days. This indicator does not say much about the resources needed and used, because each admission session can vary considerably in length. This indicator also makes it very difficult for the hospital managers to follow up on the situation, and it provides little information that is relevant for the dimensioning of resources and capacity. To make sure the admitted patients receive effective health care and improve the economic figures, it is important to plan the discharge already when the patients are admitted. Implicitly, this means that the hospital gains financially if the number of days per admission is low, and the number of readmissions increases instead. However, this also increases the risk for low quality of the health care. For example, a low degree of readmissions of frail elderly patients shows that the hospital has successfully transitioned the patients to other health-care providers close to patients' homes. It is unclear what measure to use to estimate resource use: number of admissions, or average admission time. This will in turn affect how long time the patients are admitted at the hospital, and the number of admissions says very little about the need for, and use of, resources.

A lot of work is dedicated to decreasing the number of visits at the emergency department. The work involves a lot of collaboration with the ambulance department, the primary health care, the municipality health care, the special department for frail elderly people, and the department for guidance of people via telephone. All those collaborative activities are believed to affect the inflow of patients in a positive way, so that fewer patients seek emergency health care at the hospital.

Information from the registered data is used by politicians and decision makers at the hospital. Politicians use the data to get knowledge about the health care in order to allocate money to different areas and specialties. The health-care departments use the data to get knowledge about their performance, in order to develop and improve the health-care activities and processes. Especially, readmissions of patient are followed up, in order to get knowledge about to what extent the transitions to other health-care providers are successful. At the moment, the departments follow up their performance once a week, and the politicians intend to follow up once a month.

The top managers at the hospital scan the statistics a few times per year. Once per year, they decide how many patients each department could take care of, for each group of disease. In that way, they choose how to distribute the health-care beds and other resources among different patient groups. The total number of patients who are considered in the dimensioning is decided at the regional level, as a total budget for production of health care at each hospital. This dimensioning then affects whether patients need to be rejected or referred to another health-care provider, maybe to another health-care level, such as a primary health-care center.

4.4 Summary of results

For each of the three themes, the key results are presented in **Table 1**.

Learning and knowledge creation culture	• Desire to have performance data in real time to learn and take appropriate actions • Planning and following up part of the daily management • Ad hoc-based recognitions of changes can cause a reassessment of resources • Staff adapts to changes in admission of patients • A fixed number of operating rooms based on availability determines the resources • A need to manage required resources better • Historical numbers are used for the emergency department • Coordinators from different departments meet to exchange knowledge about inflow and outflow • A fixed set of standardized test packages is sometimes used for each group of disease • Desire to have knowledge to match the staff schedule with the needs • Knowledge of shortcomings could save a lot of resources • Oral analyses at departments on a daily basis to learn and create knowledge • No formally structured learning and knowledge creation
Organizational architecture for adaptive and exaptive capacity	• Many different IT systems are used in parallel and in different ways • Excel sheets are used for scheduling and planning • Manual routines that the staff have to be reminded of • Difficult to rely on manually registered data • Dimensioning of resources is needed for planned visits • Scheduling needs to be done by experienced staff • Follow-ups on an occasionally basis • Analyses and follow-ups are not documented • No routines for documenting work performance • Number of admissions in the care agreement, but each admission can vary in number of days • Collaboration with other health-care providers is needed for frail elderly patients • Monitoring of inflow and outflow of in-patients • Professionals share knowledge about the status of patients at daily meetings • Brainstorming for improvements is conducted at department meetings
Business model for knowledge capitalization	• Middle managers follow up on the statistics at least on a monthly basis • Department managers follow up on a daily and weekly basis • In the case of warnings, follow-ups and analyses are conducted • Analyses guide the dimensioning of capacity, especially to avoid queuing of patients • An ordered amount of health care should be produced • The number of admissions says very little about the resources needed • Unclear how to estimate resource use: based on number of admissions or average admission time • One goal is to decrease the number of visits at the emergency department • Politicians use statistics to allocate money to different health-care areas • The total number of patients is decided at the regional level • Departments use statistics to improve their activities and processes • Knowledge is not diffused in a predefined way

Table 1.
The three themes analyzed in the study and the results.

5. Discussion

Different professionals have roles and work with management, creation, and sharing of knowledge within the planning of production and capacity, as part of the operations at the hospital. A complete and overall articulated architecture for this work would be fruitful. As each department now has its own way of how to conduct the planning and knowledge work, a common architecture could better support the collaboration and coordination between departments. Obviously, different departments have their own characteristics that should be considered, but an architectural base could be designed and implemented at the hospital. The departments could then adapt the architecture to the extent that is needed at each department or health-care service unit based on their specific requirements. KM and OM could in this way be smoother and more effective, and better support the overall processes that support health-care work. Even though KM processes in operations is examined in the literature, see for example Fugate, et al. [18], the infrastructure for KM in health care does not seem to be fully analyzed in existing research, except in studies related to infrastructure for health-care systems and telecommunication, see for example Von Lubitz, Patricelli and Palma [41]. KM depends on the availability of both work routines and staff, to have defined processes that could fit into the knowledge-based architecture. Now, there are variations in the planning among the different departments from ad hoc-based and resource-based KM and OM to fixed plans and experience-based planning [20].

With an appropriate KM architecture in place, the OM can be smoother and more effective, as the planning of resources and capacity can be adjusted closer in time to when different health-care processes are conducted, and more detailed and correct information can be used to plan the capacity. This will create conditions for more flexibility and give better compliance between supply and demand of health care [7]. It is also important to establish routines for learning and knowledge creation within and between the departments and the health-care service units. The emphasis on cooperation, collaboration interaction, and work within groups of individuals can be seen as crucial for creation of knowledge [42].

It will be challenging to design and implement an overall infrastructure that could function as a basis for making changes in the learning and knowledge creation processes and support the OM. Changing the planning from a scheduling perspective, to a perspective based on the patients', often emergent, needs of health care, requires an OM based on "just-in-time." This also requires coordinated planning on a horizontal level, where knowledge can be integrated and shared, in order to plan a smooth overall health care. The managers, with the responsibility for planning and managing the health care, need to have the right information and knowledge and access to staff with competence and experience in production planning in order to design the health-care processes and related activities in detail. The practical and managerial implications are summarized in **Table 2**.

Practical implications	• Formal structures for learning and knowledge creation could save resources • Documenting analyses could support future learning • Documenting follow-ups could improve the planning and dimensioning of resources • Following up on the appropriateness of activities can make health care more effective • Integrating IT systems could increase data quality
Managerial implications	• Performance data in real time could make the planning smoother • Use of more reliable data could improve the planning and dimensioning of capacity and staff

Table 2.
Practical and managerial implications.

6. Conclusions

This chapter contributes with a qualitative analysis of infrastructural KM mechanisms that have significant impacts on the OM in a health-care practice at a hospital.

A learning and knowledge creation culture is important for the use of knowledge that is prevalent in the health-care organization. Such a culture could improve the real-time planning processes and better guide how the professionals act. Deviations could be documented, and learning and knowledge creation could be based on these experiences. The organizational architecture for adaptive and exaptive capacity supports the structure for information systems, technologies, and practices. Integration of these parts could improve the planning of capacity and make it smoother and more effective, as real-time information could also be used. The regular improvement work could also be supported by changing practice to increase the use of information systems for dimensioning of resources. A business model for knowledge capitalization is important in order to derive benefits from the knowledge in the health-care organization. Knowledge diffusion and collaboration in planning with other health-care providers will impact the overall dimensions of health care. This could also provide smoother and more effective health care in the society, which does not use more resources than necessary and is of high quality.

For future research, we suggest to further study the OM in health care and to analyze the complex system of processes and resources that is needed. It is of vital importance to have detailed and reliable data that could be aggregated to make better decisions in the OM. To develop methods for OM is important under normal conditions and even more important in pandemic situations, when the capacity of health care is under great pressure. In this complex system of OM, AI can also be an interesting topic to study.

Author details

Ann Svensson[1]* and Eva Hedman[2]

1 University West, Trollhättan, Sweden

2 NU Healthcare, Trollhättan, Sweden

*Address all correspondence to: ann.svensson@hv.se

References

[1] Hans EW, van Houdenhoven M, Hulshof PJ. A framework for health care planning and control. In: Handbook of Health Care System Scheduling. US: Springer; 2012. pp. 303-320

[2] Ekdahl AW, Andersson L, Friedrichsen M. "They do what they think is the best for me." frail elderly patients' preferences for participation in their care during hospitalization. Patient Education and Counseling. 2010;**80**(2): 233-240

[3] Svensson A. Identifying motives for implementing eHealth by using activity theory. Sustainability. 2020;**12**(4):1298

[4] Brandeau ML, Sainfort F, Pierskalla WP, editors. Operations research and health care: A handbook of methods and applications. In: International Series in Operations Research & Management Science. Vol. 70. Dordrecht, The Netherlands: Kluwer Academic Publishers; 2004

[5] Gemmel P, van Dierdock R. Admission scheduling in acute care hospitals: Does the practice fit with the theory? International Journal of Operations & Production Management. 1999;**19**(9):863-878

[6] Vissers J, Beech R, editors. Health Operations Management: Patient Flow Logistics in Health Care. London and New York: Routledge, Taylor & Francis Group; 2005

[7] Hulshof PJ, Kortbeek N, Boucherie RJ, Hans EW, Bakker PJ. Taxonomic classification of planning decisions in health care: A structured review of the state of the art in OR/MS. Health systems. 2012;**1**(2):129-175

[8] Alvekrans A-L, Lantz B, Rosén P, Siljemyr L, Snygg J. From knowledge to decision—A case study of sales and operation planning in health care. Production Planning and Control. 2016; **27**(12):1019-1026

[9] Allder S, Silverster K, Walley P. Managing capacity and demand across the patient journey. Clinical Medicine. 2010;**10**(1):13-15

[10] Walley P. Does the public sector need a more demand-driven approach to capacity management? Production Planning and Control. 2013;**24**(10–11): 877-890

[11] Walley P, Silvester K, Steyn R. Knowledge and behavior for a sustainable improvement culture. Healthcare Papers. 2007;**7**(1):26-33

[12] Eriksson H. Reducing queues: Demand and capacity variations. International Journal of Health Care Quality Assurance. 2011;**24**(8):592-600

[13] Allder S, Silverster K, Walley P. Understanding the current state of patient flow in a hospital. Clinical Medicine. 2010;**10**(5):441-444

[14] Silvester K, Harriman P, Walley P, Burley G. Does process flow make a difference to mortality and cost? An observational study. International Journal of Health Care Quality Assurance. 2014;**27**(7):616-632

[15] Massaro M, Dumay J, Garlatti A. Public sector knowledge management: A structured literature review. Journal of Knowledge Management. 2015;**19**(3): 530-558

[16] Ayatollahi H, Zeraatkar K. Factors influencing the success of knowledge management process in health care organisations: A literature review. Health Information and Libraries Journal. 2020;**37**(2):98-117

[17] Nicolini D, Powell J, Conville P, Martinez-Solano L. Managing

knowledge in the healthcare sector. A review. International Journal of Management Reviews. 2008;**10**(3): 245-263

[18] Fugate BS, Stank TP, Mentzer JT. Linking improved knowledge management to operational and organizational performance. Journal of Operations Management. 2009;**27**: 247-264

[19] Karltun A, Sanne JM, Aase K, Anderson JE, Fernandes A, Fulop NJ, et al. Knowledge management infrastructure to support quality improvement: A qualitative study of maternity services in four European hospitals. Health Policy. 2020;**124**(2): 205-215

[20] Svensson A, Hedman E. Knowledge management for operations management within health care. In: Proceedings of the 19th European Conference on Knowledge Management. Padova, Italy: Academic Conferences International: Sept 5-7. 2018

[21] Aboelmaged MG. Linking operations performance to knowledge management capability: The mediating role of innovation performance. Production Planning and Control. 2014; **25**(1):44-58

[22] Cardoen B, Demeulemeester E, Beliën J. Operating room planning and scheduling: A literature review. European Journal of Operational Research. 2010;**201**(3):921-932

[23] Dexter F. Bibliography of Operating Room Management Articles. 2012. Available from: http://www.franklindexter.com/ [Retrieved: 10 August 2020]

[24] Noon CE, Hankins CT, Côté MJ, Lieb M. Understanding the impact of variation in the delivery of healthcare services. Journal of Healthcare Management. 2003;**48**(2):82-98

[25] Vanberkel PT, Boucherie RJ, Hans EW, Hurink JL, Litvak N. A survey of health care models that encompass multiple departments. International Journal of Health Management and Information. 2010;**1**(1):37-69

[26] Svensson A. Kunskapsintegrering med informationssystem i professionsorienterade praktiker [doctoral thesis]. Göteborg university. 2012

[27] Uhl-Bien M, Marion R, McKelvey B. Complexity Leadership Theory: Shifting Leadership from the Industrial RAge to the Knowledge Era. Lincoln: Leadership Institute Faculty Publications, University of Nebraska; 2007

[28] Liang TY. Innovative sustainability and highly intelligent human organization (CAS): The new management and leadership perspective. International Journal of Complexity in Leadership and Management. 2010;**1**(1):83-101

[29] Gold AH, Malhotra A, Segars A. Knowledge management: An organizational capabilities perspectives. Journal of Management Information Systems. 2001;**18**(1):185-214

[30] Chuang H-S. A resource-based perspective on knowledge management capability and competitive advantage: An empirical investigation. Expert Systems with Applications. 2004;**27**(3):459-465

[31] Svensson A. Challenges in using IT Systems for collaboration in healthcare services. International Journal of Environmental Research and Public Health. 2019;**16**(10):1773-1784

[32] DeSanctis G, Poole MS. Capturing the complexity in advanced technology use: Adaptive structuration theory. Organization Science. 1994;**5**(2):121-147

[33] Loon M. Knowledge management practice system: Theorizing from an

international meta-standard. Journal of Business Research. 2019;**94**:432-441

[34] Merriam SB. Fallstudien som forskningsmetod. Lund: Studentlitteratur; 1994

[35] Yin RK. Case Study Research: Design and Methods. London: Sage Publication Inc; 2014

[36] Myers MD. Investigating information systems with ethnographic research. Communication of the AIS. 1999;**2**:23

[37] Myers MD, Newman M. The qualitative interview in IS research: Examining the craft. Information and Organization. 2007;**17**:2-26

[38] Flyvbjerg B. Five misunderstandings about case-study research. Qualitative Inquiry. 2006; **12**(2):219-245

[39] Pope C, Mays N. Qualitative Methods in Health Research. Oxford: Blackwell Publishing Ltd; 2006

[40] Braun V, Clarke V. Using thematic analysis in psychology. Qualitative Research in Psychology. 2006;**3**(2): 77-101

[41] Von Lubitz D, Patricelli F, Palma P. Telecommunications infrastructure for worldwide networkcentric healthcare operations and the associated information system. International Journal of Business and Systems Research. 2008;**2**(1):34-51

[42] Newell S, Robertson M, Scarbrough H, Swan J. Managing Knowledge Work. London: Palgrave Macmillan; 2002

Chapter 13

Artificial Intelligence and Bank Soundness: A Done Deal? - Part 1

Charmele Ayadurai and Sina Joneidy

Abstract

Banks soundness plays a crucial role in determining economic prosperity. As such, banks are under intense scrutiny to make wise decisions that enhances bank stability. Artificial Intelligence (AI) plays a significant role in changing the way banks operate and service their customers. Banks are becoming more modern and relevant in people's life as a result. The most significant contribution of AI is it provides a lifeline for bank's survival. The chapter provides a taxonomy of bank soundness in the face of AI through the lens of CAMELS where C (Capital), A(Asset), M(Management), E(Earnings), L(Liquidity), S(Sensitivity). The taxonomy partitions opportunities from the main strand of CAMELS into distinct categories of 1 (C), 6(A), 17(M), 16 (E), 3(L), 6(S). It is highly evident that banks will soon extinct if they do not embed AI into their operations. As such, AI is a done deal for banks. Yet will AI contribute to bank soundness remains to be seen.

Keywords: bank, bank soundness, financial sector, Artificial Intelligence (AI), CAMELS

1. Introduction

The Global Financial Crisis (GFC) showcased that banks driven solely by profit, earnings and share price maximisation alone would fail. Although, success or failure of banks is highly dependent on the bank's ability to make money, it is not the only determinant of bank soundness. Amongst the equally important success factors lies in adequate capital and liquidity holdings, quality assets and making sound management decisions, that leads to the creation of value. As such, Capital (C), Asset (A), Management (M), Earning (E), Liquidity (L) and Sensitivity (S) (CAMELS) are important determinants of bank's health and wellbeing [1–3].

AI is expected to deliver additional global economic output of $13 trillion a year [4]; contribute $15.7tr. to the global economy by 2030 [5] and is expected to increase productivity gains by 20–40% [6]. Several initiatives have emerged as a result where approximately $1 trillion in costs is expected to be exposed to AI transformation in financial services sectors by 2030, out of which $450 million of this is in banking [6]. While, European Commission has increased its annual investments in AI by 70%. AI market is expected to be worth $16.06 billion by 2022, growing at 62.9% compound annual growth rate [7].

The numerous efforts and initiatives in AI investment suggests that AI is here, and it is here to stay. As such, the chapter looks to critically assess how able are banks to effectively deploy AI into their daily operations to improve CAMELS from a bank's perspective. The chapter discusses opportunities proposed by AI that could influence bank soundness.

The chapter contributes to literature in several ways. The earlier researchers have emphasised on the application of AI on the financial sector as a whole [6, 8, 9] or comparative analysis of AI applications in specific areas as service providers such as credit evaluation, portfolio management, financial prediction and planning [10–13] or by examining customer experience [5, 6, 8]. Therefore, these studies are not sufficient to understand the opportunities proposed by AI from a bank's sole perspective. To fill this gap the chapter has taken a holistic approach in scrutinising the opportunities relished by banks solely in deploying AI. By doing this the chapter provides a significant insight into the important opportunities that AI technology can offer the banking industry to ensure its survival. The chapter also further considers bank soundness with the application of AI from various aspects of Capital (C), Asset (A), Management (M), Earning (E), Liquidity (L) and Sensitivity (S) (CAMELS) determinants of bank soundness. To the best of our knowledge, this chapter is the first reviewing deployment of AI in banking operation in light of CAMELS. Earlier research [1–3] has only emphasised on bank soundness from the CAMELS perspective. The chapter also more specifically focuses from both the service provider and customer end, providing further insight from a holistic perspective. Most importantly, the intention is to examine through the lens of CAMELS how sound are banks having applied AI into their processes.

The chapter is organised as follows: the next section presents a brief theoretical discussion on the importance of embracing AI. Section three introduces the literature gathering and research method. Section four presents result and discussions on the opportunities relished by banks on application of AI from CAMELS perspective. The last section concludes the chapter and highlights insight on further research.

2. Literature review

The banking sector is the heartbeat of the economy. Yet, despite central banks efforts to keep banks afloat by recapitalization, cash injections amongst other measures banks are still underperforming, failing, with one or two microfinance banks disappearing annually [14]. On top of this, The GFC worsened the situation by causing many more bank failures leading to concentration in the financial markets. To promote economic growth governments lowered entry barriers to encourage more players to enter and stimulate competition in the financial markets. In UK, new banks could enter the market with reduced capital and liquidity requirements [8]. This led to the growth of Fintech, technology-based companies that offer financial products at competitive rates. As well as Challenger banks whose competitive advantage lies in its digital technology build on Machine

Learning (ML) outperforming UK's five big banks through fierce competition and subsector domination in the field [8, 15]. Studies have confirmed the sluggishness and weaknesses in the banking industry rest in the banks' inability to tap into AI solutions. Thus, the biggest game-changer for banks lies in its rapid adoption of AI technology. AI is a competitive advantage for banks. As it not only helps banks to remain competitive, but also to fight off weak profitability [5, 16–18]. As such, AI is no more an enabler or enhancer of productivity but a necessity that ensures survival and sustainability.

The incoming and future customers of banks are Millennials and Generation Z. These generations are more in tuned to technology-based services, and thus, demand more choices, flexibility, and control over banking. As such, banks need to embed AI into their operation to cater for the 21st century customers' expectations offering a range of services, in seconds, 24/7.

Banks also work with large volumes of data. As such, it is inhumanly impossible to process, find patterns, make fast and accurate decisions in a timely manner. AI on the other hand, is capable and has the capacity to conduct the job effortlessly in real time with increased data storage at a lower cost. The constant advancement in AI technology is also enhancing AI capabilities and capacities making it limitless [9] enabling banks to offer extraordinary services to its customers.

The GFC and the opaqueness of the banking industry has led to increased scrutiny and regulation on banks. This makes digital platforms a necessity. Digital platforms ensure all data are consistently and systematically recorded in a logical and meaningful manner, making processes more transparent, increasing the reliability and confidence in the banking system [5, 12].

It is apparent that banks cannot exist without the help of AI in moving forward.

3. Research methods

The research is conducted as a conceptual chapter with the aim to provide a deeper understanding of the opportunities parted by AI from a service provider and customer perspective. To answer the research question on how able banks are to effectively deploy AI into their daily operations to improve CAMELS from a bank's perspective, a systematic review of the literature and objective observations were undertaken to examine banks through the lens of bank soundness determinants of CAMELS. The observations found in existing literature are gathered to assemble a framework categorized by CAMELS **Figure 1**. The literature was gathered through the Scopus database as a main source of finding existing literature. The database offers a wide range of management and business-related studies relevant for the topic of research. In addition, other databases such as Google Scholar, Social Science Research Network (SSRN), SpringerLink and IEEE Xplore were also examined. Journal articles since the period 2000–2020 were extracted using the prescribed keywords of Bank, Bank soundness, Financial Sector, Artificial Intelligence (AI), CAMELS. Only articles that were available in full text, published in scholarly, peer reviewed journal were chosen to be closely examined. The search was also conducted using the backward and forward approach where reference list of articles was utilised to find further research papers.

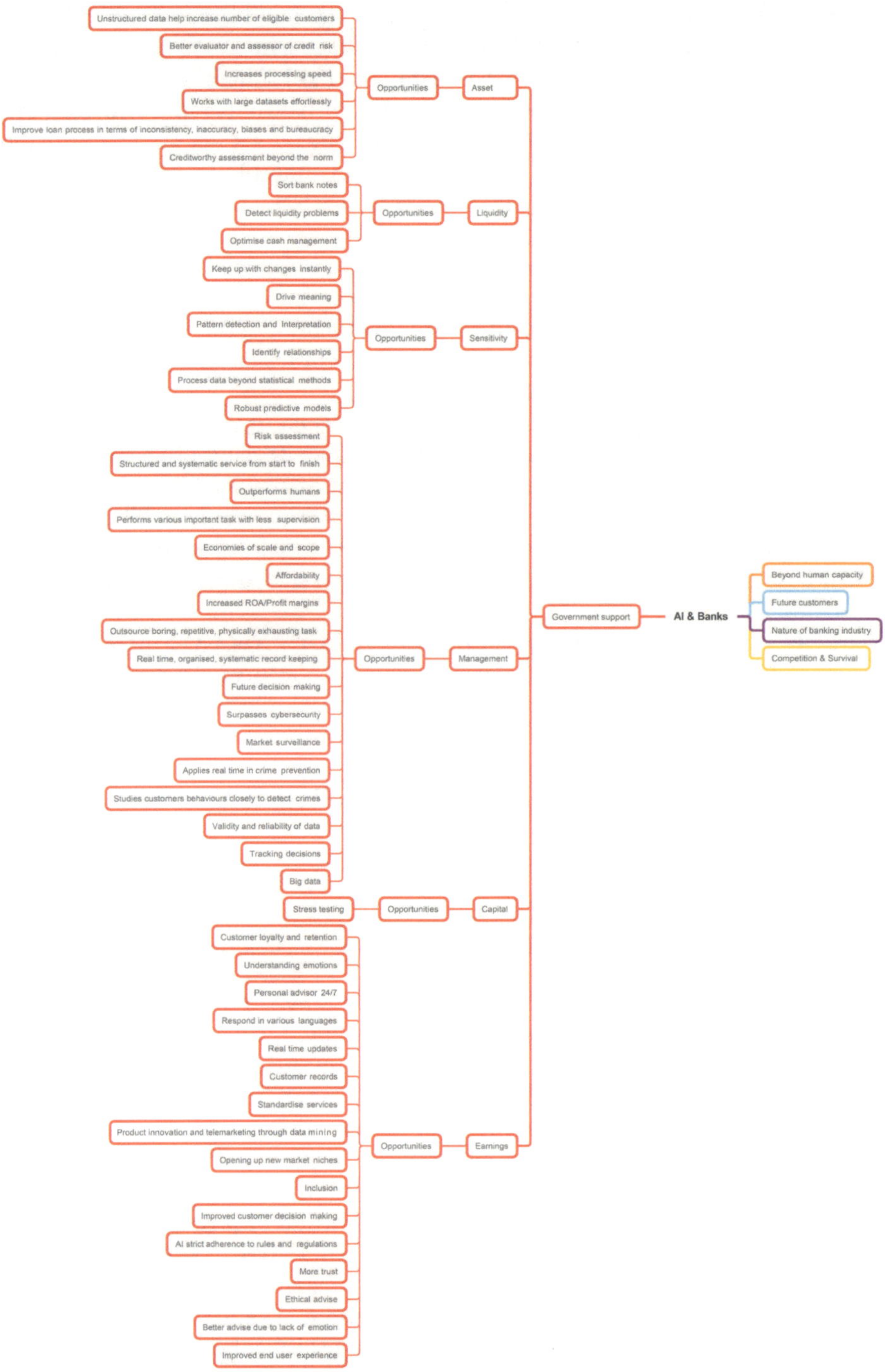

Figure 1.
Taxonomy of opportunities posed by AI on Bank soundness - a classification based on the determinants Bank soundness of CAMELS.

4. Findings and discussion

This section presents an overview of the opportunities relished by banks in deploying AI in their daily back office operations to customer services prescribed from the CAMELS perspective (see **Table 1** in the Appendix).

4.1 Capital

Bank capital acts as a core determinant for bank's survival. Capital absorbs losses during adversity and insufficient capital holdings can cause banks to collapse. AI with its limitless abilities and capabilities helps banks to hold robust capital holdings through stress testing.

Basel requires banks to demonstrate their ability to remain adequately capitalised especially during dips in the economy, stressful scenarios and most importantly during crisis. As AI works with big data, real time, real world scenarios, it is able to help banks immensely with detail capital-planning processes ensuring its robustness and forward looking. Citigroup successfully applies ML techniques to model Comprehensive Capital Analysis and Review (CCAR), thus, meeting its stress testing requirements [19–21]. ML algorithms could also project CCAR losses [21].

4.2 Asset

Asset quality is measured by the level of credit risk contained in bank's assets [22]. Therefore, a bank that can detect, measure, monitor and regulate credit risks will hold higher quality assets [23]. The GFC showcased that credit risk is the most challenging risk to manage and control as it not only absorbs profits but exposes banks to failures as well. AI helps banks to clearly assess and evaluate customers' risk, eliminating ambiguity, biasness while improving loan processes and.

As most finance related data are text heavy it can be a challenge to track data systematically and consistently. Intelligent Trading Systems screen both structured (databases, spreadsheets, etc.), unstructured (social media, news, etc.) data [12] and apply ML with improved analytics and data modelling [24, 25] to offer meaningful insights on the ability of customers to pay back their loan undertakings. Studies have also confirmed that the use of unstructured data can boost the number of eligible loan customers [9].

Banks apply credit scoring to issue loans, make investment and risk management decisions. As credit risk is evaluated through credit scoring, the accuracy of credit scoring is necessary for bank's earning as even a 1% improvement in the accuracy of prediction could lead to significant decrease in losses to financial institutions [26]. As AI supersedes traditional statistical scoring models with its ability to work with big data, nonlinear relationships, improving accuracy of prediction [26, 27] and thus, a better evaluator and predictor of credit risk, reducing significant losses from non-performing loans [26, 28–30]. ML measures credit risk detect patterns in data identifying and notifying banks on good and bad borrowers, [30, 31] detect high-risk loan applicants [32], identify customers that are being over/under charged for credit risk paying higher/lower credit risk premiums to gain market share/reduce losses. These customers can be offered lower price loans/denied loan request [15]. AI applies logical deduction to differentiate between high default risk applicants and credit applicants who are credit-worthy but do not have a comprehensive credit history [12].

Advance machine learning algorithms with an increase in number of transistors, increased computing power, improved speed of central processing units as well as increase in processing power has enabled algorithms to analyse data and process information much faster, improving processes, enabling faster loan origination, and contributing to the accuracy in decision making [9, 17, 25]. Besides, Robo-advisers and online broker community combined efforts have helped to further quickened the credit checks and loan evaluation processes [8]. JP Morgan Chase uses Contract Intelligence (COIN) to interpret commercial-loan agreement faster as it takes lawyers and loan officers 360 000 hours each year [15, 33] and is also able to extract 150 relevant attributes from commercial loan agreements quickly [21]. Bank of America and Merrill Lynch apply ML through Intelligent Receivables Solution to improve reconciliation of incoming payments that help post its receivables faster [21].

Credit scoring is evaluated based on 5Cs: the character of the consumer, capital held, collateral offered, the capacity of the customer and the economic conditions. However, the rise in number of applicants, makes it impossible to conduct the work manually for each and every customer [26]. AI in the form of ML not only can perform challenging tasks such as sifting through floods of data from repayment history to collaterals, but also simultaneously process, understand, discover useful patterns or relationships, extracting valuable information and analysing large volumes of information from large datasets, making right loan decisions based on customer's affordability, at record speeds, accurately, in a fair and objective manner [8, 9, 13, 15, 30]. Thus, improving quantity and quality of loan approvals.

Manual underwriting task requires full attention to detail, sound judgement from each and every department to produce a quality outcome. Loan officers rely on their own experiences, experiential knowledge, intuition, checklists of bank rules and conventional statistical methods to evaluate loan applications. Yet, loan officers are humans tied to relatives and friends and are subject to emotions. As loan decisions are subjective, loan judgements can be inconsistent, inaccurate, or bias. AI can train with new examples and learn from past experience to provide better outputs, reducing subjectivity [30] and biasness in decision making. For example, a loan officer who is absolute that the results will hold a linear relationship or assumes an incorrect functional form [30]. Robo-advisers and online broker community combined efforts have also known to have shortened bureaucratic processes [8]. As such, AI can improve decision making process enhancing the outcome [30, 34]. However, in exceptional circumstances, it is important to include a human in the loop [30, 34]. In summary, big data, cloud and the advancement in AI is helping banks to be more efficient, effective, effectual, and efficacious in their operations, delivering world class services to customers.

The introduction of AI has allowed banks to analyse customer's loan repayment ability beyond traditional evaluation techniques of ratios, credit history and credit scores [8]. AI tracks customer spending habits through their shopping patterns, social media activities, internet searches, as well as, customer's holidays, hobbies, interests, job related searches, connections, social activities, location consistency, network diversity...etc. [8]. This approach is particularly useful for customers with insufficient credit records. The new approach has successfully garnered 90% repayment rate [8].

4.3 Management

Banks rely heavily on management to not only generate earnings and increase profit margins [2] but also to keep banks alive [35]. AI helps banks to be more efficient, effective, effectual, and efficacious.

AI is used in risk management by designing algorithms that analyse the history of risk cases and identify early signs of potential future issues [12, 36].

AI applies structured and systematic approach in making decisions by ensuring that customers have enough understanding of investments and the mechanisms of the financial markets in general first before learning about the customer's circumstances namely number of dependencies, type of profession, assets, and liabilities etc. Then questions related to investment, namely target return and risk levels to determine the optimal risk–return level of the investment portfolio. The customer respond is then checked against the application picking up inconsistencies and gaps are filled by asking customers to provide further explanations where and when is necessary and classifying customers into different categories based on their risk–return profile and finally prescribing the right investment fund best suited for the customer. AI then provides after care service of continuous management of the portfolio while constantly touching base with the customer on the market changes and variations of asset features and its effects on the customer's financial position [12, 37]. This keeps customers up to date about their finances from start to finish.

AI takes longer to learn than humans, but it is constantly learning from its previous decisions to improve accuracy and performance overtime. As such, in the long run AI outperforms humans proving to be more effective and efficient. AI is consistent, does not easily get distracted, able to multi-task as machines can think faster, operates with higher degree of precision and accuracy, programmed to work for long hours providing non-stop services i.e., automated teller machine that able to replace the working hours of bank clerk and auto call centres that able to pick up calls at any time [8, 12, 15, 17, 37, 38]. Studies show that only 25% of human fund managers can outperform the market. Fund managers need to constantly keep up with unpredictable markets, constant flux in regulations, customers want, amongst other factors that influences decision making process [8]. While AI works with real time instantly using up-to-date information of all market activities when deciding on an investment advice [12].

AI is self-taught as it can acquire knowledge through observation and analysation, draw conclusions and take appropriate actions autonomously. AI also performs various task with little human supervision, namely ETFs, document reading through natural language processing is used to interpret written inputs, such as research documents, news stories and social media content, to create analyses that can be used to identify investment opportunities, simultaneously considering human behaviour [12, 37, 39], applying machine learning to automate data preparation, able to analyse complex task to develop insight and share [17, 25, 40]. AI is said to be revolutionising the investment industry through the application of quantitative optimisation. Where no human intervention is needed in managing customer's portfolios [12].

Although AI requires high initial investment but when the cost of developing the system has been met, the marginal cost hereon related to each new transaction is relatively low [41]. Further advancement in the field of AI have lowered the marginal cost of acquiring, managing and analysing data further [8, 15, 26, 42]. As a result, the service provider benefits from economies of scale. As more customers adopt the service, unit cost per transaction reduces considerably [15]. For example, services that require little management such as ETFs [37] have reduced management, personnel, and asset costs. The reduced price enables banks to serve a bigger customer group with a wider scale of market segment, expanding its scope globally, creating steady revenue streams [12, 37]. On top of that, AI can perform excruciating, repetitive and work for long hours without having to pay out high remunerations [8].

AI is also successfully integrated in financial institutions because of its affordability. Combined reduction in cost of data storage, cost of hard drive per gigabyte, data identification tools such as machine learning, complete with big data input has led to the success of AI [17].

Studies have shown a positive and strong correlation between higher number of AI patents registration and bank profits [17]. There are no qualms that AI has significantly increased labour productivity as this is evident in European banks with pronounced return on assets (ROA), reduced cost, combating the persistent weak profits experienced by banks prior to AI adoption [17].

AI not only perform tasks which is difficult and impossible, but it also carries out tasks that no one enjoys, repetitive tasks which are monotonous in nature, costly, laborious, low added-value tasks (e.g., replies to FAQs). Thus, reducing human error, boosting productivity, and cutting the cost of these tasks [9, 15, 17, 38].

First National Bank of Wynne in Arkansas uses AI to migrate customer account information to the acquiring bank's core system. AI possesses all the necessary skills to complete the task from its ability to learn fast, perform repetitive task quickly without mistakes, report errors to data validation. to complete the task on time effortlessly. The bank saved 70% conversion costs. AI not only helped the bank to complete the task on time and effortlessly but also with less cost.

Banks outsource repetitive task such as mass reissue of debit cards due to a breach [6] data entry, filling out forms, answering what if questions, to AI. This helps banks to reduce the demand for less-skilled labour, helping existing workers to focus on more interesting and creative tasks. 67% of nonexecutive bank employees believe that AI will improve their work-life balance while 57% expect it to expand their career prospects [6]. Banks could also put more effort to focus on the remaining bank staff by training them to improve the efficiency and effectiveness of the bank [17]. AI Debt-collectors handle loan collection work for most banks. As AI lack emotions AI can handle abusive behaviour without providing an emotional response such as getting upset, angry or misbehave, avoiding tense situations [6]. Customers can let off stim and yet not jeopardise their relationship, paying up subsequently with a patient and understanding AI.

AI ensures all information especially legal documents and requirements of collecting, processing, utilising, and categorising information is done in a systematic way and documented correctly, making it more efficient than human advisors [12, 37]. The automatic, real time record keeping ensures proper record keeping, improves transparency compared to traditional processes that are highly administrative, bureaucratic, require repetitive data entries and unnecessary paperwork [14, 17].

AI automated recorded processes helps to easily track reasons behind each decision made. AI provides consistent recommendations in a systematic, structured, and logical manner. Enabling easy understanding of the decisions made.

AI is connected to cloud. As such it can learn new things constantly and update itself but most importantly to store information cheaply in cloud for future analysis. Research predicts that increase adoption of AI in banks will allow large volumes of data to be preserved for data-driven decision making in the future [17].

Strong growth in online and mobile payments coupled with an increase in cyberattacks has forced banks to adopt AI. An increase in number, speed and complexity of the dynamic cyber-attacks carried out by intelligent agents such as computer worms and viruses show that conventional fixed algorithms are futile and only an intelligent semi-autonomous agent (such as Computational Intelligence, Neural Networks, Intelligent Agents, Artificial Immune Systems, Machine Learning, Data Mining, Pattern Recognition, Fuzzy Logic, Heuristics, etc.) is capable enough to detect, analyse, evaluate the entire process of attack before providing an appropriate respond in a timely manner to defuse the attack as well as prioritising and preventing secondary attacks [43]. On top of that, AI surpasses cybersecurity with its extraordinary safety features and abilities. While cybersecurity helps to the number of successful attacks on the system, AI wipes out attacks completely to zero [44].

AI in the form of ML techniques works with real-time in checking the credibility of credit card transactions before comparing new transactions with previous amounts and locations. Transactions that pose a risk are blocked immediately by AI. Studies have confirmed AI's accuracy in these processes and thus works well to protect customer security [9, 17, 45].

UK Banks namely Santander, Barclays, RBS, and HSBC use voice recognitions for telephone-based customer service [8]. Studies have confirmed that banks have to introduce AI into their operations from the start at data input level to detect human involvement in cybercrimes, fraud, money laundering etc. AI seeks to chapter and understand human behaviour in identifying risk. AI being present at every stage of the process enables it to critically evaluate the data given, find unusual behaviour patterns outside a person's behavioural norm triggering alerts. AI with the aid of big data is able to analyse and filters through greater volume of customer transactions for anomalies, patterns or a series of scenarios including money laundering, illicit transactions, cybercrimes [46] and security threats faster and with more accuracy [8, 9] that would otherwise have gone unnoticed with lower number of false positives [38].

Global banks bear more than $230 billion in misconduct costs [47]. As such, ML helps to closely monitor traders' behaviour from trading patterns, email traffic, calendar items and telephone calls [45]. The U.S. Securities and Exchange Commission applies ML to extract actionable insights to better regulate market activities, facilitate automated security registration processes, and assess corporate risk. The London Stock Exchange in partnership with IBM Watson and cybersecurity firm SparkCognition apply a separate type of market surveillance [21].

AI works with real time data and constantly updates itself. It also works based on "know your customers" processes where learning and chaptering every detail of the customer. As such, AI algorithms can scan client documents to check the validity and reliability of the information provided by comparing it with information from the internet. If AI algorithms identify inconsistencies, they raise a red flag and a more detailed check by bank employees is performed [17].

Banks have the upper hand in large datasets [15], equipped with a strong back-up for ML, with the help of cloud storage, banks have a competitive advantage to make more decisions with higher accuracy.

4.4 Earnings

Banks that manage their expenses well while fully utilising their assets to generate constant revenue streams are most likely to be sound [2]. AI enables banks to offer unique selling points in products that increases customer satisfaction, boosting sales and revenue [8].

As customers in banks grow, opting for various services offered, with different periods of maturity it is difficult for banks to keep a close eye or to personally service each and every customer. On average banks handles five million transactions per day, five million individual accounts, over three million customers with hundreds of product types. As such most customers feel disconnected from their otherwise committed bank. Banks are starting to realise that reconnecting and rebuilding this relationship is essential not only to retain customers but most importantly to gather information about customers preferences to provide satisfactory product and services, garnering their loyalty [48].

AI is also said to be able to detect movement and understand human emotions while responding with simulations of emotions like joy, anger, and irritation [38]. AI is an indivisible virtual personal financial advisor that attends to individual customer's need twenty-four hours, seven days a week [8, 9, 12, 38]. Banks can offer unique, extraordinary, and personalised digital experiences with personalised products and services to their customers as a result [5, 8]. As AI monitors each and every customer's account individually, AI sends instant alerts on spending, account balance and budget alerts if geolocalisation services detect that customers are shopping. AI is also quick to act i.e., providing advice to stop customer from getting into further debt if it predicts that customer will get into financial troubles. AI is also on standby to offer advice on credit cards, ETFs (Exchange Traded Fund), stocks and bonds, fixed deposit, current and saving account, personal loan, retirement planning, mortgage, ISAs, savings accounts, insurance cover if the customer is buying a house, car or electrical household item [8, 12, 49].

Algorithms can also develop personalised portfolios based on individual customer's asset and liabilities, risk and return trade off and investment capabilities [12, 50].

Bank of America, Capital One, Société Générale, Swedbank, Royal bank of Canada together with big five Canadian banks, Danske Bank and Nordea bank in Nordic countries, Hong Leong Bank, RHB and CIMB in fast east Malaysia, Mitsubishi UFJ Financial Group Inc. and Mizuho bank in Japan are all active users of AI chat-bots [6, 8, 38]. Most sophisticated robo-advisory services originate from United States, United Kingdom and Germany [51].

Chatbots, virtual assistants and robo-advisors offer automated communication channels to customers in the form of personal financial advisory services. Provide a wider range of services, include answering simple and repetitive questions, handling lost or stolen bank cards, re-setting PIN numbers, perform transactions (e.g. order transfers, open accounts, paying bills, reload prepaid phones, checking account balance, interacting with customers, guide customers through disclosure and compliance questions more carefully than humans, collect data on customer preferences and interests to offer relevant, tailored and personalised products to customers, enable users to settle common doubts, secure product recommendations, schedules appointments via the phone according to the customer's preferred date and time, processes mortgage applications and insurance claims such as opportunity and risk alerts, push notifications for market updates, periodic portfolio reviews, alert customers within 24 hours via SMS on loan approvals, guides and support customers from start to finish through an automated financial advisory process empowers customers to independently make more informed consumption, saving, personal finance and wealth management decisions).

In other words, AI has enabled customers to access, receive guidance and advise, adjust their portfolios anytime of the day and recalibrate their investments empowering customers [8, 9, 12, 17, 37, 38, 52, 53]. Besides, Robo-advisers and online broker community combined efforts have enabling loan officers to provide a more detail guidance to each and every customer based on their individual circumstances [8].

Chatbots can interpret and respond in various languages. Thus, can resolve customers' routine questions saving human advisors' time. Increasing efficiency and effectiveness of the service provided to customers increasing client base [12, 39]. AI works with real time information and therefore informs and updates itself with the most up to date information and news from connected networks before revising customers finances, alerting, or advising them accordingly and immediately [12, 17, 37].

AI automatically records every conversation, interactions, information provided by customers. The automatic, real time record keeping on customers historical records to recent transaction made, advise given are structured and organised, useful when handling customer complaints and helpful to decision makers to make fast, prompt, and accurate decisions [12].

AI's automated processes offers a standard service to all customers, ensuring customers are treated similarly, fairly, and equally eliminating biasness [12, 37]. More than 70% of financial institutions in North America agree that big data offers them a competitive advantage over their competitors and 90% believe big data is the determining success factor for banks [17]. Large global banks are turning to data mining for insight on customers' preferences and choosing telemarketing clients to remain competitive in the digital world [17, 54]. Big data coupled with great analytical capacity of AI allows banks to track customer's spending habits, preferences, developing tailor-made products and services, and even anticipate customers' needs, improving user experience. Chatbots' interactions are also collated to customise end products and services [9]. As a result, new, innovative, and tailor-made products and services are developed that are useful, helpful, and better suited to client preferences in the future [17].

According to World Economic Forum's White Paper estimates that the global retirement savings gap, caused by longevity and insufficient savings, is projected to balloon to $400 trillion by 2050 in just eight countries with some of largest pension markets or biggest populations namely Australia, Canada, China, India, Japan, Netherlands, UK, and USA. Studies agree that AI can open a new niche for banks to explore by offering simple, cheap, and non-judgemental financial advice [8]. The emergence of AI has created new set of robo-advisors designed to follow a low budget structure [37] creating a new niche for low-budget investor class [37]. Studies have also confirmed that banks are in a strong position to venture into real estate business by adopting a virtual real-estate agent powered by AI. AI algorithms coupled with customer house preference information will narrow down searches on viable housing options to the customer. This will provide transparency of the house buying process to customers, improve customers' house buying experiences, increase customer satisfaction, reduce customer churn, reduce the banks operational costs and maintain its profits margins through mortgage loans and increase revenues from adjacent areas of sales such as insurances, personal loans, etc. [5].

The adoption of AI is allowing banks to offer wider range of services to a larger number of customers at affordable prices. Therefore, all customers are granted the same access to all financial services that might have otherwise been excluded (e.g. lending to customers with whom a bank had no previous relationship with or personalised financial advice services) promoting inclusion [9].

Lack of financial knowledge and overconfidence has led customers away from human advisors [55]. In Finland, approximately 76,6 billion euro are left in personal current accounts earning virtually no interest [12]. AIs are more open than humans offering a safe platform for customers to realistically understand their level of financial knowledge without the fear, discomfort, and embarrassment of sharing their vulnerability with human advisors [12]. In this sense, robo-advisors have helped customers who previously did not have any investments or savings before to increase their investment [12].

The GFC showcased the greediness, recklessness, and dishonesty of human decision makers in globally reputed banks namely RBS, Barclays, HBOS amongst others [8]. AI's absence of self-interest ensures rules, processes as well as procedures are strictly adhered to [8].

As AI does not work with an agenda i.e., to be promoted or self-interest, AIs offers more neutral advice to customers. As a result, customers have more trust in Chatbots than human advisors [8, 12]. Customers feel comfortable to open and disclose spending habits or retirement savings to AIs for its lack of judgement on customer's spending habits and retirement savings. 68% customers trusted robo-advice in retirement planning [8].

As human advisors are attached to their own personal agenda and interest, their advice can be biased. They might suggest products that gives them the highest incentive or commission and not necessarily helpful to the customers. In this sense, AI is ethical and product-neutral and offers advisory with zero conflict of interest [12, 55]. Besides, AIs are also more transparent in revealing all the costs customers are subjected to pay and keeps customers in the loop by alerting customers when new fees are charged [12, 56].

Humans' decisions can be influenced and driven by emotions, cognitive biases, such as overconfidence, and limited cognitive abilities [55, 57], irrationality [58]. However, robo-advisors have proven to make better decisions due to lack of emotion, strict adherence to rules, following a systemic decision-making process [8, 12, 59].

AI with the combined efforts of big data and ML (BD/ML) enables speedy, flexible, tailored, cost effective by charging reduced fee by a fraction compared to human advisors, convenient, creating wider range of services, making all services available to a broader range of users, as well as offering better services, heightening efficiency, enhancing quality, creating happy customers with raised customer satisfaction levels [8, 9, 21, 60, 61]. Robo-advisory systems simplifies customers' user experience [12, 37], translate user interface into the language of the customer's choice by welcome customers in customer's preferred language while walking through products and services in the preferred language of choice [12].

4.5 Liqudity

Banks must hold sufficient liquidity funding to ensure that it is able to meet unforeseen deposit outflows. Banks that struggle to meet its daily liquidity needs will eventually fail [2]. Central banks working on larger scales overseeing the workings of the market use AI to sort large number of bank notes and detect liquidity problems.

Central bank of Netherlands applies AI to pick out potential liquidity problems in financial institutions [9]. In Banco de España, AI has been deployed to sort fit and unfit banknotes for circulation [9].

Automatic teller machines (ATMs) are the most important cash distribution channels for banks. Yet, banks face a constant challenge to hold sufficient supply of currency to meet consumer's demand causing lost surcharge fees and increased expenses from emergency currency deliveries as overstocking currency would mean a reduced investment for banks. ATMs must work closely with the dynamic and constantly changing environment to derive greater efficiency in cash management. As such, to optimise cash management and to achieve efficient cash loads routing forecasting algorithms capture and process historical data to gain insight into the future. As the demand for cash lies more on the days i.e., holidays, weekends, starting of month, festival days etc. than time itself. Hybrid Back Propagation/Genetic Algorithm approach has proven to optimise cash management of ATMS on real time with more accuracy compared to traditional ATMs.

4.6 Senstivity

Banks are subject to market risks (i.e., interest rate risk, foreign exchange risk, price risk etc) that can have adverse effects on bank's earnings and capital. AI provides solutions to real world problems [30], through real time, enabling banks to keep up, adapt and respond to constant and dynamic changes in the environment. Thus, improving bank stability and soundness.

The constant influx in business environments, banking, credit and regulatory standards, bank lending strategies, marketing strategies of banks, investor requirements, customer demands and borrowing patterns of customers require frequent revisions [30, 60]. ML systems not only can execute rules and keep up with the change in process, but it is also able to process this information in a few milliseconds [13] as it works on real time.

Neural networks identify interconnected nodes through multi-layered data from multiple disciplines, such as statistics, computer science, biology, psychology, economics i.e., game theory, and applies multitude techniques to derive meaning [40].

ML applies algorithms to sift through hundreds of thousands of factors to proficiently detect [30, 62], decode patterns and linkages in the data by continuously updating "learning" [40]. Deep learning works on pools of high- data to identify patterns of patterns. Pattern recognition uses tools such as natural language processing to classify and interpret data [40].

ML uncovers relationships beyond causal i.e., relationships that is yet to be established by theory. Supervised ML also understands non-linear relationships while unsupervised ML reveals commonalities amongst different groups whilst highlighting outliers. ML not only unravels hidden relationships but also provides additional information about the dataset which can be further used by banks and financial supervisors alike to understand the workings of the financial markets and institutions better [15, 21].

AI algorithms work with soft-computing approaches beyond parametric statistical methods (e.g., discriminant analysis and logistic regression) and nonparametric statistical methods (e.g., k nearest neighbour and decision trees) [10]. AI also works with variables that give contradicting signs, model noisy, inconsistent, or incomplete data [63, 64], identify both linear and non-linear relationships [30], work more closely with real world non-linear applications and can handle uncertain behaviours that changes over time [10]. AI's accuracy surpasses traditional statistical models and is a crucial tool in decision making processes.

Advancement in AI methodologies have also enhanced the robustness of predictive models, and thus, its outcomes. As such enabling users to predict future outcomes and make decisions more effectively [62], efficiently and more accurately.

5. Conclusion

In this chapter we were keen to explore how able are banks to effectively deploy AI into their daily operations to improve CAMELS from a bank's perspective. It has become apparent that AI contribution is limitless, and its uses are infinite [65–67], offering significant possibilities for banks to survive. Government are in joint agreement that AI will not only help banks survive but also contribute to better functioning markets. This is evident in their continuous efforts to fund, invest and support AI related projects. The chapter has successfully portrayed bank soundness in the face of AI through the lens of CAMELS. The taxonomy partitions opportunities into distinct categories of 1 (C), 6(A), 17(M), 16 (E), 3(L), 6(S). The results re-emphasise AI's advantages as being countless and numerous in helping banks to deliver world class services to its customers through efficient and effective processes. However, future study should look to investigate further the use of AI in capital and liquidity aspects as these are the core determinants of bank survival but for now, AI allows banks to survive and evolve. As such, it is a done deal.

Appendix

Government Support	AI & Banks
	Competition and Survival Nature of Banking Industry Future customers Beyond human capacity
	Opportunities
Capital	Stress Testing
Asset	• Unstructured data help increase number of eligible customers • Better evaluator and assessor of credit risk • Increases processing speed • Works with large dataset effortlessly • Improve loan process in terms of inconsistency, inaccuracy, biasness, and bureaucracy • Creditworthy assessment beyond the norm

Management	• Risk Assessment • Structured and systematic service from start to finish • Outperforms humans • Performs various important task with less supervision • Economies of scale and scope • Affordability • Increased ROA/Profit margins • Outsource boring, repetitive, physically exhausting task • Real time, organised, systematic record keeping • Future decision making • Surpasses cybersecurity • Market surveillance • Applies real time in crime prevention • Studies customers behaviours closely to detect crimes • Validity and reliability of data • Tracking decisions • Big data
Earnings	• Customer loyalty and retention • Understand emotions • Personal Advisor 24/7 • Respond in various languages • Real Time updates • Customer records • Standardise services • Product innovation and telemarketing through data mining • Opening up new market niches • Inclusion • Improved customer decision making • AI strict adherence to rules and regulations • More trust • Ethical Advise • Better advise due to lack of emotion • Improved end user experience
Liquidity	• Sort bank notes • Detect liquidity problems • Optimise cash management
Sensitivity	• Keep up with changes instantly • Derive meaning • Pattern detection and interpretation • Identify relationships • Process data beyond statistical methods • Robust predictive models

Table 1.
Taxonomy of opportunities posed by AI on bank soundness.

Author details

Charmele Ayadurai[1]* and Sina Joneidy[2]

1 Durham University, UK

2 Teesside University, UK

*Address all correspondence to: charmele.ayadurai@durham.ac.uk

References

[1] Balasubramanyan, L., Haubrich, J., Jenkins, S., & Wallman, N. (2013). Focusing on the Future: Regional Banks and the Financial Marketplace. Federal Reserve Bank of Cleveland, 4-9.

[2] Ayadurai, C., & Eskandari, R. (2018). Bank soundness: a PLS-SEM approach. In *Partial Least Squares Structural Equation Modeling* (pp. 31-52). Springer, Cham.

[3] Zervoudi, E. K. (2019). Parallel banking system: Opportunities and Challenges. Journal of Applied Finance and Banking, *9*(4), 47-70.

[4] Bughin, J., Seong, J., Manyika, J., Chui, M., & Joshi, R. (2018). Notes from the AI frontier: Modeling the impact of AI on the world economy. *McKinsey Global Institute*.

[5] Francisco, D. F. (2019). *How artificial intelligence can help banks improve the customer experience of buying a house* (Doctoral dissertation).

[6] Crosman, P. (2018). How Artificial Intelligence is reshaping jobs in banking. American Banker, *183*(88), 1.

[7] Research and Markets. (2017). Artificial Intelligence Market: Global Forecast to 2020. Ireland: Research and Markets. Retrieved from https://www.researchandmarkets.com/reports/3979203/artificial-intelligencechipsets-market-by

[8] Lui, A., & Lamb, G. W. (2018). Artificial intelligence and augmented intelligence collaboration: regaining trust and confidence in the financial sector. Information & Communications Technology Law, *27*(3), 267-283.

[9] Fernández, A. (2019). Artificial intelligence in financial services. Banco de Espana Article, *3*, 19.

[10] Bahrammirzaee, A. (2010). A comparative survey of artificial intelligence applications in finance: artificial neural networks, expert system and hybrid intelligent systems. Neural Computing and Applications, *19*(8), 1165-1195.

[11] Fethi, M. D., & Pasiouras, F. (2010). Assessing bank efficiency and erformance with operational research and artificial intelligence techniques: A survey. European journal of operational research, *204*(2), 189-198.

[12] Hakala, K. (2019). Robo-advisors as a form of artificial intelligence in private customers' investment advisory services.

[13] Sachan, S., Yang, J. B., Xu, D. L., Benavides, D. E., & Li, Y. (2020). An explainable AI decision-support-system to automate loan underwriting. Expert Systems with Applications, *144*, 113100.

[14] Longinus, O. (2018). Artificial Intelligence System: Implication for Proper Record Keeping in Microfinance Banks in Nigeria. International Journal of Academic Research in Accounting, Finance and Management Sciences, *8*(1), 131-136.

[15] Wall, L. D. (2018). Some financial regulatory implications of artificial intelligence. Journal of Economics and Business, *100*, 55-63.

[16] Bennink, K. (2017). Do traditional banks need a Chief AI Officer?: An explorative research project that aims to evaluate the appointment of a Chief AI Officer to overcome challenges that arise when traditional banks adopt AI technologies.

[17] Kaya, O., Schildbach, J., AG, D. B., & Schneider, S. (2019). Artificial

intelligence in banking. *Artificial intelligence*.

[18] Redshaw, P., & D'Orazio, V. (2018). 2018 CIO Agenda: Banking and Investment Services Industry Insights.

[19] Arnold, M. (2016), "Market grows for 'regtech', or AI for regulation," Financial Times (October 14). Available at https://www.ft.com/content/fd80ac50-7383-11e6-bf48-b372cdb1043a.

[20] Woods, Michael, (2015), "Creating Effective Revenue Forecast Models for CCAR," Available at https://s3.amazonaws.com/cdn.ayasdi.com/wpcontent/uploads/2017/02/13111852/creating-effective-revenue-forecast-models-forccar.pdf.

[21] Jagtiani, J., Vermilyea, T., & Wall, L. D. (2018). The roles of big data and machine learning in bank supervision. *Forthcoming, Banking Perspectives*.

[22] Affes, Z., & Hentati-Kaffel, R. (2019). Predicting US banks bankruptcy: logit versus Canonical Discriminant analysis. Computational Economics, *54*(1), 199-244

[23] Christopoulos, A. G., Mylonakis, J., & Diktapanidis, P. (2011). Could Lehman Brothers' collapse be anticipated? An examination using CAMELS rating system. International Business Research, *4*(2), 11.

[24] Grimes, P. E. (2005). New insights and new therapies in vitiligo. Jama, *293*(6), 730-735.

[25] Conway, J. J. E. (2018). *Artificial intelligence and machine learning: Current applications in real estate* (Doctoral dissertation, Massachusetts Institute of Technology).

[26] Ghodselahi, A., & Amirmadhi, A. (2011). Application of artificial intelligence techniques for credit risk evaluation. International Journal of Modeling and Optimization, *1*(3), 243.

[27] Šušteršič, M., Mramor, D., & Zupan, J. (2009). Consumer credit scoring models with limited data. Expert Systems with Applications, *36*(3), 4736-4744.

[28] Malhotra, R. and D.K. Malhotra, (2003). Evaluating consumer loans using neural networks. Omega, 31: 83-96. DOI: 10.1016/S0305-0483(03)00016-1

[29] Mitchell, D. and R. Pavur, (2002). Using modular neural networks for business decisions. Manage. Dec., 40: 58-63. DOI: 10.1108/00251740210413361.

[30] Eletter, S. F., Yaseen, S. G., & Elrefae, G. A. (2010). Neuro-based artificial intelligence model for loan decisions. American Journal of Economics and Business Administration, *2*(1), 27.

[31] Byrne, M. F., Shahidi, N., & Rex, D. K. (2017). Will computer-aided detection and diagnosis revolutionize colonoscopy?. Gastroenterology, *153*(6), 1460-1464.

[32] Kingston, J. (1991). X-MATE: Creating an interpretation model for credit risk assessment. In research and development in expert systems VIII: Proceedings of 11th annual technical conference of the BCS specialist group 165-174

[33] Son, Hugh (2017) "JPMorgan Software Does in Seconds What Took Lawyers 360,000 Hours," Bloomberg (February 27). Available at https://www.bloomberg.com/news/articles/2017-02-28/jpmorgan-marshals-an-army-of-developers-to-automate-high-finance.

[34] Peterson, D. (2017). Maximizeefficiency: Howautomationca nimproveyourloanorig-inationprocess.

[35] Berger, A. N., & Bouwman, C. H. (2013). How does capital affect bank performance during financial crises? Journal of Financial Economics, *109*(1), 146-176.

[36] Bachinskiy, A. (2019) The Growing Impact of AI in Financial Services: Six Examples. https://towardsdatascience.com/the-growing-impact-of-ai-in-financial-services-sixexamples-da386c0301b2

[37] Jung, D., Glaser, F., & Köpplin, W. (2019). Robo-Advisory – Opportunities and Risks for the Future of Financial Advisory. https://doi.org/10.1007/978-3-319-95999-3

[38] Rahim, S. R. M., Mohamad, Z. Z., Bakar, J. A., Mohsin, F. H., & Isa, N. M. (2018). Artificial intelligence, smart contract and islamic finance. Asian Social Science, *14*(2), 145.

[39] Ernst & Young. (2018). The evolution of robo-advisors and advisor 2.0 model. 7. Retrieved from www.ey.com

[40] O'Halloran, S., & Nowaczyk, N. (2019). An Artificial Intelligence Approach to Regulating Systemic Risk. Front. *Artif. Intell*, *2*(7).

[41] Wong, M. M. (2015). Hungry Robo-Advisors Are Eyeing Wealth Management Assets We Believe Wealth Management Moats Can Repel the Fiber-Clad Legion. 16. Retrieved from https://www.morningstar.com/content/dam/marketing/shared/pdfs/Research/equityreserach/20150409_Hungry_RoboAdvisors_Are_Eyeing_Wealth_Manage ment_.pdf

[42] Angelini, E., di Tollo, G., & Roli, A. (2008). A neural network approach for credit risk evaluation. The quarterly review of economics and finance, *48*(4), 733-755.

[43] Dilek, Selma, Hüseyin Çakır, and Mustafa Aydın (2015). Applications of artificial intelligence techniques to combating cyber crimes: A review.

[44] Yampolskiy, R. V., & Spellchecker, M. S. (2016). Artificial intelligence safety and cybersecurity: A timeline of AI failures.

[45] Van Liebergen, B. (2017). Machine learning: A revolution in risk management and compliance? Journal of Financial Transformation, *45*, 60-67.

[46] Kingdon, J. (2004). AI fights money laundering. IEEE Intelligent Systems, *19*(3), 87-89.

[47] Carney, M. (2017a), "Remarks at the Banking Standards Board Panel "Worthy of trust? Law, ethics and culture in banking" Speech at a Bank of England Conference. Available at http://www.bankofengland.co.uk/publications/Documents/speeches/2017/speech970.pdf

[48] Kishada, Z. M. E., Wahab, N. A., & Mustapha, A. (2016). Customer loyalty assessment in malaysian islamic banking using artificial intelligence. *Journal of Theoretical and Applied Information Technology*.

[49] Ivanov, O., Snihovyi, O., & Kobets, V. (2018). Implementation of Robo-advisors tools for different risk attitude investment decisions. CEUR Workshop Proceedings, 2104, 195-206.

[50] Novick, B., Lu, B., Fortin, T., Hafizi, S., Parkes, M., & Barry, R. (2016). Digital Investment Advice: Robo Advisors Come Of Age. Retrieved from https://goo.gl/Cucbx4

[51] Huikko, I., & Hykkönen, M. (2017). The Finnish perspective on robo-advisory. 16. Retrieved from https://www2.deloitte.com/content/dam/Deloitte/fi/Documents/financialservices/Robo-advisory.pdf

[52] Kaya, O., & Schildbach, J. (2017). EU Monitor Global financial markets

Robo-advice – a true innovation in asset management Robo-advice – a true innovation in asset management.

[53] Ligaya, A. (2017). Financial Post. Retrieved from Rise of the robot: How banks are using artificial intelligence upfront and behind the scenes. Retrieved from http://business.financialpost.com/news/fp-street/rise-of-the-robot-banks-using-artificial-intelligence-upfront-and-behind-the-scenes

[54] Moro, S., Cortez, P. & Rita, P. (2014). A data-driven approach to predict the success of bank telemarketing. Decision Support Systems. 62, 22-31

[55] Lewis, D. (2018). Computers may not make mistakes, but many consumers do (Vol. 10923). https://doi.org/10.1007/978-3-319-91716-0

[56] Lopez, J. C., Babcic, S., & De La Ossa, A. (2015). Advice goes virtual: how new digital investment services are changing the wealth management landscape. The Journal of Financial Perspectives: Fintech, 3(3), 1-21. Retrieved from https://papers.ssrn.com/sol3/papers.cfm?abstract_id=3084057

[57] Gervais, S., & Odean, T. (2001). Learning to be overconfident. The Review of Financial Studies, *14*(1), 1-27.

[58] Kahneman, D., & Tversky, A. (1973). On the psychology of prediction. Psychological review, *80*(4), 237.

[59] Dhar, V. (2015) 'Should You Trust Your Money to a Robot?' (2015) Big Data June, 3(2): 55-58 DOI: 10.1089/big.2015.28999.vda

[60] Krovvidy, S. (2008). Custom DU: A web based business user driven automated under writing system. AI Magazine, 29(1), 41-41.

[61] Gates, S.W., Perry, V.G., & Zorn, P.M. (2002). Automated underwriting in mortgage lending: Good news for the underserved. Housing Policy Debate, 13(2), 369-391.

[62] Chen, V. (n.d.) Creating a Better Future with Ethical AI: Key Challenges and Solutions.

[63] Tafti, M.H.A. and E. Nikbakht, 1993. Neural networks and expert systems: new horizons in business finance applications. Inform. Manage. Comput. Secur., 1: 22-28. DOI: 10.1108/096852293/0026502.

[64] Peel, M. and N. Wilson, 1996. Neural network simulation: A new approach to risk assessment and business forecasting. Manage. Res. News, 19: 50-54. DOI: 10.1108/eb028477

[65] Neti, V. (2016). Ubiquitous and Cognitive AI will Redefine our Everyday Experiences. [online] Infosys. Available at: https://www.infosys.com/insights/aiautomation/Pages/redefine-everyday-experiences.aspx

[66] Intel AI (2018). The Rise In Computing Power: Why Ubiquitous Artificial Intelligence Is Now A Reality. Forbes. Available at: https://www.forbes.com/sites/intelai/2018/07/17/the-rise-in-computing-powerwhy-ubiquitous-artificial-intelligence-is-now-a-reality/#6c3fc39e1d3f

[67] MIT Technology Review. (2018). On-Device AI. Available at: https://www.technologyreview.com/hub/ubiquitous-on-device-ai

Chapter 14

Artificial Intelligence and Bank Soundness: Between the Devil and the Deep Blue Sea - Part 2

Charmele Ayadurai and Sina Joneidy

Abstract

Banks have experienced chronic weaknesses as well as frequent crisis over the years. As bank failures are costly and affect global economies, banks are constantly under intense scrutiny by regulators. This makes banks the most highly regulated industry in the world today. As banks grow into the 21st century framework, banks are in need to embrace Artificial Intelligence (AI) to not only to provide personalized world class service to its large database of customers but most importantly to survive. The chapter provides a taxonomy of bank soundness in the face of AI through the lens of CAMELS where C (Capital), A(Asset), M(Management), E(Earnings), L(Liquidity), S(Sensitivity). The taxonomy partitions challenges from the main strand of CAMELS into distinct categories of AI into 1(C), 4(A), 17(M), 8 (E), 1(L), 2(S) categories that banks and regulatory teams need to consider in evaluating AI use in banks. Although AI offers numerous opportunities to enable banks to operate more efficiently and effectively, at the same time banks also need to give assurance that AI 'do no harm' to stakeholders. Posing many unresolved questions, it seems that banks are trapped between the devil and the deep blue sea for now.

Keywords: bank, bank soundness, financial sector, Artificial Intelligence (AI), CAMELS

1. Introduction

The Global Financial Crisis (GFC) showcased that even banks that are well established, operating in robust markets and governed by tough and forceful regulatory frameworks can fail. Over the years banks have grown larger in size, have become more complex and complicated in its nature of operation making it opaque and incomprehensible. Bank supervisors are still struggling to demystify the risk undertaken by banks during the GFC. The unknown risk posed by yet another Blackbox in the name of Artificial Intelligence (AI) could pose identical challenges of increased systemic fragility, bank failures and freezing up capital markets evident during the GFC. As such, bank supervisors are critical of banks adopting AI. To move forward banks need to give assurance that banks will "do no harm". As such, it is important to consider the lesson from the GFC which calls into question the soundness of regulations to capture the risk but most importantly to critically evaluate the process that is in place, in this case the efficacy of the implementation of AI. It is necessary to know if AI promotes safety and soundness in the financial system or adds undue burdens to the markets [1]. As such, the study looks to critically assess the challenges posed by AI that could influence bank soundness from the light of Capital (C), Asset (A), Management (M), Earning (E), Liquidity (L) and Sensitivity (S) (CAMELS), determinants of bank bank's health and wellbeing [2–4] while "doing no harm" to individuals, corporations and society as a whole.

The chapter contributes to literature in several ways. Earlier research has either focused on AI application on the entire financial sector covering banks, fintech companies, mortgage lenders, security companies amongst others [5–7] or have evaluated AI applications in the form of Machine Learning (ML), Neural Networks (NN) Artificial Neural Networks (ANN) in specific areas such as credit evaluation, portfolio management, financial prediction and planning [2, 8–10] or by examining user friendly experiences of end users [5, 6, 11]. Therefore, these studies are not sufficient to understand the challenges proposed by AI from solely a bank's perspective. As such, the chapter fills this gap by taking a holistic approach in scrutinizing the challenges faced by banks solely by deploying AI. By doing this the chapter provides a significant insight into the significant challenges that AI technology can pose on the banking industry depleting its chances of survival. The chapter also further considers bank soundness with the application of AI from various aspects of Capital (C), Asset (A), Management (M), Earning (E), Liquidity (L) and Sensitivity (S) (CAMELS) determinants of bank soundness. This chapter is the first to review the challenges of deploying AI in banking operation in light of CAMELS. Earlier research [2–4] has only focused on bank soundness from the CAMELS perspective. The chapter also more specifically focuses from both the service provider and customer end, providing further insight to regulators on what they need to look into. Most importantly, the intention is to examine through the lens of CAMELS how sound are banks having applied AI into their processes.

The chapter is organized as follows: the next section presents a brief theoretical discussion on the application of AI in different sections of the bank from front, middle to back office operation. Section three introduces the literature gathering and research method. Section four presents result and discussions on the challenges posed by banks on application of AI from CAMELS perspective. The last section concludes the chapter and highlights insight on further research.

2. Literature review

Central banks worldwide have actively embedded AI into their daily operations from microprudential and macroprudential supervisions to information management, forecasting and detecting fraudulent activities. Monetary Authority of Singapore applies AI to scrutinise suspicious transactions, while the central bank of Austria has developed a prototype for data validation. The central bank of Italy uses AI techniques to predict price moves on the real estate market [7].

Banks also have deployed AI from front office through middle-office to back office operations in different subsets of the banks [7, 12]. AI is not only widely used and applied in conventional banks it has been actively embraced in Islamic banks as well [13, 14]. AI in the form of Neural Networks (NN) has been used in risk management, forecasting i.e. inflation [15–18], identify complex patterns [16, 19, 20], predict future stock behavior, market trends, market response, real estate evaluations, financial crises, bank failures, bankruptcy, exchange rates, detecting credit card fraud [9, 20–22].

Artificial Neural Networks (ANN) has been used to analyse relationships of bonds and stocks between economic and financial phenomenon, futures and financial markets, loan application and overdraft check, loan scoring, credit scoring, credit worthiness, mortgage choice decision, portfolio optimization, portfolio management, asset value forecasting, index movement prediction, exchange rate forecasting, global stock index forecasting, portfolio selection, portfolio resource allocation, forecasting, planning, generating time series, credit evaluations, bankruptcy prediction, predicting thrift failures, financial distress prediction, asset value forecasting, and decision making [22–26].

Backpropagation neural networks (BPNN) is applied to classify loan applications from good to bad [20]. Decision Tree (DT) is applied in credit risk classification [26]. SVM is applied in corporate credit rating [26].

Machine Learning (ML), a subfield of AI is used in customer services such as search engines, offering product recommendations [27], manage customer online queries, perform voice recognition, predictive analysis, provide financial advice, analyse risks, manage assets and engage in algorithmic trading [6]. ML is deployed for call-centre optimization, mortgage decision making, relationship management, treasury-management initiatives, customer-credit decisions and equity trading [27, 28] where algorithmic trading is used to pick stocks and is able to fulfil the job specification of a portfolio manager [29, 30]. JP Morgan uses ML to execute trades in equity markets [31].

Big Data and Machine Learning (BD/ML) in the form of Robo-advisers use algorithms to deliver stock recommendations, analysing incoming information for investors, providing investment advice and financial planning services, make credit decisions [32]. Goldman Sachs uses ML platform Kensho to offer automated analysis of breaking news and compiling the information into regular summaries. Wells Fargo on the other hand, uses AIERA (Artificial Intelligent Equity Research Analyst), to issue buy and sell call options on stocks. While bank officers offer recommendations through this platform [32]. Several studies have worked with various ML models for credit scoring namely ensemble classifier [33–39], support vector machine [40–46], neural network [45, 47–52], genetic programming [53–55], Bayesian network [56–58] and decision tree [50, 59–61]. Automated trading where systems make independent decisions without human intervention or oversight is evident in stock markets. Nasdaq runs on autonomous trading by 75% [25].

Development in AI is moving in the direction of hybrid models where two or more Artificial Intelligent systems are combined to enhance performance namely with the application of intelligent systems which is able to integrate intelligent techniques to problem-solving namely the combined efforts of neural network and fuzzy system [26].

3. Methodology

The research is conducted as a conceptual chapter with the aim to provide a deeper understanding of the opportunities parted by AI from a service provider and customer perspective. To answer the research question on how able banks are to effectively deploy AI into their daily operations to improve CAMELS from a bank's perspective, a systematic review of the literature and objective observations were undertaken to examine banks through the lens of bank soundness determinants of CAMELS. The observations found in existing literature are gathered to assemble a framework categorized by CAMELS (**Figure 1**). The literature was gathered through the Scopus database as a main source of finding existing literature. The database offers a wide range of management and business-related studies relevant for the topic of research. In addition, other databases such as Google Scholar, Social Science Research Network (SSRN), SpringerLink and IEEE Xplore were also examined. Journal articles since the period 2000–2020 were extracted using the prescribed keywords of Bank, Bank soundness, Financial Sector, Artificial

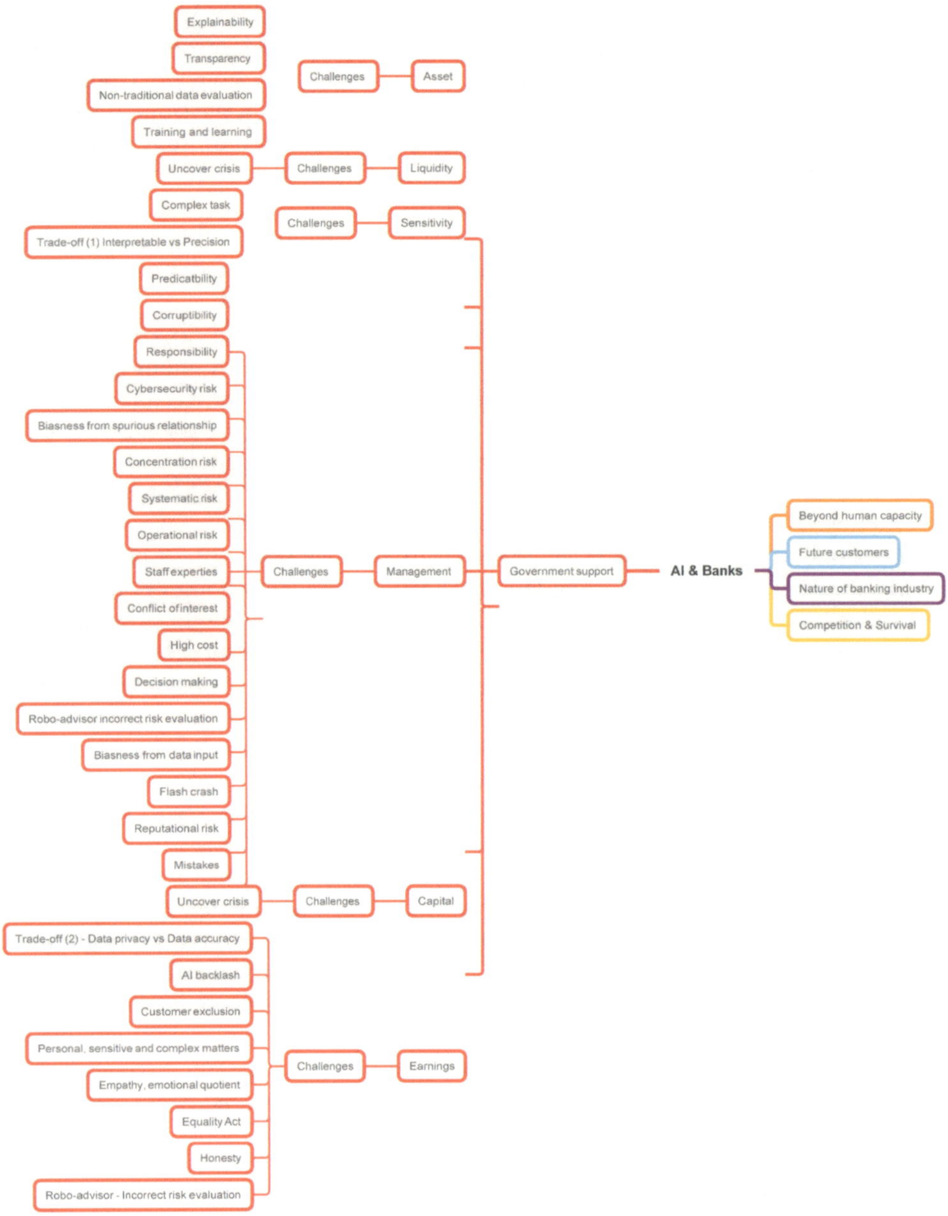

Figure 1.
Taxonomy of challenges posed by AI on Bank Soundness - A classification based on the determinants Bank Soundness of CAMELS.

Intelligence (AI), CAMELS. Only articles that were available in full text, published in scholarly, peer reviewed journal were chosen to be closely examined. The search was also conducted using the backward and forward approach where reference list of articles was utilized to find further research papers.

4. Findings and discussion

This section presents an overview of the challenges posed by banks in deploying AI in their daily front, middle and back office operations prescribed from the CAMELS perspective (see **Table 1** in the Appendix).

4.1 Capital and liquidity

Bank capital acts as a core determinant for bank's survival. Capital absorbs losses during adversity and insufficient capital holdings can cause banks to collapse. AI with its limitless abilities and capabilities helps banks to hold robust capital holdings through stress testing.

Banks have to hold sufficient liquidity funding to ensure that it is able to meet unforeseen deposit outflows. Banks that struggle to meet its daily liquidity needs will eventually fail [3]. Central banks working on larger scales overseeing the workings of the market use AI to sort large number of bank notes and detect liquidity problems.

AI's ability to detect or uncover crisis depends on the quantity and quality of data provided and used to train the algorithms. As such if the dataset lacks important conditions such as economic crashes, as normal periods exist more than crisis periods, limited crisis data could reduce AI's predictive abilities and the output will have limited use in measuring or projecting future risk under stress [7, 32]. As such, will have little value for bank's setting their minimum prescribed capital/liquidity holding (Basel accords) to remain solvent while lending through recessions [27]. Banks have little choice but to rely on theory of distribution of losses and parametric statistical structure to link normal times data to determine large losses that causes instability. Yet, a more accurate prediction will yield from data of distribution of losses itself [27].

4.2 Asset

Asset quality is measured by the level of credit risk contained in bank's assets [62]. Therefore, a bank that can detect, measure, monitor and regulate credit risks will hold higher quality assets [63]. The GFC showcased that credit risk is the most challenging risk to manage and control as it not only absorbs profits but exposes banks to failures as well. AI helps banks to clearly assess and evaluate customers' risk, eliminating ambiguity, biasness while improving loan processes and.

Banks are accountable for each decision that they make. As such employ verification and checks at several levels to weed out incorrect or weak decisions. As such, Loan officers should be able to provide a logical explanation on what grounds a loan has been accepted or rejected to their superiors, compliance officers, auditors, regulators and customers [5, 7, 10, 12, 64, 65]. The working logic of AI decision has to be traceable backwards. Customers need to understand the reasons why their loan application has been rejected or why AI has recommended a particular product before acting on it. Keeping customers in the dark without proper justification will cut short the chances of them determining the real cause behind the rejection, finding solutions to their problems and improving their circumstances or proving an identity theft if it happened to them. In short, AI adverse decision will have a permanent detrimental effect on someone's future [28, 64, 66–68].

Transparency is also important to fully trust the system through validating the decision made by AI, by not only detecting anomalies in the decision process such as biasness, mistakes, manipulations of data, deficiencies, compliance to rules i.e. GDPR, cybersecurity crimes linked to work processes such as dataset poisoning, internal network manipulation, and side-channel attacks [69] but also to detect clearly and precisely at which step the anomalies occurred and what information AI fed itself [10, 12, 64, 66–68, 70–72].

Although AI can assess customers from various angles namely with non-traditional data such as customers connection, internet searches, network diversity, etc., yet how reliable is this information to make an informed decision about a person's repayment ability, thus future. Does the credit score of a person increase if they socialise with those who are creditworthy? Borrowers may also be judged based on how they behave online or their dishonesty in disclosing financial data, forming biases and being judged unfairly [73]. Also, are customers aware that non-traditional data is used in the evaluation process to assess their loan repayment ability? [6].

AI that are trained based on supervised learning where both inputs and output are fed into the system have zero chance of biasness unless the data fed itself is biased. Data used to train ML algorithm must be representative of a wide range of customers that will apply for loans representing namely a whole population [27, 72]. If a population is underrepresented or there are rare cases such as women, race, ethnicity, marital status, zero credit history and this information is used to train AI, AI will deliver biased results if data is highly correlated in these categories [7, 28, 72–74].

In unsupervised learning, AI trains itself to make independent decisions. As such, based on what AI trains itself with, decisions can be bias. In reinforced learning AI takes uses its own initiative to combine various decision to make an ultimate decision where biasness can form as well. In checking creditworthiness AI can discriminate based on gender if more men are in professions or earning higher salaries, race if more discount stores are located near ethnic minorities, spelling mistakes in internet searches etc. Statistics reveal, algorithms accept white applicants and reject black applicants evident from the gradual reduction in black applicants' loan approval in banks [64].

According to Janssen algorithms can systematically introduce inadvertent bias, reinforce historical discrimination, favor a political orientation or reinforce undesired practices [75]. Standard affordability rules such as defaults, loan-to-value, and loan-to-income may not be applicable to all groups of borrowers [76] causing low-income borrowers to be marginalised. Looking from a different perspective, the contribution of one's data could contribute to a whole set of minority, race, gender, marital status or society to be judged in a certain way, forming biases, causing more harm than intended. For example, algorithms picking up 20 black female who are constantly delinquent on their loans as a representative of the whole black female population. AI could also link financially vulnerable customers to mental health issues [12]. Banks could utilize this information to turn down loan applications causing more harm to the society than intended [12]. Yet, to train AI systems to replicate human decision-making skills is a challenge. As it is difficult to transform various algorithmic concepts into training data to solve every problem for a range of lending products [10, 77].

4.3 Management

Banks rely heavily on management to not only generate earnings and increase profit margins [3] but also to keep banks alive [78]. AI helps banks to be more efficient, effective, effectual and efficacious.

The legal profession requires predictability in its approach i.e. contracts are written in a way knowing how it will be executed. As such, the legal system offers a predictable environment where customers can improve their lives [64]. Therefore, AI needs to be predictable to customers.

The GFC is the outcome of human greed, manipulation and corruption. As such, AI algorithms need to be robust against exploitation [64]. Discontented employees or external foes may learn the inner workings of the AI model to easily corrupt algorithms or use AI application in malfeasant ways [28]. This will strike of a worst catastrophe than the GFC, as the involvement of AI increases complexity and opaqueness of the financial systems making it is difficult to configure a solution.

When an AI system fails at its assigned task, it is difficult to pin down who is to take the blame for its actions. As the AI ecosystem comprises of wide range of stakeholders from the philosopher, the AI researcher, the data scientist, the data provider, the developer, the library author, the hardware manufacturer, the OS provider, programmers, etc. Each staff has established procedures to part with AI and their responsibilities have been distributed widely amongst them. As such, when a catastrophe strikes, it would be difficult to assign liability and could be a perfect cover for mistakes, manipulation and exploitation [64, 72, 79]. In the pursuit of accumulating big data, banks could cross the boundaries to incorporate customers private information. As such, when any loss results from the use of AI should the scientists who work to tune the experience to the needs of consumers, employees write the content the chatbots or the algorithm provider be liable? [5, 7].

As AI systems are interconnected, hackers or malicious software can manipulate bank's data by hacking client's financial details, creating false identities, flooding systems with fabricated data resulting in misclassification or bad clusters causing incorrect decisions facing consumer backlash and regulatory repercussions [28, 72].

Algorithms are constantly looking to improve predictive power. As such are in the constant look out for correlations producing spurious relationships which eventually leads to bias conclusions [80].

Literature have pointed out the potential for AIs to act on biased data [81–92]. scientists have realised that ML can discriminate customers based on race and gender. One such example is the 'white guy' syndrome where men are picked over women. Input data is directly linked to the outcome. As every individual have their own biases, norms and ethics it is difficult to establish that biases will not exist even after AI has gone through training data [84, 93]. Also to perfect the existing biases under the Fair Lending Act, and to improve processes and innovation, more data from people with disability, colour, age, gender and creed could be incorporated into the system. Only if customers feel comfortable to share [12, 32].

As developing and operating AI requires extensive resources and big data, only large banks can be players in this field. This encourages concentration affecting healthy competition in the market [7]. Banks have to rely heavily on technology companies for AI's critical tools and infrastructure. As such, increasing operational risk [7].

As there are only a few players in the market, operational risk could easily feed into systemic risk. On top of that, the widespread use of AI in similar functions such as provision of credit or trading of financial assets and the uniformity in data, training and methodology employed to develop the algorithms could spark off herding and procyclical behaviour [7].

Banks that are extensively working with AI need staff that have expertise not only the field of finance but also have formal training in computer science, cybersecurity, cryptography, decision theory, machine learning, formal verification, computer forensics, steganography, ethics, mathematics, network security, psychology and other relevant fields. The challenge would be to find sufficient number of staff to fit this role.

AI in the form of robo-advisory services incur high development, marketing as well as advertising costs. As such a single client acquisition costs ranges between \$300–\$1,000 with clients at the lower end only generating \$100 in annual revenues [9, 94]. Robo-advisors' slim operating margin and low average account size would eat up the profits garnered quickly taking banks a decade or more to cover the \$10 to \$100 million in marketing costs [9, 95].

Some studies have pointed out that ML is only able to act on the primary stages of decision making such as data processing and forming predictions. However, at higher levels of judgement, action and task, requires special skills such as empathy, lateral thinking as well risk evaluations where AI is unable to muster [6].

Algorithmic trading through ML could also facilitate trading errors. A single failure in the AI system could lead to devasting catastrophes without a chance for recovery resulting in flash crashes [96, 97], causing "excess volatility or increase pro-cyclicality as a result of herding" [98]. Besides, major financial institutions returned compliance software that stopped detecting trading issues from excluded customer trades [28]. Developers prescribe that new intelligent features embedded into AIs could pose unexpected and unknown risks creating new points of vulnerability leaving loopholes for hackers to exploit [28, 32].

If humans make mistakes, manipulated the system they can be fired instantly. However, if AI makes mistakes or corrupts, customers will lose hope and trust in the bank and its systems [5]. Robo-advisors work with several parties namely clearing firms, custodians, affiliated brokers, and other companies in the industry to offer its services to customers. While Lewis suggest that robo-advisors resolve conflict of interest amongst the parties [99], Jung et al. suggest conflicts remains costing customers [100]. If company uses brokers for an example, this cost is transferred to customers increasing the price of service while the robo-advisor makes profit as the middleman [101]. In other scenarios, robo-advisors could receive fee for order flow in exchange for routing trades to a clearing firm or have an equal interest in securities that customers are looking into [101].

Scripting errors, lapses in data management, misjudgments in model-training data can compromise fairness, privacy, security, as well as compliance [28]. As the volume of data being sorted, linked, ingested is large, further complicated with unstructured data, mistakes such as revealing sensitive information could take place. i.e. client's name might be redacted from the section that is used by an AI but present in the stock broker's notes section of the record thus breaching European Union's General Data Protection Regulation (GDPR) or the California Consumer Privacy Act (CCPA) [28].

4.4 Earnings

Banks that manage their expenses well while fully utilising their assets to generate constant revenue streams are most likely to be sound [3]. AI enables banks to offer unique selling points in products that increases customer satisfaction, boosting sales and revenue [6].

Studies have recorded chatbot controversy and backlash [5, 6]. AI chatbot Microsoft's Tay, tweeted racist, sexist, xenophobic, and prejudiced opinions through learning them through tweets it read and interaction with younger demographics on Twitter upsetting customers [7, 102–104].

To increase market share and competitive positions, improving the predictive power of algorithms and ensuring AIs are trained properly to avoid bias decisions, banks need a large set of quality and diverse data. In the pursuit and pressure to achieve this goal, banks might share customer's private data without their consent when customers trusted the bank to keep it confidential [7, 9, 12]. Privacy is important to not only for customers but also banks to allow banks to remain in its competitive position [27].

Training AIs to exclude certain segments of customers in sales could also lead to discrimination and biasness [28]. AI could also weave zip codes and income of individuals to create target offering causing discrimination against other classes and groups of people [28]. Robo-advisors can provide incorrect risk evaluation if it is not equipped with all aspects of an investor's financial condition to accurately access the overall risk. When customised questions are unable to capture unique life experiences such as receiving large inheritance customers are better off with human advisors [9].

Customers are more likely to rely on human advisers than chatbots and robo-advisors to assist when it comes to more personal and sensitive matters. One example is when large sums of money is involved either through wealthy customers or due to death and illness. Another is when there is market volatility. Customers are less inclined to trust new technologies and would prefer humans to handle such transactions for accountability purposes [6, 9, 105, 106]. Customers are also more confident to gain insight from human advisors when it comes to complex financial products such as derivatives, discussing complicated matters or making complains [6].

AI lacks emotion quotient. As such is unable to connect, understand or react to human at deeper level to comprehend their emotion and to empathise, rejoice or sympathize with them [5, 6]. As such, some prefer front-desk receptionist to a chatbot or an electronic menu that needs to be navigated.

Although the equality act, that oversees the violation against race and gender acknowledges inequality when "a person" treats "another person" favourably. It does not recognise discrimination by AI's but it recognises discrimination by other "non-humans" such as government agency or regulator. As such, AI cannot be taken to court [6].

Banks may not disclose their use of AI to customers to either benefit the bank or to avoid "fear factor of AI" amongst customers. Banks have to be transparent with their customers revealing if they are working with AI or human advisors. Although Swedbank and Société Générale agree it is best to be honest with customers others may not agree as it disadvantages the banks in many ways. As such AI are being trained to offer seamless interaction through training to very closely mimic humans. [6].

4.5 Senstivity

Banks are subject to market risks (i.e. interest rate risk, foreign exchange risk, price risk etc) that can have adverse effects on bank's earnings and capital. AI provides solutions to real world problems [20], through real time, enabling banks to keep up, adapt and respond to constant and dynamic changes in the environment. Thus, improving bank stability and soundness.

Unsupervised ML techniques are the only ones that can be used to detect frauds as it is able to identify unusual transaction characteristics to then investigate and test further to prove the authenticity of the transaction [27, 32]. Besides, unsupervised ML can also be used to closely monitor traders' behavior enabling auditing [107]. Yet, as unsupervised ML is linked to blackbox decision making, it is difficult to point out if the decision was made fairly.

ML in the form of decision tree and logistic regression models are interpretable but lack accuracies [10]. AI in the form of Deep Neural Networks (DNNs) and ensemble models such as random-forest, XGboost and Adaboost have strong higher predictive power and accuracy as it works with multiple layers of hundreds and thousands of parameters and neurons while applying nested non-linear structure which makes them opaque and a complex blackbox [10, 27, 32, 71]. Various stakeholders have labelled the effort of allowing black boxes to make decisions as irresponsible as decisions created by these blackboxes are not justifiable, not interpretable, trackable, legitimate, trustworthy as the decision lacks detailed explanations [71].

5. Conclusion

Investment in AI is one of the most essential and a core element in bank survival. Therefore, it is vital for banks to continue to deploy AI into their operations. Yet, AI suffers from a series of limitations that must be considered in assessing its use. Many studies have raised concerns about AI biasness, discrimination, violating privacy, manipulating political systems compromising national security, economic stability, political stability, digital safety, and financial safety causing disastrous consequences from reputational damage, revenue losses, regulatory backlash to criminal investigation, ignores equality and fair treatment, difficult to evaluate decisions due to poor explainability, transparency, resulting in lack of trustworthiness, accountability and reliability [12, 20, 28, 108]. The chapter has successfully portrayed bank soundness in the face of AI through the lens of CAMELS. The taxonomy partitions challenges posed by AI into 1(C), 4(A), 17(M), 8 (E), 1(L), 2(S) distinct categories. Ironically, both AI and banks are opaque in nature and have diminished public trust in them. Governments will be held accountable once again by taxpayers if markets come to a standstill as a result of AI. As such, banks need to provide answers on how well they are protecting customer's privacy and security with a range of protocols, controls and measures. If a silver bullet is not found than either banks will have to disappear, or the world will witness yet another catastrophe created by banks but this time with the help of AI. As such, trapping banks further into the conundrum of being in between the devil and the deep blue sea.

Appendix

Government Support	AI & Banks
	Competition and Survival Nature of Banking Industry Future customers Beyond human capacity
	Challenges
Capital	• Uncover crisis
Asset	• Explainability • Transparency • Non-traditional data evaluation • Training and learning
Management	• Predictability • Corruptibility • Responsibility • Cybersecurity risk • Biasness from spurious relationship • Concentration risk • Systemic risk • Operational risk • Staff expertise • Conflict of interest • High cost • Decision making • Robo-advisor incorrect risk evaluation • Biasness from data input • Flash crash • Reputational risk • Mistakes

Government Support	AI & Banks
Earnings	• Trade-off 2- Data privacy vs. accuracy • AI backlash • customer exclusion • Personal, sensitive and complex matters • Empathy, emotional quotient • Equality Act • Honesty • Robo-advisor-incorrect risk evaluation
Liquidity	• Uncover crisis
Sensitivity	• Complex task • Trade-off 1-Interpretable vs. precision

Table 1.
Taxonomy of challenges posed by AI on bank soundness.

Author details

Charmele Ayadurai[1]* and Sina Joneidy[2]

1 Durham University, UK

2 Teesside University, UK

*Address all correspondence to: charmele.ayadurai@durham.ac.uk

References

[1] O'Halloran, S., & Nowaczyk, N. (2019). An Artificial Intelligence Approach to Regulating Systemic Risk. Front. *Artif. Intell*, *2*(7).

[2] Balasubramanyan, L., Haubrich, J., Jenkins, S., & Wallman, N. (2013). Focusing on the Future: Regional Banks and the Financial Marketplace. Federal Reserve Bank of Cleveland, 4-9.

[3] Ayadurai, C., & Eskandari, R. (2018). Bank soundness: a PLS-SEM approach. In *Partial Least Squares Structural Equation Modeling* (pp. 31-52). Springer, Cham.

[4] Zervoudi, E. K. (2019). Parallel banking system: Opportunities and Challenges. Journal of Applied Finance and Banking, *9*(4), 47-70.

[5] Crosman, P. (2018). How Artificial Intelligence is reshaping jobs in banking. American Banker, *183*(88), 1.

[6] Lui, A., & Lamb, G. W. (2018). Artificial intelligence and augmented intelligence collaboration: regaining trust and confidence in the financial sector. Information & Communications Technology Law, *27*(3), 267-283.

[7] Fernández, A. (2019). Artificial intelligence in financial services. Banco de Espana Article, *3*, 19.

[8] Fethi, M. D., & Pasiouras, F. (2010). Assessing bank efficiency and erformance with operational research and artificial intelligence techniques: A survey. *European journal of operational research*, *204*(2), 189-198.

[9] Hakala, K. (2019). Robo-advisors as a form of artificial intelligence in private customers' investment advisory services.

[10] Sachan, S., Yang, J. B., Xu, D. L., Benavides, D. E., & Li, Y. (2020). An explainable AI decision-support-system to automate loan underwriting. Expert Systems with Applications, *144*, 113100.

[11] Francisco, D. F. (2019). *How artificial intelligence can help banks improve the customer experience of buying a house* (Doctoral dissertation).

[12] Burkhardt, R., Hohn, N., & Wigley, C. (2019). Leading your organization to responsible AI. *McKinsey Analytics*.

[13] Abdullah, O. (2017). Digitalization in Islamic Finance. Retrieved from http://kliff.com.my/wpcontent/uploads/2016/09/Sesi-3-Digitalization-of-Islami-Finance-Othman.pdf.

[14] Khan, N. (2017, December 13). The Impact of Financial Technology. Retrieved from https://www.linkedin.com/pulse/impact-financial-technology-nida-khan

[15] Huntley, D.G., (1991). Neural nets: An approach to the forecasting of time series. Soc. Sci. Comput. Rev., 9: 27-38. DOI: 10.1177/089443939100900104.

[16] Huang, Z., Chen, H., Hsu, C.-J., Chen, W.-H., Wu, S., (2004). Credit rating analysis with support vector machines and neural networks: A market comparative study. Decision Support Systems 37, 543-558

[17] Limsombunchai, V., C. Gan and M. Lee. (2005). An analysis of credit scoring for agricultural loans in Thailand. Am. J. Applied Sci., 2:1198-1205.

[18] Binner, J. M., Gazely, A. M., & Kendall, G. (2009). An evaluation of UK risky money: an artificial intelligence approach. Global Business and Economics Review, *11*(1), 1-18.

[19] Zhang, G.P., (2004). Neural Networks in Business Forecasting. 1st Edn., Idea Group Inc., ISBN: 1-59140-176-3, pp: 1-41.

[20] Eletter, S. F., Yaseen, S. G., & Elrefae, G. A. (2010). Neuro-based artificial intelligence model for loan decisions. American Journal of Economics and Business Administration, *2*(1), 27.

[21] Conway, J. J. E. (2018). *Artificial intelligence and machine learning: Current applications in real estate* (Doctoral dissertation, Massachusetts Institute of Technology).

[22] Rossini, P. (2000). Using expert systems and artificial intelligence for real estate forecasting. In *Sixth Annual Pacific-Rim Real Estate Society Conference, Sydney, Australia* (pp. 24-27).

[23] Shachmurove, Y., 2002. Applying artificial neural networks to business, economics and finance. http://ideas.repec.org/p/cla/penntw/5ecbb5c20d3d547f357aa130654099f3.html

[24] Nag, A. K., & Mitra, A. (2002). Forecasting daily foreign exchange rates using genetically optimized neural networks. Journal of Forecasting, *21*(7), 501-511.

[25] Rao, A. (2017). A strategist's guide to artificial intelligence. *Strategy+ Business*, *87*, 46-50.

[26] Bahrammirzaee, A. (2010). A comparative survey of artificial intelligence applications in finance: artificial neural networks, expert system and hybrid intelligent systems. Neural Computing and Applications, *19*(8), 1165-1195.

[27] Wall, L. D. (2018). Some financial regulatory implications of artificial intelligence. Journal of Economics and Business, *100*, 55-63.

[28] Cheatham, B., Javanmardian, K., & Samandari, H. (2019). Confronting the risks of artificial intelligence. McKinsey Quarterly, 1-9.

[29] Reuters (2017), "BlackRock Is Cutting Jobs and Banking on Robots to Beat the Stock Market" Fortune (March 29). Available at http://fortune.com/2017/03/29/blackrock-robots-stockpicking/.

[30] Siegal (2016), "BlackRock Is Making Big Data Bigger" Institutional Investor. Available at http://www.institutionalinvestor.com/article/3598029/asset-managementfixed-income/blackrock-is-making-big-data-bigger.html#/.WahmfsiGOr1.

[31] Noonan, Laura (2017), "JP Morgan Develops Robot to Execute Trades," Financial Times (July 31).

[32] Jagtiani, J., Vermilyea, T., & Wall, L. D. (2018). The roles of big data and machine learning in bank supervision. *Forthcoming, Banking Perspectives*.

[33] Xu, D., Zhang, X., & Feng, H. (2018). Generalized fuzzy soft sets theory based novel hybrid ensemble credit scoring model. International Journal of Finance & Economics, 24(2), 903-921.

[34] Abellán, J., & Castellano, J.G. (2017). A comparative study on base classifiers in ensemble methods for credit scoring. Expert Systems with Applications, 73, 1-10.

[35] Xiao, H., Xiao, Z., & Wang, Y. (2016). Ensemble classification based on supervised clustering for credit scoring. Applied Soft Computing, 43,73-86.

[36] Marqués, A.I., García, V., & Sánchez, J.S. (2012). Exploring the behaviour of base classifiers in credit scoring ensembles. Expert Systems with Applications, 39(11), 10244-10250.

[37] Wang, G., Ma, J., Huang, L., & Xu, K. (2012). Two credit scoring models based on dual strategy ensemble trees. Knowledge Based Systems, 26, 61-68.

[38] Hung, C., & Chen, J.H. (2009). A selective ensemble based on expected probabilities for bankruptcy prediction. Expert systems with applications, 36(3), 5297-5303.

[39] Nanni, L., & Lumini, A. (2009). An experimental comparison of ensemble of classifiers for bankruptcy prediction and credit scoring. Expert systems with applications, *36*(2), 3028-3033.

[40] Harris, T. (2015). Credit scoring using the clustered support vector machine. Expert Systems with Applications, 42(2), 741-750.

[41] Tomczak, J.M., & Zieba, M. (2015). Classification restricted boltz mann machine for comprehensible credit scoring model. Expert Systems with Applications, 42(4), 1789-1796.

[42] Hens, A.B., & Tiwari, M.K. (2012). Computational time reduction for credit scoring: An integrated approach based on support vector machine and stratified sampling method. Expert Systems with Applications, 39(8), 6774-6781.

[43] Chen, W., Ma, C., & Ma, L. (2009). Mining the customer credit using hybrid support vector machine technique. Expert Systems with Applications, 36(4), 7611-7616.

[44] Huang, G. B.; Ramesh, M.; Berg, T.; and Learned-Miller, E. (2007). Labeled faces in the wild: A database for studying face recognition in unconstrained environments. Technical Report 07-49, University of Massachusetts, Amherst.

[45] Li, S.T., Shiue, W., & Huang, M.H. (2006). The evaluation of consumer loans using support vector machines. Expert Systems with Applications, 30(4), 772-782.

[46] Schebesch, K.B., & Stecking, R. (2005). Support vector machines for classifying and describing credit applicants: Detecting typical and critical regions. Journal of the Operational Research Society, 56(9), 1082-1088.

[47] Luo, C., Wu, D., & Wu, D. (2017). A deep learning approach for credit scoring using credit default swaps. Engineering Applications of Artificial Intelligence, 65, 465-470.

[48] Zhao, Z., Xu, S., Kang, B.H., Kabir, M. M., Liu, Y., & Wasinger, R. (2015). Investigation and improvement of multi-layer perceptron neural networks for credit scoring. Expert Systems with Applications, 42(7), 3508-3516.

[49] Khashman, A . (2010). Neural networks for credit risk evaluation: Investigation of different neural models and learning schemes. Expert Systems with Applications, 37(9), 6233-6239;

[50] Bensic, M., Sarlija, N., & Zekic-Susac, M. (2005). Modelling small business credit scoring by using logistic regression, neural networks and decision trees. Intelligent Systems in Accounting, Finance & Management: International Journal, 13(3), 133-150.

[51] Kim, Y.S., & Sohn, S.Y. (2004). Managing loan customers using misclassification patterns of credit scoring model. Expert Systems with Applications, 26(4), 567-573.

[52] West, D. (2000). Neural network credit scoring models. Computers & Operations Research, 27 (11-12), 1131-1152.

[53] Metawa, N., Hassan, M.K., & Elhoseny, M. (2017). Genetic algorithm based model for optimizing bank lending decisions. Expert Systems with Applications, 80, 75-82.

[54] Abdou, H.A. (2009). Genetic programming for credits coring: The case of Egyptian public sector banks. Expert Systems with Applications, 36(9),11402-11417.

[55] Ong, C.S., Huang, J.J., & Tzeng, G.H. (2005). Building credit scoring models using genetic programming. Expert Systems with Applications, 29(1), 41-47.

[56] Leong, C.K. (2016). Credit risks coring with bayesian network models. Computational Economics, 47(3), 423-446

[57] Wu, W. W. (2011). Improving classification accuracy and causal knowledge for better credit decisions. International Journal of Neural Systems, 21(04), 297-309.

[58] Zhu, H., Beling, P.A., & Overstreet, G.A. (2002). A bayesian framework for the combination of classifier outputs. Journal of the Operational Research Society, 53(7), 719-727.

[59] Bijak, K., & Thomas, L.C. (2012). Does segmentation always improve model performance in credit scoring. Expert Systems with Applications, 39(3), 2433-2442.

[60] Yap, B. W., Ong, S. H., & Husain, N. H. (2011). Using data mining to improve assessment of credit worthiness via credit scoring models. Expert Systems with Applications, 38(10), 13274-13283

[61] Zhang, D., Zhou, X., Leung, S.C., & Zheng, J. (2010). Vertical bagging decision trees model for credit scoring. Expert Systems with Applications, 37(12), 7838-7843.

[62] Affes, Z., & Hentati-Kaffel, R. (2019). Predicting US banks bankruptcy: logit versus Canonical Discriminant analysis. Computational Economics, *54*(1), 199-244

[63] Christopoulos, A. G., Mylonakis, J., & Diktapanidis, P. (2011). Could Lehman Brothers' collapse be anticipated? An examination using CAMELS rating system. International Business Research, *4*(2), 11.

[64] Bostrom, N., & Yudkowsky, E. (2014). The ethics of artificial intelligence. The Cambridge handbook of artificial intelligence, *1*, 316-334.

[65] Kaya, O., Schildbach, J., AG, D. B., & Schneider, S. (2019). Artificial intelligence in banking. *Artificial intelligence*.

[66] Bary, E. (2018). How Artificial Intelligence Could Replace Credit Scores and Reshape How We Get Loans. MarketWatch. Available at: https://www.marketwatch.com/story/ai-based-credit-scores-will-soon-give-onebillion-people-access-to-banking-services-2018-10-09

[67] Hornigold, T. (2018). Life-or-Death Algorithms: Avoiding the Black Box of AI in Medicine. Singularity Hub. Available at: https://singularityhub.com/2018/12/18/life-or-death-algorithms-the-black-box-ofai-in-medicine-and-how-to-avoid-it/

[68] Gilchrist, K. (2018). Your next job interview could be with a robot. CNBC. Available at: https://www.cnbc.com/2018/10/03/future-of-jobs-your-next-jobinterview-could-be-with-a-robot.html

[69] Papernot, N., Abadi, M., Erlingsson, U., Goodfellow, I., & Talwar, K. (2016). Semi-supervised knowledge transfer for deep learning from private training data. *arXiv preprint arXiv:1610.05755*.

[70] Samek, W., & Müller, K. R. (2019). Towards explainable artificial intelligence. In *Explainable AI: interpreting, explaining and visualizing deep learning* (pp. 5-22). Springer, Cham.

[71] Arrieta, A. B., Díaz-Rodríguez, N., Del Ser, J., Bennetot, A., Tabik, S., Barbado, A., ... & Chatila, R. (2020). Explainable Artificial Intelligence (XAI): Concepts, taxonomies, opportunities and challenges toward

responsible AI. Information Fusion, *58*, 82-115.

[72] Nassar, M., Salah, K., ur Rehman, M. H., & Svetinovic, D. (2020). Blockchain for explainable and trustworthy artificial intelligence. Wiley Interdisciplinary Reviews: Data Mining and Knowledge Discovery, *10*(1), e1340.

[73] Byrne, M. F., Shahidi, N., & Rex, D. K. (2017). Will computer-aided detection and diagnosis revolutionize colonoscopy?. Gastroenterology, *153*(6), 1460-1464.

[74] Petrasic, K., Saul, B., Greig, J., Bornfreund, M., & Lamberth, K. (2017). Algorithms and bias: What lenders need to know. *White & Case*.

[75] Janssen, M., & Kuk, G. (2016). The challenges and limits of big data algorithms in technocratic governance.

[76] Gates, S.W., Perry, V.G., & Zorn, P.M. (2002). Automated underwriting in mortgage lending: Good news for the underserved. Housing Policy Debate, 13(2), 369-391.

[77] Aggour, K.S., Bonissone, P.P., Cheetham, W.E., & Messmer, R.P. (2006). Automating the underwriting of insurance applications. AI magazine, 27(3), 36-36.

[78] Berger, A. N., & Bouwman, C. H. (2013). How does capital affect bank performance during financial crises? Journal of Financial Economics, *109*(1), 146-176.

[79] Howard, P. K., (1994). The Death of Common Sense: How Law is Suffocating America. NewYork, NY: Warner Books.

[80] Research and Markets. (2017). Artificial Intelligence Market: Global Forecast to 2020. Ireland: Research and Markets. Retrieved from https://www.researchandmarkets.com/reports/3979203/artificial-intelligencechipsets-market-by

[81] IBM. (n.d.). AI and Bias. IBM. Available at: https://www.research.ibm.com/5-in5/ai-and-bias

[82] Bloomberg, J. (2018). Bias Is AI's Achilles Heel. Here's How To Fix It. Forbes. Available at: https://www.forbes.com/sites/jasonbloomberg/2018/08/13/bias-isais-achilles-heel-heres-how-to-fix-it/#2b782e676e68

[83] Eder, S. (2018). How Can We Eliminate Bias In Our Algorithms?. Forbes. Available at: https://www.forbes.com/sites/theyec/2018/06/27/how-can-weeliminate-bias-in-our-algorithms/#63f855a9337e

[84] McCullen, A. (2018). Ethical AI "Without reason, without heart, it destroys us"?. [online] Medium. Available at: https://medium.com/thethursdaythought/ethical-aiwithout-reason-without-heart-it-destroys-us-5436fb81ae1a

[85] Sears, M. (2018). AI Bias And The 'People Factor' In AI Development. [online] Forbes. Available at: https://www.forbes.com/sites/marksears1/2018/11/13/ai-bias-andthe-people-factor-in-ai-development/#f5874879134c

[86] Simpson, G. (2018). The Societal Impact of AI. CIO. Available at: https://www.cio.com/article/3273565/the-societal-impact-of-ai.html.

[87] Vanian, J. (2018). Unmasking A.I.'s Bias Problem. Fortune. Available at: http://fortune.com/longform/ai-bias-problem.

[88] Hao, K. (2019). This is How AI Bias Really Happens—and Why It's So Hard to Fix. [online] MIT Technology Review. Available at: https://www.technologyreview.com/s/612876/

this-is-how-ai-bias-reallyhappensand-why-its-so-hard-to-fix/

[89] Intel AI. (2019b). The Risks Of Dirty Data And AI. Forbes. Available at: https://www.forbes.com/sites/intelai/2019/03/27/the-risks-of-dirty-data-andai/#70b8807e2dc7

[90] Leetaru, K. (2019). Why Is AI And Machine Learning So Biased? The Answer Is Simple Economics. Forbes. Available at: https://www.forbes.com/sites/kalevleetaru/2019/01/20/why-is-ai-and-machinelearning-so-biased-the-answer-is-simple-economics/#28a8b3bf588c

[91] Marr, B. (2019). Artificial Intelligence Has A Problem With Bias, Here's How To Tackle It. [online] Forbes. Available at: https://www.forbes.com/sites/bernardmarr/2019/01/29/3-steps-to-tackle-theproblem-of-bias-in-artificial-intelligence/#691845ed7a12

[92] Pandya, J. (2019). Can Artificial Intelligence Be Biased?. Forbes. Available at: https://www.forbes.com/sites/cognitiveworld/2019/01/20/can-artificial-intelligencebe-biased/#678865bf7e7c

[93] McRaney, D. (2012). You are Not So Smart: Why You Have Too Many Friends on Facebook, why Your Memory is Mostly Fiction, and 46 Other Ways You're Deluding Yourself. New York: Gotham Books.

[94] Ernst & Young. (2018). The evolution of robo-advisors and advisor 2.0 model. 7. Retrieved from www.ey.com

[95] Wong, M. M. (2015). Hungry Robo-Advisors Are Eyeing Wealth Management Assets We Believe Wealth Management Moats Can Repel the Fiber-Clad Legion. 16. Retrieved from https://www.morningstar.com/content/dam/marketing/shared/pdfs/Research/equityreserach/20150409_Hungry_RoboAdvisors_Are_Eyeing_Wealth_Manage ment_.pdf

[96] Kirilenko, Andrei A., and Andrew W. Lo. (2013). Moore's Law versus Murphy's Law: Algorithmic Trading and Its Discontents." Journal of Economic Perspectives 27, no. 2: 51-72.

[97] Yampolskiy, R. V., & Spellchecker, M. S. (2016). Artificial intelligence safety and cybersecurity: A timeline of AI failures.

[98] Carney, M. (2017b), "The Promise of FinTech – Something New Under the Sun?" Speech at the Deutsche Bundesbank G20 conference on "Digitising finance, financial inclusion and financial literacy", Wiesbaden Germany. Available at http://www.bankofengland.co.uk/publications/Documents/speeches/2017/speech956.pdf.

[99] Lewis, D. (2018). Computers may not make mistakes, but many consumers do (Vol. 10923). https://doi.org/10.1007/978-3-319-91716-0

[100] Jung, D., Glaser, F., & Köpplin, W. (2019). Robo-Advisory – Opportunities and Risks for the Future of Financial Advisory. https://doi.org/10.1007/978-3-319-95999-3

[101] Fein, M. L. (2015). Robo-Advisors : a Closer Look By Melanie L. Fein. 1-33. https://doi.org/10.2139/ssrn.2658701

[102] Vincent, J. (2016). Twitter Taught Microsoft's AI Chatbot to be a Racist Asshole in Less Than a Day. The Verge. Available at: https://www.theverge.com/2016/3/24/11297050/tay-microsoft-chatbot-racist.

[103] Buranyi, S. (2017). Rise of the Racist Robots – How AI is Learning All Our Worst Impulses. [online] The Guardian. Available at: https://www.theguardian.com/inequality/2017/aug/08/

rise-of-the-racist-robotshow-ai-is-learning-all-our-worst-impulses

[104] Newman, D. (2017). Your Artificial Intelligence Is Not Bias-Free. [online] Forbes. Available at: https://www.forbes.com/sites/danielnewman/2017/09/12/yourartificial-intelligence-is-not-bias-free/#165b087c783a

[105] Ludden, C. (2015). The Rise of Robo-Advice: Changing the Concept of Wealth Management. Accenture, 12. https://doi.org/10.1007/978-3-319-54472-4_67

[106] Cocca, T. (2016). Potential and Limitations of Virtual Advice in Wealth Management. Journal of Financial Transformation, 44(December), 45-57. Retrieved from https://ideas.repec.org/a/ris/jofitr/1581.html

[107] Van Liebergen, B. (2017). Machine learning: A revolution in risk management and compliance? Journal of Financial Transformation, *45*, 60-67.

[108] Di Maio, P. (2020). Neurosymbolic Knowledge Representation for Explainable and Trustworthy AI.

Chapter 15

Implementation of Computerized Maintenance and Management System in Wine Factory in Ethiopia: A Case Study

Dejene Addis Nigussie and Muralidhar Avvari

Abstract

The productivity improvement is being main concern of all industries in spite of the type of product and amount of recourses or production system. One of the wine factories (Awash wine factory, Ethiopia) is also struggling to achieve its expected productivity since 1970s because of its poor maintenance management system. So far, productivity has been achieved by implementing various methodologies exclusively computerized maintenance and management system. Accordingly, the main aim of the study is to enhance productivity through smart maintenance management system. The current chapter emphasis on effective maintenance strategy and management system as the solitary way of improving productivity of the case company. The study used descriptive research design by applying both quantitative and qualitative research approach. Maintenance strategy followed by the company (brake down), frequent failure of critical machines, poor maintenance management system, lack of pertinent attention on maintenance, inadequate maintenance budget, considering maintenance as a cost center rather than a business center were the foundations for the problems. Arising from these problems, this study proposed smart maintenance strategy (Preventive Maintenance) and management system tool (CMMS) that improves reliability of machines reduce frequency of equipment failure, reduce breakdown time, decrease cost of maintenance, and then enhance productivity.

Keywords: CMMS, maintenance, management, wine factory, preventive maintenance, work flow

1. Introduction

The production system maintenance has evolved to be one of the most important areas in the business environment since few decades [1]. The growth of global competition caused remarkable changes in the way of production operation and process. These changes have been affected maintenance management and made its role even more essential in business success. Implementing maintenance management concept is one of the significant changes in production companies [2]. Consequently, maintenance management is a significant part of activities that impacts on production performance.

As automation and mechanization came into picture to change the production processes are converted from manual to machines in high end [3]. Accordingly, the global market is working in one-faced industrialization and to be competitive. Hence, every industry is striving to improve the productivity through different types of tools, techniques, systems, and philosophies. Improved maintenance management system is the best but unseen way that should be used by different manufacturing industries. The role of maintenance management must change to support the growing worldwide competition. It can no longer limit its role to immediate reaction to emergencies and overpower problems with more bodies and excessive overtime. In other words, maintenance should not always be reactive rather it should be proactive. In this regard, there is a need of executing maintenance more effectively.

The industries minimized the losses by locating right systems, infrastructure, process of the product development, and all these executed in a well manner. Hence, the operation will become stable to maximize the production output. This type of improvement is also known as maintenance excellence. It can redefine the roles, responsibilities of traditional, and even maintenance process is also necessary to improve to enhance the asset useful life and cost as well [4]. It can run like any other for-profit business and expected to meet its critical contribution to a fully integrated plant organization. All of these results are significant reductions in maintenance spending. It is not unusual for organizations to experience as much as a 50% reduction in maintenance cost as a result of moving from a reactive style of management to a preventive approach [5]. Therefore, appropriate maintenance resembles the improved management system in industries. Accordingly, preventive maintenance management has been introduced to repair or replacement of components, accomplishment of servicing with information management system. All of them are organized and implemented clearly to exploit the convenience of operational tasks in the industry.

However, productivity should be linked with maintenance management system and then only the possible improvements are occurring effectively. Productivity is associated with the outcomes established from the organization by what extent the inputs are accessing into the system [6]. The link between productivity improvement and maintenance management is well established. Proper maintenance management maximizes the performance and availability of machinery, which leads to increased productivity. Equipment effectiveness is no longer restricted to availability, but involves other factors, such as quality and efficiency. The impact of maintenance management on business performance aspect such as productivity and profitability has increased indefinitely in recent times due to its role in ensuring and improving machine availability, performance efficiency, product quality and swift delivery, environmental and safety requirements [7]. Therefore, one of the main areas for productivity improvement is maintenance management and so the research focuses on how to improve the productivity of Awash Wine Share Company through effective maintenance management system.

Awash Wine Factory is Ethiopia's longest established wine maker and supplier starting from 1936EC. Wine is the end product of the complete/partial alcoholic fermentation of grapes juice in many processes. Different machines and equipment are used in the process of wine production. In the course of time, these processing machines and equipment may fail regularly and resulted in reduction of productivity. Failures decrease production as well as erode profits off. The reduction in productivity can arise either through maintenance cost or increasing in production time. Higher plant-reliability leads to reductions in the frequency of equipment failure and wastages of energy. This study focuses on the investigation of the existing maintenance problems of the case factory and proposed the best solution that increase the reliability of the machinery and then enhance productivity. Consideration is focused like, maintenance management system to improve the output, machinery condition, operation standards, and reasons for failure, while simultaneously enhance productivity through reducing the downtime of machinery and overall maintenance cost. Hence, possible solutions have been found to resolve the maintenance system and its strategy to minimize the machine downtimes and the cost for maintenance. All these can be controlled with the preventive problems at the beginning by observing the failure root causes and also, proper leadership over the operators, testers, and helpers in the system. Moreover, the chapter tries to find the behavior of the industry to enhance the productivity by implementing smart maintenance strategy and management system, which is suitable to the factory based on its objective.

1.1 Maintenance management

Accorsi et al. [8] stated maintenance management as a business concept that describes the successful and efficient management of maintenance issues involved in the upkeep, operation and productivity of a factory, manufacturing facility or plant.

Maintenance management is considered as the collective actions like administrative, managerial, and technical tools, and over the product is to be produced without failures in the factory. Also, maintenance management helps to change and understands the state of functions or combination of its actions of the product to perform effectively. Accordingly, maintenance management is regulating the objectives which are given to produce a part and makes sure that the described targets are necessary to assign carefully from the department. To fulfill the given objectives, management methods such as strategies and responsibilities are to be planned and controlled by supervising economical aspects in the factory. This is closely aligned to other such notions found in modern maintenance literature. Further definitions consider maintenance management as the management of all assets owned by a company, based on maximizing the return on investment in the asset.

Maintenance management may be described as the function of providing policy guidance for maintenance activities, in addition to exercising technical and management control of maintenance programs [9]. Maintenance management system consists evaluation and analysis of indicative control variables such as performance, quality, equipment history and cost, and systematic implementation of programs like planned preventive maintenance, computerized maintenance management system, reliability improvement, cost reduction, and skill development schemes.

An effective maintenance management might be characterized as the product of prudence. Good maintenance programs and the efficient management systems behind them are essential for economically viable and operationally safe machinery. But most of the factories including awash wine factory lack appropriate machine/equipment maintenance strategy and management system.

A good maintenance management system helps to accomplish minimal downtime [10]. The management of maintenance is equally important with performing maintenance. A good maintenance management system makes equipment and facilities available. If the required equipment or service is down, or if the machine stops short of completing a job, time and money are wasted.

Literature on maintenance management has so far been very limited [11]. Basic steps of maintenance management stated that are request, approval, plan, schedule, performing work, recording data accounting for costs, developing management information, updating equipment history and providing management control reports [10]. Generally, the size of the maintenance group is determined in the company for achieving superior management and their control. Previously, the size could be varied 5–10% to the operating force. Whereas, now the trend has been changed like the size of the maintenance efforts that are increased significantly when compared with operating force. The increment of the size is because of advanced technologies introducing in all companies. It shows the maintenance requirements are greater than the necessity of manual operators.

The main goal of the maintenance management is to increase productivity and also profits from the effective operations by which many areas are to be considered and focused in the organization. It is very important to maintain the equipment over a prescribed schedule and plan in a company. In addition, practice of preventive maintenance guidelines and warranties enhances the productivity by saving company's budget in maintenance costs and acquired the overall safety of the crews.

Besides, inventory management is also necessary to control the effective business in the organization. It reduces the overhead and physical space on the stock parts and goods for better utilization. Definitely, the company's efficiency is increased professionally by managing the inventory. However, maintenance department acts at where the company is consuming bulk of its overhead costs. Consequently, business philosophy and approaches have been adjusting according to the management costs. Recent trends that have been developed in maintenance technology would convince to enhance the company's margins.

In other side, maintenance system could be seen as input to the output. Here, manpower, equipment, management, etc. are considered as input. Whereas in output, well equipment configuration and working conditions to operate effectively. Similarly, approaches and philosophies are to be maintained when the company needs to enhance its profit margins.

A good maintenance management system makes equipment and facilities available. If the required equipment, machineries or service is down, or if the machine stops short of completing a job, time and money are wasted. A good maintenance management system helps to accomplish minimal downtime [9].

An effective maintenance management encompasses the following:

1.2 Maintenance policy

These are rules or set of rules describing the triggering mechanism for the different maintenance actions. It is a question of what triggers maintenance actions. The common maintenance policies are: failure-based maintenance (FBM), condition-based maintenance (CBM), opportunity-based maintenance (OBM), can be sorted in this category.

1.3 Maintenance inventory/material control

Inventory is stores of goods and stocks kept to meet future demand. In manufacturing industries, items in inventory are called stock keeping items that are held at a stock (storage) point. Usually, the stock keeping items are raw materials, work-in-process, finished products, purchased parts, and supplies.

There are two types of inventories; independent and dependent demand inventories. Independent demand inventories are finished products inventory, while dependent demand inventories are inventories of raw materials and in process goods. The demand for independent items is forecasted, while the demand for dependent items should be calculated from the production requirements for independent demand items [12].

The objective of inventory management system is to make decisions regarding the approved level of inventory and changes in the level of inventory.

Various types of inventory management systems incorporate different decision rules toward inventory. Some are dependent on time and others on the levels of inventory, but the essential decisions are the same.

All inventory systems can be classified as one of the three varieties based on the above basic concept:

- Reorder point system
- Periodic review system
- Material requirement planning system

Steps such as job planning, coordinating with purchasing, coordinating with stores, coordination of issuance of materials, and reviewing the completed job can help reduce material related problems [13].

In the middle of the nights cannot be able to check the spare parts following an urgent call from the operating manager. However, many large companies have been arranged few sort of computers to control the spare parts ordering and warehousing. But, there is no alternative to check spare parts by individual knowledge like what computer can do regularly. Also, the spare parts are necessary to check whenever it is necessary to reduce the time and cost. For this, large companies have special arrangements to major equipment spare parts storage. The boxes have all the spare parts in an arrangement with size and stock numbers to make easy process. It will be consumed wherever the work has been carried out and then it will be returned with all lists of consumed parts to the warehouse.

This is also the time to check on the availability of special tools. These should be kept in a separate box, inventoried at regular intervals, and generally treated as a valuable spare part or essential resource.

1.4 Principles of spare part inventory

To facilitate the setting of inventory control policies, spare parts can be classified according to their usage rates into fast moving (where the demand is greater than, three items per year) and slow moving (demand less than three times per year).

1.5 Spare part inventory for fast moving parts

The main aim of the spare part administration activities is to control the holding stock cost against the running out cost. Inventory techniques are used to control the procedures which will reduce the cost of the following stocks such as running stock, replenishing stock, and holding stock. Two basic levels are considered to control the policies for enhancing the spares:

i. Re-order level: replenishment has been driven by the stock falling to fix the re-order levels;

ii. Re-order cycle: stocks are reviewed and replenishment has been allowed.

In the re-order level policy, the so-called “two-bin” system (**Figure 1**): the inventory policy is set in terms of a re-order level M and a re-order quantity q. Continuous monitor has been done over the stocks, and replenishment order to a fixed size q is ordered when the stock is on-hand falls to or below a fixed re-order level M. Hence, re-order levels are used as reservoir which can reduce the

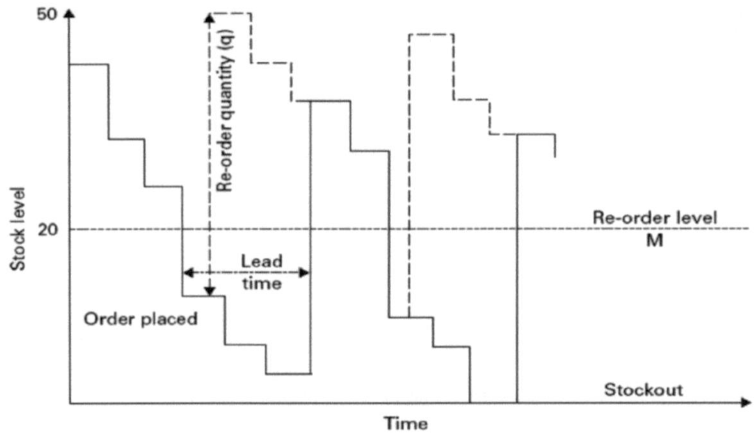

Figure 1.
Re-order level (source MTB by Alembazezwu).

possibilities of running out of stock caused from the dwell lead time and the random variability of demand. The outcome of the stock holding is as shown in **Figure 1**. Here, solid lines describe the stock held, and the dotted lines reveal about stock on-hand.

The expression for re-order quantity is:

$$q = \sqrt{2DCo}/\text{Ch} \tag{1}$$

where q = reorder quantity; C_o = ordering cost; D = mean demand per unit time; C_h = holding cost.

Assume stack is re-ordered at regular time interval.

Lead time between order and delivery is negligible.

Negative stack is not allowed by the policy.

Constant rate of demand per unit.

The reordering level M is calculated as:

$$M = DL + K\sigma\sqrt{L} \tag{2}$$

where L = lead time, σ =standard deviation of demand per unit time, k = normal standard variety.

Maintenance work order system:

A work order is approved and given the directions in a separate or group vise to carry out the tasks. In this, all maintenance activities and responsibilities are covered to control the costs and to enhance the job performance. Work order should at least contain information such as requested and planned completion dates, work description and its reasons, planned start date, labor and material costs, item or items to be affected, work category, and appropriate approval signatures.

The two main parts of work order system are:

1. The documents required to facilitate work planning, execution and control, and the work order flow process.

2. The basic documents required for the work order system include the work order, materials and tools requisition forms, job card, maintenance schedule, maintenance program, plant inventory, and equipment history files.

Detailed written instructions for any work or activities (job) to be carried out, in any component or part of a plant/equipment/machinery, must be clearly shown in the work order [14].

1.6 Equipment records

Equipment records consist the files including inventory, job performed, and maintenance cost. All these activities are well mannered which can be done while performing the jobs. Also, the maintenance costs are categorized in a historic profile manner. In general, stores and accounting departments provide the information regarding inventory. Operations, drawings, service manuals, warranties and so on are included in the files.

1.7 Maintenance job planning

Job planning is an essential element of the effective maintenance management. A number of tasks may have to be performed prior to commencement of a maintenance job; for example, procurement of parts, tools, and materials, coordination and delivery of parts, tools, and materials, identification of methods and sequencing, and coordination with other departments. Job planning is considered as a main technique to make good observations and preventive maintenance. Here, the primary job is to complete what parts to be maintained and what is the best method to enhance the performance of the jobs. Skills, talent, and good deal time are necessary to accomplish the better procedures. However, this should be performed to gain the good experience to train the future design engineers. But, writing skills are also an important qualification along with practical experience in many practices. The way of writing the language is short, concise, and clear to understand easily. All the statements should be written clearly and the following should be followed strictly:

- Title and identification number must be given for each and every procedure.
- Mention the purpose of the task.
- List out the tools, parts, and reference documents.
- Safety precautions are displayed clearly.
- Location of the job to an operator should be stated effectively.

When the procedure is completed and turned into maintenance control, the planner or scheduler should note any additional work required and see that it gets done according to priority.

1.8 Maintenance scheduling

Maintenance scheduling is as important as job planning. Schedule effectiveness is based on the reliability of the planning function. Fixed interval maintenance tasks and schedules are generally to be considered only when the option has to control the failures detected in advance. Hence, the planning for identifications and preventive levels can be finalized in days, weeks, and even months together soon to make sure that the time for production must be convenient.

1.9 Backlog control and priority system

Maintenance management effectiveness is having many determining factors in that one of the most promptly used factors is amount of backlog. These backlogs are important to manage the manpower and workload necessities in the organization. In addition, it uses in making the overtime decisions, assigning the jobs and recruitment, subcontracting, etc.

1.10 Performance measurement

Performance measures are necessary and it has been implemented in successful organizations so that measures can be controlled in various manners. Hence, it can be used as powerful tool to measure all the activities before and after the tasks to enhance the productivity of the organizations.

1.11 Maintenance strategy

Based on views, maintenance strategy includes: Corrective Maintenance (CM), Preventive Maintenance (PM), and Predictive Maintenance (PRM) [12].

1.11.1 Corrective maintenance

This type of maintenance has been used to recollect the items in satisfactory levels after it has a repair or after degraded below its performance. Planning is also to be considered to maintain the tasks and for logistics which can improve the efficiency of the company. The establishment of the maintenance and logistics requirements is accomplished through analysis of the system's designed-in maintainability and reliability features and attributes, whereas it is most expensive approach to make the maintenance. Service levels should be generally below the levels which can be acceptable. Sometimes, failures happen inappropriate times and hence, interrupt the production system severely. Definitely, maintenances are most expensive because resource planning is not been considered. But, parts need express movement, and the cost is also in effective manner. Overall, the corrective maintenance has been used in successful companies.

1.11.2 Preventive maintenance

It is "Maintenance carried out at predetermined intervals or according to prescribed criteria and intended to reduce the probability of failure or the degradation of the functioning of an item."

PM is maintenance performed to retain a system in a satisfactory operational condition by inspection, and subsequent repair or replacement, and by scheduled overhaul, lubrication, calibration, etc. Poor maintenance practice will actually degrade the condition of the system. Poorly executed preventive maintenance will result in significantly lower operational reliability, higher maintenance cost, and unnecessary downtime.

Preventive maintenance is any planned maintenance activity that is designed to improve equipment life and avoid any unplanned maintenance activity. All these levels are used to control the small failures before they arise in equipment outage. With this PM, many potential issues can be controlled before they occur.

Preventive maintenance is considered majorly in two factors. Such as, cost of the program and the utilization of the equipment.

Also, preventive maintenance can be characterized as follows:

a. **Scheduled Preventive Maintenance:** it can be carried out on size of use and the time schedule.

b. **Predetermined Preventive Maintenance/Time Based Maintenance (TBM):** it can be carried out without considerations of previous conditions with an interval times.

c. **Condition-Based Preventive Maintenance:** it is needed when the equipment is monitored repeatedly. The main aim of the maintenance evaluates and concludes the equipment that is necessity of the repairs and what conditions actions to be performed without suffering a breakdown. Many large and successful companies implemented this type of maintenance through automation techniques like PLCs (Programmable Logic Controllers). Respective signals to find out the failures are produced automatically using automation. However, the initial costs and start-ups are considered as very high when established by automation.

1.11.3 Elements of preventive maintenance

Activities of maintenance function could be either repair or replacement activities, which are necessary for an item to reach its acceptable productivity condition or these activities, should be carried out with a minimum possible cost consists of different activities in a given sequence.

Mentioned below are the most common preventive maintenance activities/elements:

Inspection: periodically inspecting materials/items to determine their serviceability by comparing their physical, electrical, mechanical, etc., characteristics (as applicable) to expected standards.

Servicing: cleaning, lubricating, charging, preservation, etc. of items/materials periodically to prevent the occurrence of incipient failures.

Calibration: periodically determining the value of characteristics of an item by comparison to a standard; it consists of the comparison of two instruments, one of which is certified standard with known accuracy, to detect, and adjust any discrepancy in the accuracy of the material/parameter being compared to the established standard value.

Testing: periodically testing or checking out to determine serviceability and detect electrical/mechanical-related degradation.

Alignment: making changes to an item's specified variable elements for the purpose of achieving optimum performance.

Adjustment: periodically adjusting specified variable elements of material for the purpose of achieving the optimum system performance.

Installation: periodic replacement of limited-life items or the items experiencing time cycle or wear degradation, to maintain the specified system.

1.11.4 Predictive maintenance

Another type of maintenance allows the forecasting of failures through analysis of the condition of the equipment. The analysis is generally conducted through some form of trending of a parameter such as vibration, temperature, or flow. Predictive maintenance allows equipment to be repaired at times that does not interfere with production schedules. This removes one of the largest factors from the downtime cost. The equipment service level will be very high under this type of maintenance. The comparison in different maintenance strategies is shown beneath in **Figure 2** and **Table 1**.

1.12 Maintenance planning, priority, scheduling, and control system

The existing system is observed and revealed that the planning of the maintenance and control system was not fully developed and carried out in a proper manner. Hence, proper planning, Scheduling, all necessary activities can be used in the organizations to minimize such kind of challenges/issues.

i. **Maintenance planning**

Planning is defined as the determination of overall all essential elements required to perform a job before its start time in advance. Good observations and preventive maintenance are necessary to find out the volume of size produced using the machines/equipment in daily, weekly, monthly, and yearly. This helps the maintenance to find the free time of the equipment and thus can save the time and cost. The following factors need to be considered in order to make a plan and schedule precise PM.

- Availability of storage information and required parts
- Availability of skilled labors
- Availability of necessary tools
- History of Machine and Equipment
- Working environment or condition
- Manufacturers recommendation

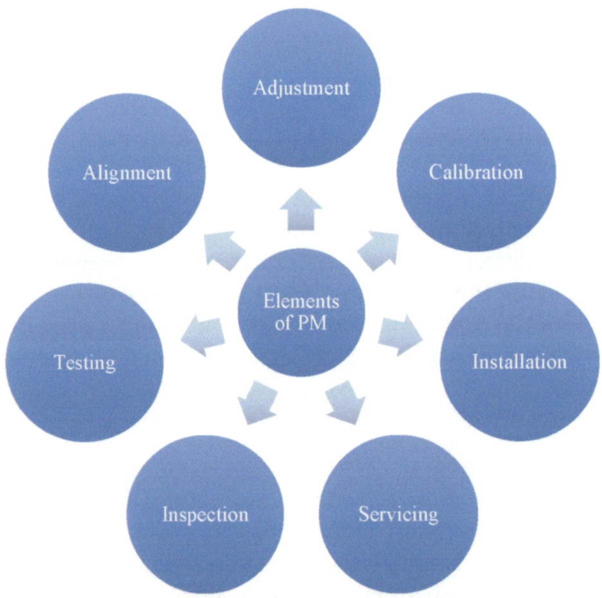

Figure 2.
The seven elements of preventive maintenance.

Strategy	Depiction	Payback	Limitation
Preventive maintenance	Schedule of planned maintenance actions aimed at the prevention of breakdowns and failures. Scheduling maintenance activities based on defined time intervals. It is assumed that equipment condition is directly related to time or use.	Prevent failure of equipment before it actually occurs. Preserve and enhance equipment reliability. More accurate and effective equipment maintenance due to application of recent technological advanced inspecting tools. Reduces reactive maintenance and provides a structure to maintenance actions. Flexible, energy savings, cost savings over reactive. Low cost, reduction in failures and breakdowns, extension of equipment life, improved trade-in/resale value of equipment, increased equipment reliability, increased plant productivity, fewer surprises, reduced cycle time, increased service level for the customer, reduction in the number of defective parts, and reduced overall maintenance	Does not totally eliminate unexpected equipment problems. Unneeded maintenance performed regardless of condition. Wastes resources/labor and results in large inventories.
Reactive maintenance	Remedial action carried out due to failure or deficiencies discovered during preventive maintenance, to repair an equipment/item to its operational state. Fix or replace a device, only after failure. Suitable for noncritical and low cost equipment.	Low cost/resources required. Little time, effort or expense for maintenance until absolutely necessary.	Unscheduled maintenance action, potential safety hazards, and increased costs due to unplanned maintenance and associated downtime, overtime, spare parts, and secondary damage.
Predictive maintenance	Assesses the equipment health through diagnostics testing and/or on-line monitoring to find and isolate the source of equipment problems.	Predicts when a device is likely to fail, minimizing the risk of random failure. Directs actions aimed at failure root causes as opposed to faults or machine wear conditions. Increased availability, quality, and safety.	High investment in diagnostic equipment and training. Results in being proactive in areas which have little effect on the plant's operation.

Table 1.
Comparison with in the three maintenance strategy.

- Criticality of machines and/or equipment
- Type of training and/or skill manpower to assist maintenance
- The task performed by the machine related to time
- Budget allocation

ii. **Maintenance Priority**

When the individual pieces of equipment have been identified for preventive maintenance, there must be a procedure for identifying the order in which they are to be done.

Awash wine factory uses first come first serve (FCFS) principle when it makes the maintenance action but this principle is not appropriate in such companies. The problem with this "first come, first served" method is that the more desirable work in friendly locations tends to get done while other equipment somehow never gets its preventive maintenance.

Need Urgency Customer Rank Equipment Criticality (NUCREC) is the recommended way of work scheduling. The NUCREC priority system helps to ensure that the most important items are done first. Also, it helps to recognize the crucial factors and enhance on the Ranking Index for Maintenance Expenditures (RIME) in many conditions. Ratings are decided based on the priority of the customer rank. This rating has been varied from one organization to another based upon the needs of particular parts produced in the company.

The rating system over the scale has been recommended from number 1–4. Generally, people will follow the numbering as number 1 is first choice to work get it done, similarly the NUCREC system does number 1 first.

Need urgency ratings include as follows:

- Rate 1: for emergency like safety hazards with potential further damage if not corrected immediately
- Rate 2: downtime of the equipment which produces less revenue than expected
- Rate 3: preventive maintenance and repetitiveness
- Rate 4: esthetic look and appropriate

The ranks of the customer are:

- Rate 1: top management
- Rate 2: production lines through direct revenue associations
- Rate 3: middle management, R&D, and frequent customers
- Rate 4: all others

The criticality ratings of the equipment are:

- Rate 1: effect of large areas includes utilities and safety systems
- Rate 2: major equipment with no back ups
- Rate 3: less impact on morale and productivity
- Rate 4: low, little use or effect on output

After overall ratings have been concluded, priority of the works would be determined and the value is varied in the range of 1–64 ($1 \times 1 \times 1$–$4 \times 4 \times 4$).

According to NUCREC principle, lowest products get the first priority. "Rate 1," is considered as high class emergency. Also, many of the factors are influenced to be selected as first priority but, they can be found using availabilities of labor and materials availability, locations, and scheduling fit. These priorities have been decided in a regular meeting of management. Initially, PM starts with rate 3 and it continuously increases the priority as the weeks passed and the rank will move to 2 and then rank 1. This can be done at the responsible time so that the developments have been found in the organizations.

iii. **Maintenance Scheduling**

Maintenance scheduling is the successive arrangements by which repairs have been carried out. It has a following stage like task to be achieved, who will conduct? Where to be carried out? When to be performed? Including necessary activities and estimated accomplishment time. Therefore, procedures of maintenance schedules make the successful preventive maintenance in an organization. Moreover, the amount of work to be determined for each and every craft workers is strictly mentioned by backlogs. This can be achieved by using the following Eq. (3).

The backlog is calculated as:

$$Backlog = \frac{total\ pland\ hour\ ready\ to\ schdule}{true\ craft\ capacity} \tag{3}$$

where:

Backlog = the amount of work that is documented as needing to be performed by the craft.

True craft capacity = the total hours scheduled for the craft minus scheduled interruptions.

Scheduled interruptions = average hours spent on emergencies, absenteeism, vacations, and routine maintenance work.

An illustrative example for a week with 6 working days 8 hour/day working time and 15 crews (permanent = 10 and contract =5) is:

Total permanent labor hours for next week (10 × 48 hours) = 480 hours.
Total contract labor for next week (5 × 48 hours) = 240 hours.
Total over time worked (average from last 3 months) = 60 hours.
Gross labor hours available = 780 hours.
Average emergency work (consider 30% from the last 3 months) = 235 hours.
Average absenteeism/week = 15 hours.
Average routine (non-back log) hours/week = 45 hours.
Average vacation hours/week/ = 05 hours.
Total deductions =300 hours.
Gross minus deduction = 780–300 = 480 hours.

These 480 hours represent what can really be expected to be completed from backlog work for the week. This number is used to calculate the true backlog as follows. Presume a planned hour to schedule is 3000 hours; the backlog is determined using Eq. (5.1).

$$\text{backlog1} = \frac{3000}{780} = 3.85\ weeks\ and\ backlog2 = \frac{3000}{480} = 6.25\ weeks$$

The illustrative example shows, it seems that 3.85 weeks are enough to complete the back log but in reality considering the constraints the back log needs 6.25 weeks to accomplish.

iv. **Maintenance Control**

Maintenance control refers to coordination and allocation of the set of maintenance resources, activities, tools, and procedures utilized to achieve the objectives of the maintenance system. For the successful accomplishment of planned maintenance, it is necessary to establish the mechanism by which each planned maintenance input and output activities are controlled. Maintenance is controlled based on the following major input and output; these are:

- Cost incurred
- Work control
- Stock control
- Quality and process control
- Performance
- Safety required
- Reporting and feedback system

Cost controlling: the most neglected part in maintenance activities conducted in AWF is the cost incurred due to improper maintenance management. Operation department does not stop the work for preventive maintenance purpose for the sake of time saving and short-term profit, respectively. Routine maintenance is not emphasized in the department rather break down maintenance is done. These problems resulted in complex maintenance work, high machine down time, and increased maintenance costs. Maintenance costs to be controlled include:

direct maintenance cost: labor cost and material cost and indirect maintenance costs: down time cost and overhead costs (**Figure 3**).

Successful cost reduction program requires:

- Approval of maintenance work before it is performed
- Planning of maintenance work according to its scope, criticality, cost, etc.
- Prioritized planned work

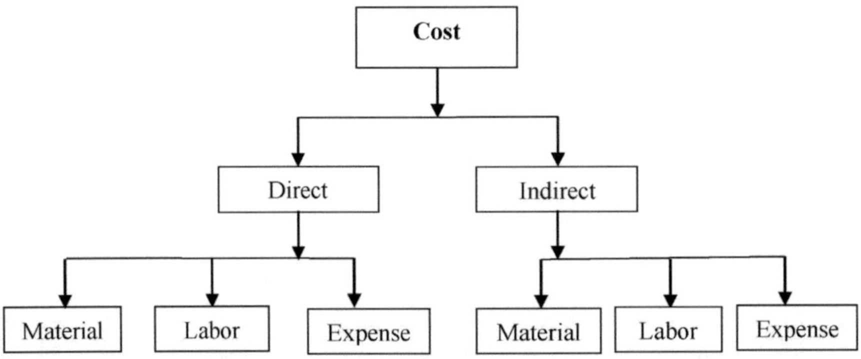

Figure 3.
Cost category in service and/or production sector (cost center based).

- Scheduling of planned and prioritized work
- Scrutinization of manpower; performance to be measured, and any questionable use of labor being justified by maintenance supervision

Work control: deals with monitoring the work status and the accomplished work to investigate if the work is done according to standards (quality and time). To do so, supervisors take responsibility for auditing the maintenance works either the routine maintenance or break down maintenance. To achieve this type of control, it is assumed that the maintenance control system includes standards that are assigned in advance of performing actual maintenance work. Proper planning of man-hours required is necessary to make use of optimum available human resources and implement it accordingly. To increase availability of production machine, use priority order as mentioned above under work order priority analysis.

Stock control: stock policies are considered as prior role to reduce the downtime over the selection of materials and parts. Planned maintenance activities are used to enable the giving orders of spare and consumable parts. Having the economic re-order quantities can minimize the inventory, labor cost, and also, plant downtime based upon the usage of the materials. Stock control based on fast moving items and slow moving items in the factory maintenance section is essential under maintenance schedule analysis as discussed in the sub-topic spare part supply.

Performance control: performance control can be seen from the technical personnel assigned to conduct the maintenance activities and the equipment maintained point of view. Therefore, it is essential to evaluate the whole system which includes the productivity and service quality of every technician and entire workshop. Moreover, the maintenance management department must consider and include the activities as follows:

Maintenance costs: the maintenance cost is inversely proportional to the efficiency of the organization.

Yearly down time: the efficiency is reduced when the downtime increases.

Service life: if the service life increases, efficiency is also increases.

The overall performance of the organization has been evaluated by cumulating of all mentioned parameters so that the productivity and quality can act accordingly to give promotions and incentives.

Quality and process control: the quality of the product has a direct link to the maintenance. Accordingly, effective maintenance gives minimal scrap and also increases the process capabilities. Scrap and repeated jobs are required to record and repot by monthly basis to enable for determining the quality of the machines. The corrective action will be taking place after identifying the repaired machines.

Reporting and feedback system: maintenance activities need follow-up, reviewing, monitoring, and streamlining of practice (corrective actions) and making continuous improvement to become a genuine part of organizational culture. Maintenance report should be produced every week by maintenance supervisor. The maintenance activities within a week or a couple of weeks should be evaluated by a committee of maintenance workers, maintenance foreman, planners, and production foreman and give solution to the problems and feedback goes to the planner for improvement. It encourages everyone involved in the maintenance process to be responsible for their performance efforts and accomplishments.

2. Maintenance program/cycle

Maintenance program/cycle is the maintenance type taking place in chronological order. In other words, maintenance cycle is the period between two successive major overhauls. Each preventive maintenance program conducted on each machine/equipment comprises maintenance cycles of the machine/equipment. Maintenance programs are inputs for preparation of maintenance plan. Maintenance program is the set of maintenance activities that are planned to be accomplished in a definite interval based on the following factors:

- Manufacture recommendations
- Age of the machinery/equipment/service life of major components
- Operation conditions
- Criticality of the machine or equipment
- Maintenance history of the machine or reported data

The maintenance program/cycle for the washing machine is illustrated as follows:
Type of machine: **BONY Washer Machine**.
Machine made: **Turkey**.
Machine model: **Mbt0213**.
Purchased date: **May 2013**.
Monthly running/operating hour: 8000 hours.
Total hours to be covered before the first overhaul: 288,000 (5 years).
Preventive maintenance interval: 6000 hours.

Interval of preventive maintenance conducted in month = $\frac{\textit{Interval of Preventiv Maintenance}}{\textit{Monthly running hour}}$

$$= \frac{6000}{8000} = 0.75 month = 23 days$$

Number of maintenance program/cycle = $\frac{\textit{total hour in a complet cycle}}{\textit{Preventive Maintenance Interval}} = \frac{288,000}{6,000} = 48.$

Thus, the 48 preventive maintenance programs/cycles for the washer machine to its first overhaul time is **MP1-MP2-MP1-MP2-MP1-MP2-MP1-MP3-MP1-MP2-MP1-MP2-MP1-MP2-MP1-MP3-MP1-MP2-MP1-MP2-MP1-MP2-MP1-MP4-MP1-MP2-MP1-MP2 MP1-MP2-MP1-MP3-MP1-MP2-MP1-MP2-MP1-MP2-MP1-MP3-MP1-MP2-MP1-MP2-MP1-MP2-MP1-MP5.**

The common activities to be performed on each maintenance program (PM_n) are listed in **Table 2**.

The 1 year preventive maintenance program proposed to the washer machine operates for *6200 hours and the last maintenance type performed is MP1 at the first position in the cycle as on April 30/2019* is discussed below (**Table 3**).

Determination of maintenance programs, maintenance cycles, and maintenance schedules for each machine or equipment simplify the task of maintenance planning. Maintenance planning can easily be determined (derived) from maintenance programs and maintenance cycles.

Maintenance planning may be prepared for the life of the machine/equipment or working budget year. Annual maintenance plan is recommended to Awash Wine Factory. The annual maintenance plan has to be broken down to quarterly, monthly, weekly, and daily maintenance works to be done. The above maintenance planning samples help the maintenance department to prepare and adopt their annual preventive maintenance plan for other machineries/equipment and instruments according to the criteria already discussed.

Maintenance program	Types of maintenance activities	Period
MP_1	Clean parts, inspection parts, lubricate necessary parts, clean burner chamber, etc.	Every 6000 hour or 3 weeks whichever comes first
MP_2	Perform all activities in MP1, check and adjust pump drive belts, inspect and adjust the main chain, etc.	12,000 hour or 1 and half months whichever comes first
`MP_3	Perform all activities in MP2, check and repair: carriers, roller rails, pipes, pump impellers, electric motors	Every 48,000 hour or 10 months whichever comes first
MP_4	Perform all activities in MP3, check and change: bearings, mechanical seals, bushings	Every 144,000 hour or 2 and half years whichever comes first
MP_5	Perform PM4, check and change the main chain, check and change the injector, check and change the electric control panels, etc.	288,000 hour or 5 year whichever comes first

Table 2.
Maintenance activities in a maintenance program.

PM conducting date	Running hours	PM program	Position/cycle
May 23, 2019	12,000	MP2	2
June 16, 2019	18,000	MP1	3
July 9, 2019	24,000	MP2	4
August 2, 2019	30,000	MP1	5
September 25, 2019	36,000	MP2	6
October 18, 2019	42,000	MP1	7
November 11, 2019	48,000	MP3	8
December 4, 2019	54,000	MP1	9
December 27, 2019	60,000	MP2	10
January 20, 2020	66,000	MP1	11
February 13, 2020	72,000	MP2	12
March 6, 2020	78,000	MP1	13
March 29, 2020	84,000	MP2	14
April 22, 2020	90,000	MP1	15
May 15, 2020	96,000	MP3	16

Table 3.
Annual maintenance schedule for the washer machine with PMI 6000 hours.

Maintenance program for other machineries and equipment is left to the maintenance staffs of the factory.

2.1 Maintenance work flow

The success of maintenance organization relays highly on workflow system. For successful accomplishment of maintenance activities, it is necessary to develop and implement a well understood workflow system. This simplifies the task of maintenance and enhances productivity. Thus, for the case company, the suitable work flow system that should be followed by maintenance planner is proposed in **Figure 4** beneath.

Work orders raised from different departments will be filled on formal work order request form and approved by department manager and submitted to maintenance planner. Therefore, the planner will organize work order from its request formats. Based on the demand, works can move quickly in terms of allocating manpower, arrangement of spare parts to complete the given task effectively. Next work orders released to executers during execution maintenance planners has to follow the progress and quality of work. After completion, the maintenance work should be validated or confirmed for successfulness. This can measure by reworks and/or number of defects. Record keeping is the last stage. It helps to keep histories of machine, measuring performance, and prop up decision making. The

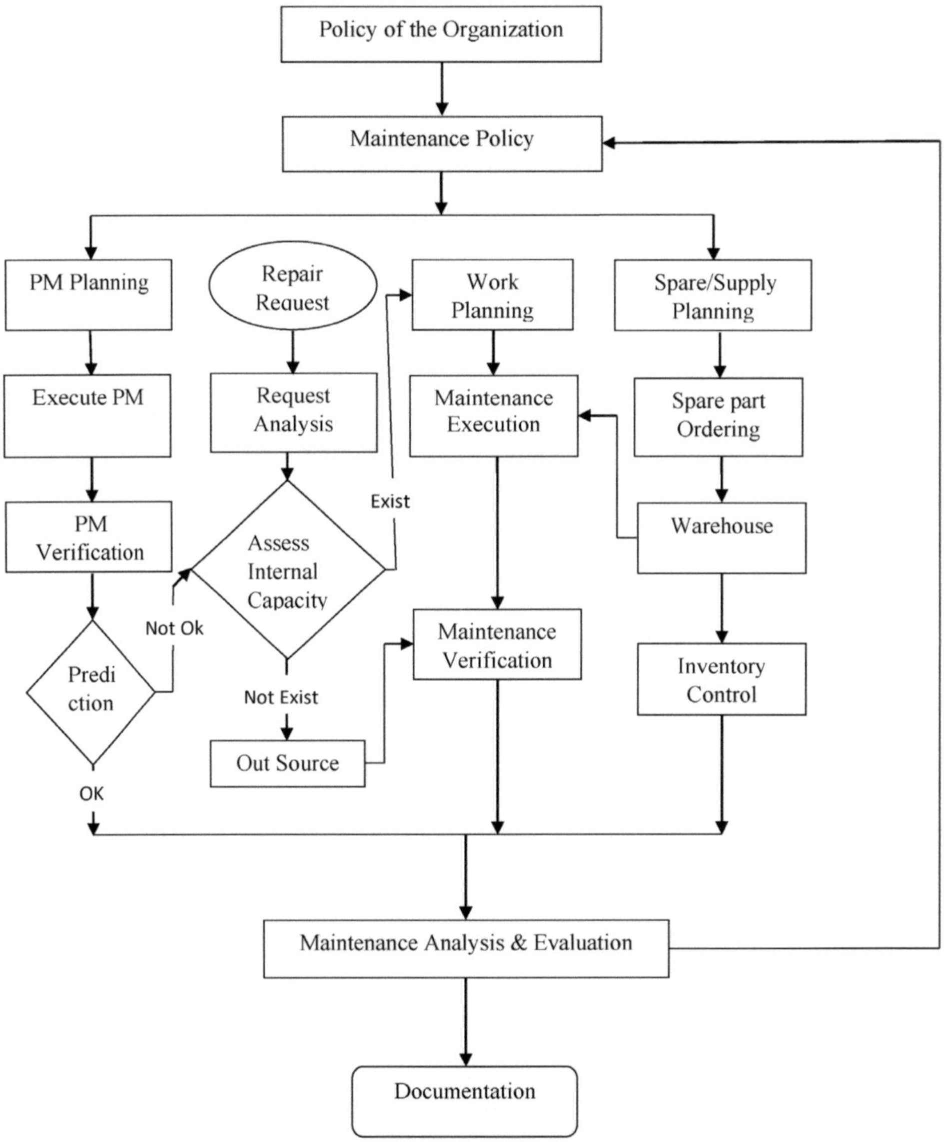

Figure 4.
Recommended maintenance work flow algorithm.

recommended workflow model has considered all the requirements of an effective & efficient recent workflow system.

2.2 Description

Awash wine factory uses breakdown maintenance strategy with fixed time preventive maintenance management system. That means the machines are maintained according to the maintenance program planned once in a year. But these types of maintenance system is not efficient for such batch production type process, thus preventive maintenance strategy with condition-based maintenance policy is advisable for such industries. The recommended work flow model consists both preventive and reactive work flow system, for the following basic reasons: (1) it is ideal to make an accident free production system, (2) it is difficult to make a complete paradigm shifting of the existing maintenance culture of the plant from breakdown to preventive once, thus breakdown maintenance should be incorporated in the recommended system and gradually changed into pure preventive strategy.

Regularly, preventive maintenance activities have been planned and implemented to each and every machine and equipment in the organization. As a result, the condition of the machine or equipment could be found and hence, the maintenance is decided whether it should be small, medium or high and even some times the maintenance is not required. Accordingly, the feedback is considered to make a plan to work out on good and repaired machineries together. The selection of a part and quantity of the spare parts are to be ordered concurrently to minimize the time to wait for spare parts. Finally, the observations and feedback has been analyzed for the future maintenance and stored safely. The model consists of basic activities to be performed so as to reduce the breakdown time and maintenance cost.

Preventive maintenance: at the beginning of every fiscal year, preventive maintenance plans are prepared by maintenance planner discussing with maintenance supervisor, and finally approved by engineering and maintenance manager. Based on the annual plan, preventive maintenance is initiated by assigning an expert/technician. The assigned maintenance crew conducts PM activity on machineries/equipment and identifies the necessary services to be undertaken. Based on the assessment of the assigned maintenance crew, engineering and maintenance manager decides whether internal capacity exists or not. If internal capacity exists for the periodical service, then the assigned maintenance crew will service the machine. If internal capacity does not exist, then engineering and maintenance manager/commercial department head will outsource the service.

Verification is conducted to ensure that the maintenance job is done to prevent the occurrence of potential failures. Maintenance report is prepared indicating the type of service undertaken and the responsibilities allocated at various maintenance stages.

Corrective maintenance: Assigned expert/technician conducts inspection on the machine/equipment requested for maintenance and records the findings including:

- The nature of failure
- In-house capability
- Resources needed to undertake the maintenance if internal capability exists
- Availability of spare parts in store and in the market
- If there is a need for modification and other related activates

Based on the assessment of the assigned expert/technician, depending on the machine to be maintained, engineering and maintenance manager decides on whether internal capacity exists or not and other related information with maintenance. If internal capacity exists for the break down maintenance, then the maintenance work is planned and executed. If internal capacity does not exist, then engineering and maintenance manager/commercial department head will outsource the service. Verification is conducted to ensure that the maintenance job is done to prevent the occurrence of potential failures. Maintenance report is prepared indicating the type of maintenance undertaken and the responsibilities allocated at various maintenance stages.

3. Computerized maintenance management system

A Computerized Maintenance Management System (CMMS) is a computer software program designed to assist in the planning, management, and administrative functions required for effective maintenance.

The main challenge of the management maintenance system in the large organization process is complexity, and it has a substantial impact over the entire business. In addition, the process is impossible to maintain or manage effectively without the computerization. Therefore, alternative management program is necessary to achieve successful implementation of the systems in organization.

Functions of computerized system include inputs, methods, outputs, and improvement activities as shown in **Figure 5**.

These functions perform the generating, planning, and reporting of work orders; the development of a traceable history; and the recording of parts transactions. CMMS and their integration into pre-existing organizations have been proven as an excellent platform to promote communications while improving coordination between different functions in the organization. Companies of all sizes can benefit from maintenance software, whether they have a maintenance team of five or five hundred. Users of CMMs can generate the following benefits/advantages:

- **Less work outages:** it is easy to do preventive maintenance/less surprise breakdowns

- **Better accountability**: quick check on one time work done and completion alert

- **Less overtime:** better scheduling; even work distribution; no sitting idle/ working overtime

- **Information capture:** recorded problems and solutions, used as information for next task

- **Savings on purchases:** replacement of acceleration time to purchase the spare parts and pricing from the inventory planning features.

- **Certification and analysis:** managers can understand and analyze the total recorded data for the energy usage and utilization of maintenance planning in the company.

Considering the benefits listed above and believing that it is vital to improve the maintenance management in the company, the CMMS software user friendly

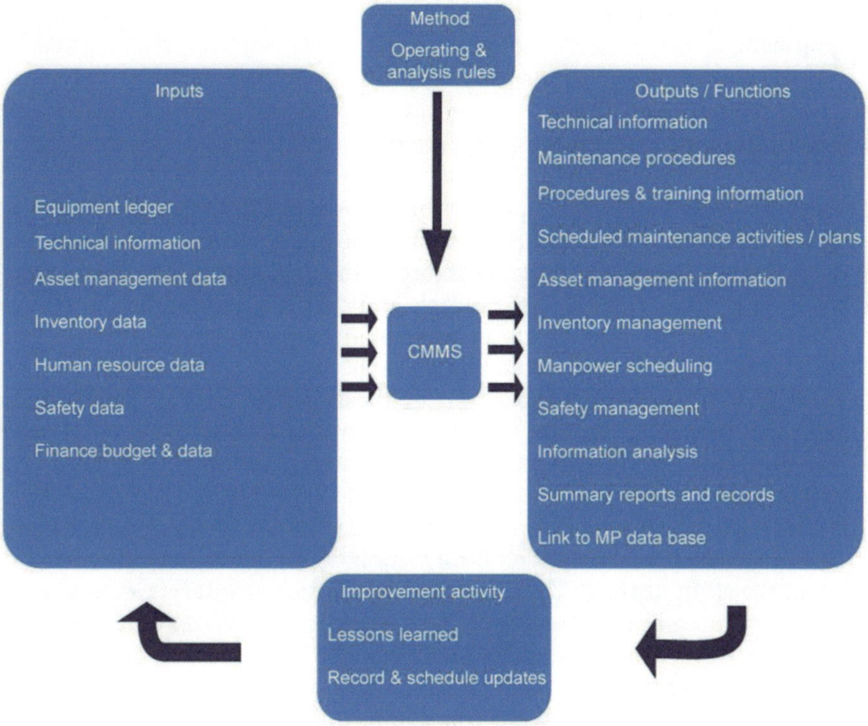

Figure 5.
Process in CMMS (source SMMT industry forum ltd. 2680 king court).

program with seven entities (*Search Menu, department menu, employees menu, equipment/machinery menu, work requisition menu, task/maintenance work menu and materials or spare parts menu*) is developed to the company.

4. Objective

The foremost intention of this study was to enhance productivity of Awash Wine Factory through smart maintenance management system (strategy and management tool). The study focused on the investigation of the existing maintenance problems of the factory and proposed the best solution that increases the reliability of the machinery and then enhances productivity. Consideration is focused on: the maintenance management system to improve the output, machinery condition, operation standards, and reasons for failure, while simultaneously enhances productivity through reducing the downtime of machinery and overall maintenance cost. Hence, in the current chapter took a challenge to develop a model or strategy that will minimize the downtime of the equipment and cost of maintenance. Alternatively, smart maintenance strategy and management system has been developed to enhance the factory's productivity effectively.

The specific objectives of the study were:

- To assess the existing maintenance system
- To explore and identify the problems in maintenance management system
- To assess the maintenance inventory control mechanism

- To develop preventive maintenance program for critical machinery and/ equipment

5. Research design and target population

The research methodology adopted in this study was a descriptive survey type since this research method accurately describes the relation between variables, increase fairness, and maximize the reliability of the data. The method provides straightforward summaries about the sample and the observations that have been made. This ensured that appropriate answers are obtained for the research questions.

The data collecting techniques employed in this work is: subject matter literature surveys using: journal articles, books, related thesis papers; survey questionnaires; unstructured interview questions; visit and assessment of secondary sources such as factory maintenance log book, annual finance reports, published documents, and work order sheets.

The data used in this study have been collected from primary and secondary sources. Primary sources focused on information that helps to have entire understanding of the study through different techniques such as interviews, questionnaires, observations, and discussions, whereas secondary sources of information related to the study have detail understanding with the supportive documents and reports. Both quantitative and qualitative data were collected.

Non-random sampling which is purposive technique was selected in this study because this type of sampling is extremely functional to construct reality, describe a phenomenon or build up an impressive about a universe using specific knowledge. This sampling strategy is feasible to quantitative researches.

The sample for this study is taken from the target population of maintenance department, production department, and management parts of AWF at the two sites (Lideta and Mekanisa)

Population variability, availability of participants, and the suitability to the required information are factors considered when the sample size is determined.

Eqs. (4) and (5) are used to determine the sample size since the population is finite.

$$\mathrm{N} = Z^2 P(1 - \mathrm{P})/\mathrm{C}^2 \tag{4}$$

$$\mathrm{Ns} = \frac{N}{1 + \frac{N-1}{No}} \tag{5}$$

where N = sample unit (initial sample size)
Z = value of level of confidence (consider 90% confidence level, Z is 1.645)
P = percentage picking a choice expressed as decimal 0.5 used for sample size needed
C = confidence initial expressed as decimal 0.08 = +/− 8%
Ns = size of targeted sample
No = size of the population (214 in this study)
Applying Eqs. (3.1) and (3.2), the target sample size is determined as:
N = 105 and Ns = 70

6. Summary

The first part of the questionnaires presented on the six parameters was discussed in primary data analysis, and the result is summarized as follows:

Regarding the worth given to maintenance, most of the respondents agreed that the attention given to maintenance department is less and it is not considered as a business center rather as a cost center. The budget for maintenance is negligible when compared with other departments. The maintenance management system is outdated, and the scheduling operation plan of maintenance department is poor. The common maintenance action takes place in the company is breakdown based which contributes much for breakdown time increment.

Concerning the intense of the company toward preventive maintenance, the respondents assure that there is a fixed time based preventive maintenance that takes place once in a year regardless of the condition of machines/equipment. However, priority is for reactive maintenance than preventive. While coming to maintenance planning and scheduling, most of the respondents disagreed on the presence of planned and scheduled maintenance. The back log control system, recording and documentation system of the company is said to be meager. The maintenance work is not prioritized technically. The maintenance work recording and documentation system is loyal to mistake and not used as a source for next work. The repair work analysis habit is so poor and needs to be changed. The company puts much effort toward training, and the communication system in the company is committed. The inventory management mechanism is so poor that is why much cost is elapsed when compared with the other maintenance expenses in the company.

The second part of the questioner shows that the frequently failed and bottle-neck machine in the company is washer machine with the production rate of 64.17%. The main causes of the losses in production are stated as mechanical failure in washing section which covers the largest portion (35%), maintenance strategy followed by the plant (27%), maintenance management program (25%), and high cost of maintenance inventory control, and other (13%).

The company has got problems of highest machine breakdown time resulted in a loss of 1,188,950; 1,126,600; 1,174,975 bottles of wine in 2018, 2017, 2016 production year, respectively. In production year 2018, the total product is given by (2336–116.5) hour $\times$ 4300 bo/hour = 9,543,850 bottles. The loss in production in this year is (116.5 + 160) hour $\times$ 4300bot/hour = 1,188,950 bottles which is 12.46% of the total product. In addition to production loss, high amount of money is elapsed in maintenance in the stated production years due to the lack of proper maintenance strategy and management system.

7. Conclusion

The aim of this study was to enhance productivity of Awash Wine Factory through improved maintenance management system. The basic maintenance problems of the factory were investigated, and smart maintenance system that decrease machine down time and reduce maintenance cost is recommended. Data collected through questionnaires, unstructured interview, visit and document analyses were used to identify the major causes of the problem. Case and effect diagram and failure mode effect and criticality analysis were used as a tool to differentiate the root cause of the problems. The collected data were analyzed by SPSS software, and the following results were obtained. The company has got problems of highest machine breakdown time and high yearly maintenance cost due to wrong maintenance strategy, poor maintenance management system, and repeated failure of washer machine. As a result, the company has lost more than 12% of its yearly product and incurred much money as cost of maintenance.

After the analysis, this study has suggested ways to enhance productivity and increase competitiveness of the factory in the modern market. The recommended maintenance system incorporates preventive maintenance as maintenance strategy and CMMS as a maintenance management tool with their respective merits explained in the fifth chapter of the study. New annual maintenance program for critical machines like washer machine is developed. The suggested smart maintenance system helps to decrease machine downtimes, increase availability of the machinery, and reduce maintenance costs through the creation of a sense of ownership in each of the plant-equipment operators, maintainers, and support staff so as to encourage “a prevention of problems at source” attitude. Implementing the new maintenance system, the factory will generate many benefits. For instance, considering the present efficiency of the company leaving other factors, the yearly product loss due to machine break down will be reduced from 12 to 6%.

Author details

Dejene Addis Nigussie[1] and Muralidhar Avvari[2]*

1 Awash Wine Factory, Addis Ababa, Ethiopia

2 Faculty of Mechanical and Production Engineering, Arba Minch Institute of Technology, Arba Minch University, Arba Minch, Ethiopia

*Address all correspondence to: seemurali@gmail.com

References

[1] Rivera-Gomez H, Oscar Montano-Arango O, Corona-Armenta JR, Garnica-Gonzalez J, Hernandez-Gress ES, Barragan-Vite I. Production and maintenance planning for a deteriorating system with operation-dependent defectives. Applied Sciences. 2018;**8**:165

[2] Cohena T, Baretichb MF, Gentlesc WM. Chapter 3 & 4 - computerized maintenance management systems. In: Clinical Engineering Handbook (Second Edition). Elsevier; 2019. pp. 208-218

[3] He M, Li H, Duan Y. Research on railway intelligent operation and maintenance and its system architecture. In: 2019 6th International Conference on Dependable Systems and Their Applications (DSA), Harbin, China. 2020. pp. 488-490

[4] Vilarinho S, Lopes I, Oliveira JA. Preventive maintenance decisions through maintenance optimization models: A case study. Procedia Manufacturing. 2017;**11**:1170-1177

[5] Duan C, Deng C, Wang B. Multi-phase sequential preventive maintenance scheduling for deteriorating repairable systems. Journal of Intelligent Manufacturing. 2019;**30**: 1779-1793

[6] Lofsten H. Management of industrial maintenance-economic evaluation of maintenance policies. International Journal of Operations & Production Management. 1999;**19**:716-737

[7] Swanson L. Linking maintenance strategies to performance. International Journal of Production Economics. 2001; **70**:237-244

[8] Accorsia R, Galloa A, Tufanoa A, Bortolinia M, Penazzia S, Manzinia R. A tailored maintenance management system to control spare parts life cycle. Procedia Manufacturing. 2019;**38**:92-99

[9] On the Definition of Maintenance Management. The Maintenance Management Framework. London: Springer Series in Reliability Engineering. Springer; 2007. pp. 3-10

[10] Budai G, Dekker R, Nicolai RP. Maintenance and production: A review of planning models. In: Complex System Maintenance Handbook. London: Springer Series in Reliability Engineering. Springer; 2008. pp. 321-344

[11] Goyal RK, Maheshwari K. Maintenance management practices: A retrospective and literature review. International Journal of Advances in Engineering Research. 2012;**3**:1-18

[12] De Salvo MC, Gaudenzi R, Begalli B. Risk management strategies and residual risk perception in the wine industry: A spatial analysis in Northeast Italy. Land Use Policy. 2019;**83**:47-62

[13] Mahilet M. Enhancing Productivity through Improved Maintenance System (Case Study in Dashen Brewery Share Company). Thesis of AUU Institutional Repository; 2007

[14] Olechowskia A, Oehmenb J, Seeringa W, Ben-Dayac M. The professionalization of risk management: What role can the ISO 31000 risk management principles play? International Journal of Project Management. 2019;**34**:1568-1578

Chapter 16

Maintenance Decision Method Based on Risk Level

Yang Tang, Xin Yang and Guorong Wang

Abstract

Maintenance decision method of the mechanical equipment still has some deficiencies and shortcomings, including unreasonable maintenance methods, surplus or insufficient repair, and unscientific inspection and repair intervals, especially for equipment with different risks. As a consequence, more frequent failures and higher maintenance costs of the mechanical equipment occur along with some major safety accidents and economic losses in the process of production. To overcome these problems, a framework for maintenance decision method based on risk level is presented for mechanical equipment in the petrochemical industry. First, 10 evaluation indexes and a set of scoring criteria quantifying the subjective evaluation are defined to evaluate the risk level of each mechanical equipment. Based on the analytic hierarchy process (AHP), the weight of the evaluation indexes and the evaluation model of the risk levels were established. Then, the subjective effects in the scoring process were removed using the Monte Carlo simulation (MCS) approach. Next, maintenance decision-making trees of the mechanical equipment were formulated by referencing the logic decision tree of reliability-centered maintenance. Finally, the feasibility of the framework was effectively verified by testing a well-control system in the oil field, for which the risk level and maintenance methods were obtained.

Keywords: maintenance decision method, mechanical equipment, risk level, petrochemical industry, maintenance decision-making tree

1. Introduction

Through extensive research, we found that nearly half of all major accidents and economic losses were caused by failure of the equipments that are inherently high risk in the petrochemical industry. While most equipment failures were attributed to the current maintenance decision method in the petrochemical industry that was backward and unscientific. Moreover, the conventional maintenance methods could not guide the maintenance personnel to carry out timely maintenance for reliability and safety of the mechanical equipment. As we all known, some maintenance decision models and maintenance methods are widely applied in different industries, such as aerospace, electricity generation, and transportation. But there were no effective decision-making methods or scientific theoretical models to satisfy the special mechanical equipment in the petrochemical industry. So that there are some negative outcomes, including surplus repair, insufficient repair, unreasonable repair intervals, higher maintenance costs, and so on, which were brought about in their maintenance work [1]. At present, based on the different attention focuses in each industry and the different model and method choices in the maintenance and management process, many research results are related to the maintenance decision method. Bertolini and Bevilacqua [2] proposed a new maintenance decision method to adopt a modified FEMCA analysis and a type of Monte Carlo simulation (MCS) approach based on different important levels of the power plant equipment. Bertolini and Bevilacqua presented a new maintenance decision technique to determine the better maintenance strategies for the critical centrifugal pumps in an oil refinery [3]. A maintenance decision method of a multi-criteria classification of equipment was proposed by Gómez de León Hijes and Cartagena by the analytic hierarchy process (AHP), and oil pipeline projects were effectively evaluated by Dey with a multiple attribute decision-making technique [4, 5]. Chang et al. applied a new maintenance decision model to estimate the production availability in offshore installations [6]. In the study above, some mathematical models, including AHP and MCS, are often used for making maintenance decisions. However, through research and investigation, very few applications of both AHP and MCS exist for making maintenance decisions of the mechanical equipment based on their different risk levels. Moreover, there are some differences between the mechanical equipment in the petrochemical industry and the ones in other industries, such as types and distribution of the failure, the methods and costs of the maintenance, and requirements for reliability and safety, due to the factors of harsh construction environments, complicated working conditions, and extremely high safety requirements in the production process [7].

Thus, these existing maintenance decision models and the maintenance strategies applied to the equipment in other industries are not directly suitable for the mechanical equipment in the petrochemical industry [8]. Therefore, it is necessary to study the maintenance decision method belonging to the mechanical equipment in the petrochemical production process by focusing on the features of its high risks and hazards [9]. A new framework is put forward for making maintenance decisions based on the different risk levels of the mechanical equipment. Finally, through the framework of maintenance decision making, a more reasonable and more effective maintenance strategy can be devised for the mechanical equipment to guarantee the reliability and security of the production operation.

The remainder of this chapter is organized as follows. In Section 2, the influence factors of the risk level of the mechanical equipment are defined, and their scoring criteria are formulated. In Section 3, an evaluation model for the risk level of the mechanical equipment is established using AHP, and then, the MCS approach is applied to reduce the subjective influences in the scoring process. Then, three MDMETs for the mechanical equipment are obtained based on their categories of different risk levels in Section 4. Finally, Section 5 provides some discussion and conclusions.

2. Maintenance decision method based on risk level

2.1 Definition of evaluation indexes and scoring criteria of risk level

Through the FMECA of the mechanical equipment in the petrochemical industry, from the four aspects of reliability, economics, monitorability, and maintainability, 10 influencing factors directly related to its risk level were analyzed, as shown in **Table 1** [10].

In order to ensure that the evaluation of the risk level of machinery and equipment is not too complicated, and that the accuracy of evaluation is balanced, and

Index	Serial number	Factors of affecting risk Level
Reliability factor	1	Personnel safety (PS)
	2	Environment and health (EH)
	3	System functions (SF)
	4	Failure frequency (FF)
Economic factor	5	Production loss (PL)
	6	Maintenance costs (MC)
Monitorability factor	7	Inspectability (IN)
Maintainability factor	8	Downtime (DT)
	9	Maintenance difficulties (MD)
Other factor	10	Service length (SL)

Table 1.
Influencing factors of risk level about the mechanical equipment.

Serial number	Casualty	Grading
1	No impact at all	0
2	Minor injury	1–3
3	Seriously injured	4–7
4	One death	8–9
5	Mass casualties	10

Table 2.
Scoring criteria of PS.

the degree of influence of each factor is more effectively quantified. Therefore, by experts, professional maintenance personnel and field operators through the review pointed out that the risk level of influencing factors according to the situation divided into 3–5 levels, using a 10-point system for scoring. The scoring standards for the 10 influencing factors related to the risk level of mechanical equipment are as follows [11].

2.1.1 Influence of failure on personnel safety (PS)

For petroleum and petrochemical companies, ensuring production safety is the most important issue. Among·the petroleum and petrochemical companies' mechanical equipment, some object failures will cause casualties to the platform personnel. The PS indicator is divided into five levels, and the scoring standards are formulated as shown in **Table 2**.

2.1.2 Influence of failure on environment and health (EH)

Whether object failures have an impact on the environment and health is receiving much more attention from enterprises, which will cause social public opinion and bring disaster to enterprises. Therefore, considering the degree of the impact of object failures on the environment, the determination of the EH-scoring standard is mainly based on the country and the enterprise environmental safety system and requirements. The EH index is divided into five levels, and the scoring standards are shown in **Table 3**.

Serial number	Pollution degree	Grading
1	No pollution	0
2	Slight pollution	1–3
3	Local pollution	4–7
4	Severe pollution	8–9
5	Major pollution	10

Table 3.
Scoring criteria of EH.

Serial number	Influence level	Grading	
		Have spare	No spare
1	Total loss of system function	6	10
2	Basic loss of system function	5	8–9
3	Significant decrease in system function	3–4	6–7
4	Reduced system functionality	1–2	1–5
5	System function has no effect	0	0

Table 4.
Scoring criteria of SF.

2.1.3 Influence of failure on system function (SF)

The SF index mainly considers the impact of object failure on the entire production system or the function of mechanical equipment. In order to ensure continuous production operations, the important mechanical equipment in petrochemical companies generally has certain spare. During the production operation, once such mechanical equipment failure, you can immediately switch the standby equipment and quickly resume operations, which can effectively control downtime and avoid large economic losses. Therefore, the degree of impact of object failures with spare parts on system functions will be reduced accordingly. Therefore, in evaluating the object failures on the system, it needs to be divided into two cases: the object has standby and no standby. Based on the actual production process, the level of the SF index is determined, and its scoring standards are established, as shown in **Table 4**.

2.1.4 Influence of failure frequency (FF)

The FF index is related to the trouble-free working time (MTBF) of the object. The MTBF value of the mechanical equipment of petroleum and petrochemical enterprises that has been in operation for a period of time can generally be obtained by analyzing the historical records of the operation of mechanical equipment. Relevant mechanical equipment reliability data are obtained through on-site operation and maintenance personnel's correction. Therefore, the FF index is divided into six levels by the MAXIMO system, the SAP management system, and the historical maintenance records of mechanical equipment during on-site maintenance. The specific scoring criteria are shown in **Table 5**.

Serial number	MTBF (hours)	Grading
1	>10,000	0
2	5000–10,000	1–2
3	3000–5000	3–4
4	1500–3000	5–6
5	500–1500	7–8
6	<500	9–10

Table 5.
Scoring criteria of FF.

Serial number	Outage loss	Score
1	No loss	0
2	Little loss	1–2
3	Loss	3–4
4	Big loss	5–7
5	Great loss	8–10

Table 6.
Scoring criteria of PL.

Serial number	Maintenance cost (Yuan)	Score
1	<10,000	0–2
2	10,000–50,000	3–5
3	50,000–150,000	6–8
4	>150,000	9–10

Table 7.
Scoring criteria of MC.

2.1.5 Influence of failure on production loss (PL)

The PL indicator is due to the failure of components, subsystems, or systems, which will cause changes in product operation, cause standby products to be put into operation, and cause production problems such as drilling stoppages, which will cause economic losses in the enterprise. In combination with the production of offshore oil and gas, in the case of production loss caused by failure, the PL index is divided into six levels, and the specific scoring standards are shown in **Table 6**.

2.1.6 Influence of maintenance cost (MC)

The MC index needs to consider the complexity of the maintenance object, the labor cost of maintenance, the cost of spare parts, and the inventory cost. Therefore, considering the above aspects of the mechanical equipment of petroleum and petrochemical enterprises, MC is divided into four levels, and the specific scoring standards are shown in **Table 7**.

2.1.7 Influence of inspectability (IN)

The IN indicator mainly considers whether the object can be monitored, the number of monitoring parameters, the monitoring cost, and the technical level of the monitoring personnel. The maintenance of the mechanical equipment in petrochemical enterprise cannot be separated from the monitoring of its important subsystems and components. A comprehensive analysis of the existing monitoring technology of mechanical equipment in petroleum and petrochemical enterprises has determined the difficulty of monitoring the object, that is, the IN index is divided into four levels, and the specific scoring standards are shown in **Table 8**.

2.1.8 Influence of downtime (DT)

The outage time includes the time required for the outage, maintenance, and start-up of mechanical equipment, so the DT index is related to factors such as the structure type, power, temperature, and pressure of the object. According to the distribution of outage time caused by object failure in offshore oil and gas production, the DT index is divided into four levels according to the working hours. The scoring standards are shown in **Table 9**.

2.1.9 Influence of maintenance difficulties (MDs)

From the perspective of object maintenance, MD is an important indicator that affects the level of object risk, which is related to the difficulty of the object (including height and surrounding environment), the complexity of the object, and the supply of spare parts. Based on a comprehensive consideration of on-site maintenance of mechanical equipment in petroleum and petrochemical enterprises, the MD index is divided into four levels, and its scoring standards are shown in **Table 10**.

Serial number	Ease of monitoring	Score
1	Low	0–2
2	Medium	3–5
3	High	6–8
4	Very high	9–10

Table 8.
Scoring criteria of IN.

Serial number	Working hours (h)	Score
1	<2	0–2
2	2–8	3–5
3	8–12	6–8
4	>12	9–10

Table 9.
Scoring criteria of DT.

Serial number	Ease of maintenance	Score
1	Easily	0–2
2	General	3–5
3	More difficult	6–8
4	Difficult	9–10

Table 10.
Scoring criteria of MD.

Serial number	Length of service (year)	Score
1	<3	0–2
2	3–8	3–5
3	8–12	6–8
4	>12	9–10

Table 11.
Scoring criteria of SL.

2.1.10 Influence of service length (SL)

From the perspective of mechanical equipment maintenance management, object service age (or operating time) and frequency of use are also indicators that affect its risk level. The SL index is related to the service life of the mechanical equipment, the operating environment and working conditions, the frequency of use (continuous operation, interval operation, and temporary use), the past failures, and the maintenance of the mechanical equipment. Therefore, through discussions with field operations and maintenance personnel, it was determined that the SL index of mechanical equipment in petroleum and petrochemical enterprises is divided into four levels, and its scoring standards are shown in **Table 11**.

2.2 Research on risk assessment methods of mechanical equipment

2.2.1 Establishment of risk assessment model for mechanical equipment

After completing the scoring according to each evaluation index, it is necessary to comprehensively evaluate the risk level of the object. Therefore, it is necessary to comprehensively consider the rating vector V of its evaluation index and its corresponding weight value vector W [12]. The object's risk level evaluation index can be expressed as

$$Index(V, W) = F[v_1 w_1, v_2 w_2, \dots, v_n w_n]. \tag{1}$$

Among them, $F[\cdot]$ is a comprehensive evaluation function that reflects the degree of influence of each evaluation index on the object risk level and can be a function of various forms. Using a simpler linear weighted model, the formula for calculating the object risk level evaluation index, namely, *Index*, is as follows:

$$Index = \sum_{i=1}^{n} v_i w_i \tag{2}$$

where n is the number of evaluation indicators; v_i is the rating value of the ith evaluation index of the evaluated object; and w_i is the weight value of the ith evaluation index of the evaluated object.

Therefore, the magnitude of evaluation value *Index* indicates the risk level, so that the objects can be sorted and screened based on their different risk levels.

From Eq. (2), we can know that the weight w_i of the influencing factors will have a great influence on the final value of the risk level evaluation index *Index*. Therefore, the AHP method is used for this calculation above. The specific calculation steps are as follows:

Step 1: Constructing a judgment matrix D through pairwise comparisons among the evaluation indexes,

$$D = \begin{bmatrix} u_{11} & u_{12} & \cdots\cdots & u_{1n} \\ u_{21} & u_{22} & \cdots\cdots & u_{2n} \\ \cdots\cdots & \cdots\cdots & \cdots\cdots & \cdots\cdots \\ u_{n1} & u_{n2} & \cdots\cdots & u_{nn} \end{bmatrix} \tag{3}$$

where u_{ij} is a relative risk level value that of the ith evaluation index compared with the$j$th evaluation index and $u_{ji}u_{ji}$ is the relative risk level value that of the jth evaluation index compared with the ith evaluation index.

Thus, the value of u_{ji} is the reciprocal value of u_{ij}, namely $u_{ji} \times u_{ij} = 1$. The definition and fundamental scale of the relative risk level are shown in **Table 12**.

Step 2: Calculating the maximum eigenvalue λ_{max} of the judgment matrix D using the system of homogeneous linear equations as follows:

$$\begin{cases} (u_{11} - \lambda)\omega_1 + u_{12}\omega_2 + \cdots + u_{1n}\omega_n = 0 \\ u_{21}\omega_1 + (u_{22} - \lambda)\omega_2 + \cdots + u_{2n}\omega_n = 0 \\ \cdots\cdots\cdots\cdots \\ u_{n1}\omega_1 + u_{n2}\omega_2 + \cdots + (u_{nn} - \lambda)\omega_n = 0 \end{cases} \tag{4}$$

Step 3: Determining the eigenvector relative to the maximum eigenvalue λ_{max}, which is given by

$$W = (\omega_1, \omega_2, \cdots, \omega_n) \tag{5}$$

Step 4: Checking the consistency of the judgment matrix D,

$$CR = CI/RI \tag{6}$$

Important scale	Definition
1	Equally important
3	Moderately important
5	Strongly more important
7	Very strongly important
9	Extremely more important
2, 4, 6, 8	Situation between the above levels

Table 12.
The definition and fundamental scale of the relative risk level.

$$CI = (\lambda_{\max} - n)/(\text{n-1}) \quad (7)$$

where CR is the random consistency ratio of the judgment matrix; CI is the general consistency index of the judgment matrix; and RI is the mean random consistency index of the judgment matrix.

For 2 to 9th-order judgment matrix, the value of RI is shown in **Table 13**.

Step 5: Performing consistency adjustment and weight ordering.

If CR < 0.01, the consistency of the judgment matrix D is satisfactory, which means that the weight apportionment of each evaluation index is reasonable; if not, the judgment matrix D should be adjusted until the consistency meets the above requirement. At this time, the maximum eigenvector of the judgment matrix D corresponds to the weight value of each factor. The priority of mechanical equipment can be determined according to the weight of each factor.

2.2.2 Analysis of eliminating the subjective factors based on the MCS

Because the scoring process of the influencing factors of the risk level of mechanical equipment has subjective factors and differences among individual experts, based on the AHP analysis method to determine the ranking of the influencing factors of each level of risk, the Monte Carlo simulation method is used for calculation [13]. In the calculation process of the Monte Carlo method, the weight of each evaluation factor can be changed by generating random numbers, so that the robustness of the risk level ranking of mechanical equipment is enhanced, and the ranking results are less affected by subjective factors. The logic block diagram of the Monte Carlo simulation is shown in **Figure 1** [12].

As shown in **Figure 1**, a certain random numbers in [0, 1] are generated in the calculation process. The random numbers are regarded as the weight value of certain evaluation indexes and assigned with the priority order obtained in the previous calculation process [14]. In other words, for any group of random numbers, the largest random number will be assigned to the top priority, the smallest one will be assigned to the lowest priority, and the rest of random numbers will be assigned to the other evaluation indexes in order of priority from large to small. Then, in an MCS computation, the total score of all evaluation indexes can be calculated using Eq. (1), and the risk level of the mechanical equipment will be obtained and ranked according to the calculated *Index*. Through N times simulation calculations in the MCS, a number of ranking values are obtained based on different risk levels of the same mechanical equipment. Then, the risk level of a single equipment can be displayed from their sequence of cumulative frequency reaching 1, namely, the faster cumulative frequency of one mechanical equipment reaches "1," so that it will be a higher risk level.

n	2	3	4	5
RI	0.00	0.58	0.9	1.12
n	6	7	8	9
RI	1.24	1.32	1.41	1.45

Table 13.
RI values of the 2 to 9th-order judgment matrix.

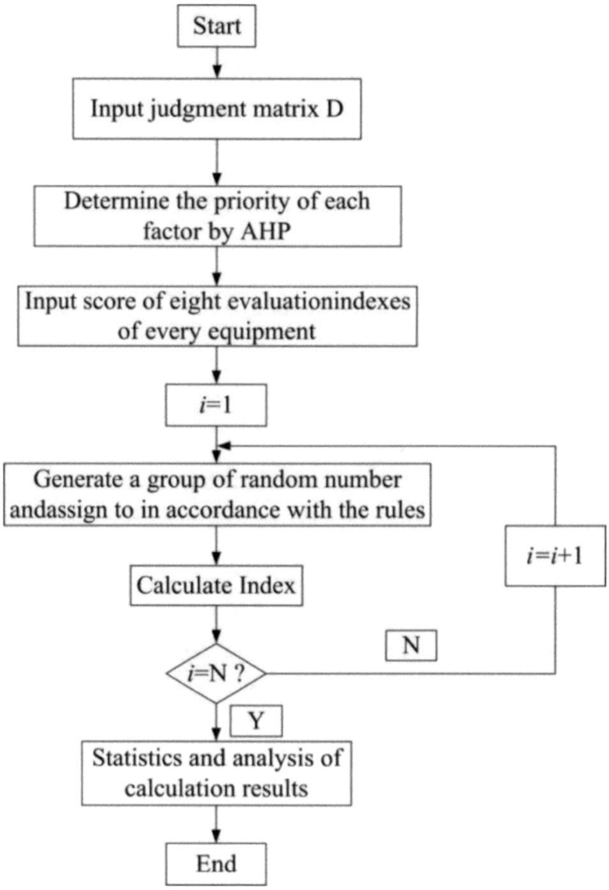

Figure 1.
Logic block diagram of the Monte Carlo simulation.

2.3 Research on maintenance decision-making methods based on the risk level of mechanical equipment

According to statistical data about the priority orders of the mechanical equipment from their evaluation risk levels in the previous step, their cumulative frequency can be plotted with a curve chart. Based on the principle of establishing a cumulative frequency curve chart, the percentage of the area on the right side of the curve from the total area can be taken as another representation evaluating the risk level of the mechanical equipment [15]. A larger area percentage is indicating that one mechanical equipment has a higher risk level. According to their different area percentages, namely, the different risk levels among them, the mechanical equipment can be divided into three categories, including Classes A, B, and C. Class A is an area percentage of 0–30% of mechanical equipment. Class B is an area percentage of 30–80% of mechanical equipment. Class C is an area percentage of 80–100% of mechanical equipment [16, 17].

According to different own characteristics and failure modes of the mechanical equipment in the petrochemical industry, the existing maintenance methods include Lubrication, Service, Corrective Maintenance, Time-based Maintenance, Hidden Failure Detection, and Condition-based Maintenance. In order to reasonably establish maintenance decisions of mechanical equipment tree (MDMET) and effectively implement the existing maintenance methods above, the failure mode, failure effect, and failure cause of the three categories of mechanical equipment are determined based on the FMEA method. Then, the MDMETs on the mechanical equipment of the petrochemical industry are established by referencing the logic decision diagram in the reliability centered maintenance (RCM) theory. The MDMETs are shown in detail as follows [18–20]:

1. The failure consequence of Class A of mechanical equipment has little or no influence on the function of the whole system or causes lower maintenance costs. Increasing the spare part inventory or decreasing the failure frequency for Class A of mechanical equipment cannot affect the production process. A MDMET of Class A of mechanical equipment is shown in **Figure 2**.

2. When Class B of mechanical equipment has been failed, it might result in severe failure consequences, but it usually does not influence personnel safety and environment. The failure frequency of Class B of mechanical equipment could be reduced through reasonable maintenance strategies, so that failure consequences could be decreased as well. But these maintenance strategies might cause higher maintenance costs. A MDMET of Class B of mechanical equipment is shown in **Figure 3**, which includes corrective maintenance, time-based maintenance, and hidden failure detection.

3. The failure of Class C of mechanical equipment might endanger the personnel safety, pollute the environment, and cause the significant economic consequences. In order to ensure the operation reliability and maintenance economy of Class C of mechanical equipment, some special maintenance

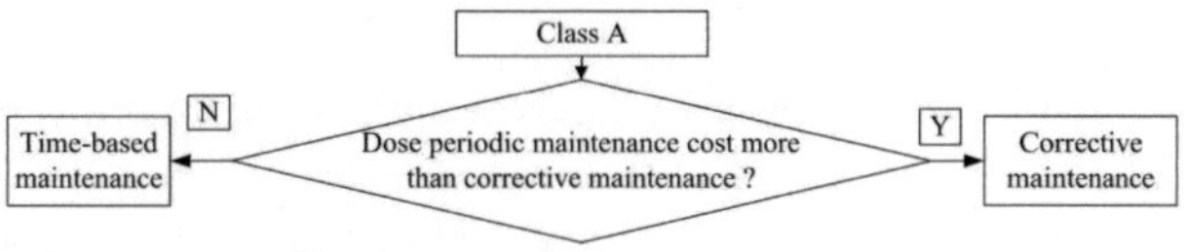

Figure 2.
MDMET of Class A of mechanical equipment

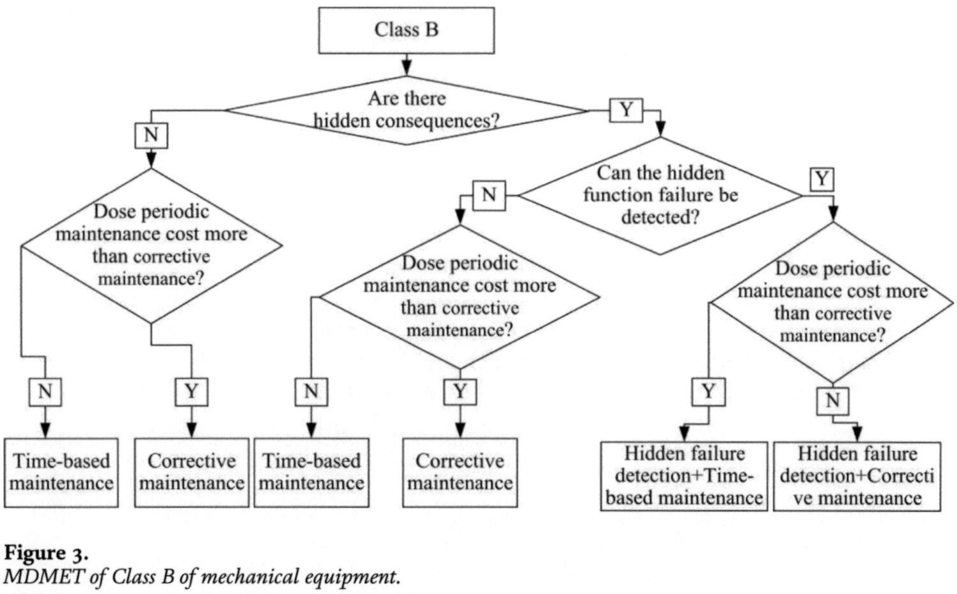

Figure 3.
MDMET of Class B of mechanical equipment.

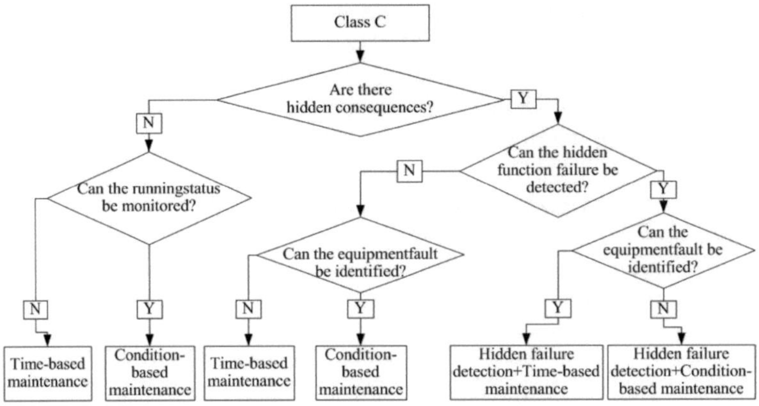

Figure 4.
MDMET of Class C of mechanical equipment.

strategies should be applied, and the failure frequency should be reduced by increasing the maintenance costs with advanced maintenance methods. A MDMET of Class C of mechanical equipment is shown in **Figure 4**, which includes corrective maintenance, time-based maintenance, hidden failure detection, and condition-based maintenance.

3. Conclusions

The aim of this study was to present the framework for making maintenance decisions for mechanical equipment to improve their reliability and security and to decrease safety accidents and economic losses. The framework adopts the idea of different equipment classifications based on their own risk level with different MDMETs. We summarized 10 evaluation indexes for evaluating the risk level of the mechanical equipment and defined a set of scoring criteria to quantify the subjective evaluation in the petrochemical industry. The evaluation model of the risk level was established based on the AHP and the MCS, which calculate the weight of the evaluation indexes and remove the subjective effects in the scoring process. Moreover, we divided the mechanical equipment into three categories based on their risk level and established their MDMETs.

From the discussion above, we can conclude that the framework for making maintenance decisions, combining the quantitative and qualitative methods with the mathematical model and decision-making theory, can be effectively applied to mechanical equipment and provides them with reasonable and scientific maintenance strategies. The framework is suitable not only for mechanical equipment but also for other similar equipment on the premise that the evaluation index and scoring criteria are revised based on their respective features. This framework also includes three analysis methods for the risk level of the mechanical equipment, which can be put into use for other equipment or systems to filter out the unimportant parts. Moreover, the analyzed methods and results in this study will help the study of asset integrity management (AIM) in oil companies.

Author details

Yang Tang*, Xin Yang and Guorong Wang
School of Mechatronic Engineering, Southwest Petroleum University, Chengdu, China

*Address all correspondence to: tangyanggreat@126.com

References

[1] Doostparast M, Kolahan F, Doostparast M. A reliability-based approach to optimize preventive maintenance scheduling for coherent systems. Reliability Engineering & System Safety. 2014;**126**:98-106. DOI: 10.1016/j.ress.2014.01.010

[2] Bertolini M, Bevilacqua M. A combined goal programming—AHP approach to maintenance selection problem. Reliability Engineering & System Safety. 2006;**91**(7):839-848

[3] Dey PK. Analytic hierarchy process helps evaluate project in Indian oil pipelines industry. International Journal of Operations & Production Management. 2004;**24**(6):588-604. DOI: 10.1108/01443570410538122

[4] Bevilacqua M, Braglia M, Gabbrielli R. Monte Carlo simulation approach for a modified FMECA in a power plant. Quality and Reliability Engineering International. 2000;**16**(4): 313-324

[5] Chang KP, Chang D, Zio E. Application of Monte Carlo simulation for the estimation of production availability in offshore installations. In: Simulation Methods for Reliability and Availability of Complex Systems. London: Springer; 2010. pp. 233-252

[6] Gómez de León Hijes FC, Cartagena JJR. Maintenance strategy based on a multicriterion classification of equipments. Reliability Engineering & System Safety. 2006;**91**(4):444-451. DOI: 10.5267/j.msl.2014.3.028

[7] Perrons RK, Richards MG. Applying maintenance strategies from the space and satellite sector to the upstream oil and gas industry: A research agenda. Energy Policy. 2013;**61**:60-64. DOI: 10.1016/j.enpol.2013.05.081

[8] Hmida JB, Gaspard AJ, Lee J. TQM-based equipment maintenance in oilfield service industries. Global Perspectives on Engineering Management. 2013;**2**(2):60-69

[9] Triantaphyllou E et al. Determining the most important criteria in maintenance decision making. Journal of Quality in Maintenance Engineering. 1997;**3**(1):16-28

[10] Gang D, Ren MP, Yu YF, Li XF, et al. Risk ranking and well control measures of drilling blowout in Xinjiang oilfield. Advanced Materials Research. 2013;**616**:844-849

[11] Tang Y, Liu Q, Jing J, et al. A framework for identification of maintenance significant items in reliability centered maintenance. Energy. 2017;**118**:1295-1303. DOI: 10.1016/j.energy.2016.11.011

[12] Saaty TL. Decision making with the analytic hierarchy process. International Journal of Services Sciences. 2008;**1**(1): 83-98

[13] Arunraj NS, Maiti J. Risk-based maintenance policy selection using AHP and goal programming. Safety Science. 2010;**48**(2):238-247. DOI: 10.1016/j.ssci.2009.09.005

[14] Azizi A, Fathi K. Selection of optimum maintenance strategies based on a fuzzy analytic hierarchy process. Management Science Letters. 2014;**4**(5): 893-898. DOI: 10.5267/j.msl.2014.3.028

[15] Marseguerra M, Zio E. Optimizing maintenance and repair policies via a combination of genetic algorithms and Monte Carlo simulation. Reliability Engineering & System Safety. 2000; **68**(1):69-83. DOI: 10.1016/S0951-8320(00)00007-7

[16] Eti MC, Ogaji SOT, Probert SD. Reducing the cost of preventive maintenance (PM) through adopting a proactive reliability-focused culture. Applied Energy. 2006;**83**(11):1235-1248

[17] Kyriakidis EG, Dimitrakos TD. Optimal preventive maintenance of a production system with an intermediate buffer. European Journal of Operational Research. 2006;**168**(1):86-99

[18] Tang Y, Zou Z, Jing J, et al. A framework for making maintenance decisions for oil and gas drilling and production equipment. Journal of Natural Gas Science and Engineering. 2015;**26**:1050-1058. DOI: 10.1016/j.jngse.2015.07.038

[19] Murthy DNP, Atrens A, Eccleston JA. Strategic maintenance management. Journal of Quality in Maintenance Engineering. 2002;**8**(4):287-305

[20] Shen Q, Lo K-K, Wang Q. Priority setting in maintenance management: A modified multi-attribute approach using analytic hierarchy process. Construction Management and Economics. 1998;**16**(6):693-702. DOI: 10.1080/014461998371980

Chapter 17

Exploring Constituents of Short Food Supply Chains

Sahand Ashtab, Yang Xing and Cheng Zheng

Abstract

By deploying a systematic review approach, this chapter provides a holistic exploration of AFNs which contributes to further mobilization of locally produced products. This chapter explores the constituents of AFNs by studying food citizenship, sustainability and food democracy, food safety and quality, embeddedness and social capital, the relationship between the level of participation in AFNs and consumers' demographics, consumers' motivations to engage in buy-local activities, vendors' perspective on selling products in farmers' market, and the development of short food supply chains in the Canadian context. Specifically, the social interaction aspect of buying local, for example, engaging with vendors and other consumers, has been cited as a factor that motivates consumers to buy local food products from the farmers' market; however, consumers had to deploy online ordering channels with door delivery option during COVID-19 pandemic to access locally produced products safely. To capture one aspect of the potential impacts of COVID-19 pandemic on AFNs, future research can explore whether social interaction is still an influential factor in consumers decision to buy local, or the importance of the social interaction aspect of buying local will be replaced by the convenience of receiving the fresh, locally produced food products at consumers' doorstep via online ordering process.

Keywords: sustainable food production, short food supply chains, alternative food networks, local food buyers, environmental sustainability, food citizenship, food democracy, food quality, food safety, consumers' motivations, Covid-19 pandemic

1. Introduction

For decades, global and centralized food supply chains and supermarkets have served as a one-stop shopping outlet and provided the consumers with the convenience of accessing a variety of products all year round. Due to globalization, advanced technologies, and economies of scale in the transportation of mass volumes of products, consumers across the globe can access a variety of products all year round. The global food production and distribution system brought forward some concerns for the society, economy, and environment for acceleration of the disappearance of local agriculture [1]. Supermarkets had taken control of about 80% of the food consumption market in Britain [2]. Conventional food systems were challenged, and consumers questioned the food quality provided by conventional food networks. Also, the sustainability of industrialized farming was also called in question [3]. With little knowledge of the place of origin and methods of production, some consumers of the globalized network found it uncomfortable to deal with [3].

Recent years have shown increasing interest in the local food system. In both North America and Europe, the number of farmers market has grown rapidly since the 2000s. For instance, in 2008, there were 508 farmers' market registered across Canada, while in 2014, this number increased to 653 [4]. The US has even more farmers' market than Canada. The number of farmers' market in the US in 2004 increased 53% compared with the number in 1994 [5]. In the UK, Miller [6] reported that the number of farmers' market was about 800.

Many European countries started to find alternatives for regional food supply solutions [7]. Local food networks, along with fair trade, forms part of alternative food networks (AFNs) [8]. Johnson and Endres [9] described the local food movement as a push back against the globalization of food system. Contrary to global food supply chains, short food supply chains, also referred to AFNs, have facilitated the mobilization of locally produced fresh products [10] by connecting suppliers, e.g., farmers, with consumers [11].

AFNs emerged as a result of political, cultural, and historical processes [12]. It was the uneven development of participants in a commercial food supply network that gives rise to the nurturing of AFNs. These chains can be categorized into three kinds: direct sales by individuals, e.g., U-picks and farm gate sales; collective direct sales, e.g., farmers market; and partnerships, e.g., Community Supported Agriculture (CSAs) [13]. At the very beginning of the twenty-first century, CSA was not as commonly known as today in the US. CSA was still a growing social movement at its earliest stage when industrialized farming was dominating the market [3]. CSA is a form of partnership which secures the amount demanded by customers, and therefore, the risks and benefits of production are shared by both farmers and consumers [13]. With a contractual agreement between producers and consumers, shareholders and stakeholders at the same time, CSAs are an ideal solution to build a long-term relationship with mutual trust [3].

In the US, Departments including Agriculture, Commerce, Health and Human Services, and Urban development have initiatives to promote the development of farmer's market [14]. Food cooperatives and farmers' markets are vibrant constituents enabling the whole domestic food system to prosper. Direct farmer-to-consumer food distribution channels, regardless of type, provide short food miles, reduce the number of intermediaries in the food supply chain, and connect producers with consumers. The elimination of numerous intermediaries secures the farmers' fairer share and ensures the traceability and higher quality standards of the products. Also, the firm support from consumers reflects the position of short food supply chains. In a conventional chain, the imbalance between bargaining powers of farmers and distributors made many small farmers hard to make a profit [13]. While large chain stores and farmers' markets are the top choices for grocery shopping [15], the level of competition between these two venues is not clear.

Supports from the state and consumer's willingness facilitate the growth of the local food economy, and more local food communities and systems are created with increased link between consumers and producers. Farmers' markets are important channels of direct marketing for locally grown foods in Canada, the US, and Europe. What made the farmers' market stand out of the crowd is the quality of food, either product-based or process-based [16]. From food safety perspective, during an era of increasing globalization, the origin of fertilizers, seeds, and feeds used by small farms is likely to be from all over the globe, and contamination is liable to come from these sources [17]. Opportunities lie in the support of cooperatives and society since more resources were invested in food safety education and better system and management [17]. Also, food suppliers have more opportunities to communicate directly to customers for any feedbacks.

Farmers' markets are self-organized and locally controlled. Their size can depend on the local demand and supply, and they are formed by local values and culture [18]. In Canada, farmers' market is prevalent and has gained a great success; it has a strong consumer base with the demands of fresh, healthy fruits, vegetables, and baked foods. Farmers' markets in Canada provide a large set of diverse choices to consumers, and consumers are highly motivated to buy local foods. They visit farmers' markets regularly which helps those markets to thrive.

Numerous scholarly papers have focused on specific aspects of short food supply chains. By deploying a systematic review approach, this chapter provides a holistic exploration of alternate food networks which contributes to further mobilization of locally produced products. Specifically, the constituents of short food supply chains, namely, food citizenship (e.g., [19]), sustainability and food democracy (e.g., [20]), embeddedness social, and human capital (e.g., [21]), food quality and safety concerns (e.g., [22]), defining local food (e.g., [15]), investigating the relationship between level of participation in AFNs and consumers' demographics (e.g., [23]), consumers' motivations to engage in buy-local activities (e.g., [24]), vendors' perspective on selling products in farmers' market (e.g., [25]), and the development of short food supply chain in the Canadian context (e.g., [26]) are studied in this chapter.

1.1 Localism

Local food does not have a universally accepted definition. The notion of local food is identified by the geographic dimension, which means the distance between food producer and consumer [27]. However, people hold different opinions on defining the distance of local food. Based on the survey, people's perceptions on local are varied; some of them believe local to be in the same province or state, some believe local is the region within 100 kilometers, and some others cannot describe precisely and just think local is nearby counties [28]. Although there are some controversies that exist on defining local food, the movements of local food are emerging and become one of the most prevalent topics in food sector.

Despite that different European countries have their respective understanding of short food supply chain, the need for reasonable comparability forced EU to come to a publically recognized definition. On the other hand, local food system is much harder to be generalized since the term local food is relatively subjective [13]. Also, there are different ranges for the definitions of local in different Canadian provinces. Defining local food can be tricky for the suppliers at farmers' markets, which makes the consumers question the products' authenticity [15].

1.2 Food citizenship

Current theoretical frameworks such as solidarity buying groups, consumer co-ops, or collective urban gardening initiatives are still not a perfect solution for problems incurred by conventional global food networks [19]. Although AFNs lack a set of clear standards, and a boundary between them and conventional food networks is ill-defined, the emergence and development of AFNs are a good interpretation of a new type of producer-consumer cooperation focusing on the perspective of sustainable consumption apart from sustainable rural development.

Consumer co-operatives and buying groups in some European countries, for instance, are more than channels for direct-selling and producer-initiated activities

[19]. The transition from passive end-consumers to proactive citizen-consumers clearly represents a new type of relationship between the producers and consumers of food. For these producers, this means the transition from supplying the market with food in bulk to directly supply the community with more sustainable and organic food.

Unlike the uncommon existence in the European literature, the concept of food citizenship occurs quite often in North America. It was introduced into Canada to criticize the fading food skills of citizens and corporate control. Also, compared with conventional food networks, the existence of power relationship is less evident in these civic food networks since the disappearance of the intermediaries between producers and consumers [29].

Renting et al. [19] argues that society-based and civil forms of governance is an important concept to understand to better promote food citizenship, especially in times of several political and economic crises [19]. Wittman et al. [15] also suggest that the rebuilding of linkages between civil society and markets, and the creation of new connections can be interpreted as the governance mechanisms.

1.3 Sustainability and food democracy

The significance of AFNs is not limited to providing communities with access to locally sourced fresh and organic food, but also allowing consumers to contribute to the pillars of sustainability, e.g., economic and environmental, by supporting sustainable food production and distribution systems [30]. The environmental dimension of sustainability is particularly urged by many stakeholders, and the changes these businesses about to make are challenging [31]. The dominant philosophy of such changes can be hard for the senior management of a firm to accept. The public and the companies perceive the domain of sustainability differently. Also, the customers are aware of sustainability, but they think the companies are responsible for the necessary expenditure [31].

Consumers are becoming more conscious about environmentally friendly production and distribution systems [32], and are interested in finding out about the origin of their food [16, 33]. The transparent and domestic origin of the food products distinguishes them from their counterparts sold in conventional supermarkets [16]. Researchers believed that this kind of desire derives from the decline of consumer's trust on the standardized foods, and they have more concerns about the production methods and ethical issues of those imported foods [34, 35]. This kind of mistrust can be eliminated by locally produced food, because locally produced food builds a closer connection between buyers and producers. Consumers can easily access the whole process of local food from the farm to their table, and that reduces their ethical and environmental concerns of the food production. In terms of environmental benefits, local food economy shortens the distance of food and reduces the carbon emissions incurred in the transportation process. Meanwhile, it also encourages farmers to adopt organic farming, which is more environmentally friendly and promotes the consumer's preference of organic food [1]. However, Brown [22] argued that the relationship between farmers' market and the growth of organic food market was not documented, but she also agreed that farmers' market is the major source of organic and exotic foods. In another study, M. D'amico et al. [36] pointed out that participants, who were selected from three Italian regions to better represent the target consumers, favor the food items circulated through short food supply chain because of the quality and lower environmental impact. This study tried to figure out the main features and aspects of the direct selling of wine through analyzing the main factors that directly influence consumers' choice and purchasing behavior.

Hamprecht et al. [20] demonstrated how enterprises could control the sustainability through managing the economic, social, and environmental performance of the supply chain [20]. In the process of tracking supplies of the production, a controlling framework was used to assess the economic, social, and environmental performance aspects of milk and cereal factories [20]. Another initiative was a wave of organic farming activities which took place in Czech Republic, and it was aimed at reducing environmental impacts and improving alternative food quality [16]. Contrary to the agro-industrial model, short food supply chain is an ideal alternative in reconnecting production and consumption with quality and values, which also promotes sustainable agriculture [8].

As direct channels of distributing food, AFNs drew these participants closer, and they enabled the redistribution of power across the food chain [29]. Critics doubt these networks' ability to bring structural changes on a greater scale [29]. Either Food democracy or civic agriculture represents a certain academic perspective in understanding all these food networks [29]. With the direct involvement in distributing and selling the food produced, the AFNs in France and Brazil have greater control over the economic actions downstream. Initiatives in both countries matter to the food democracy, as their members share the decision-making power [29].

1.4 Embeddedness, social and human capital

Other than farmers' market which has brought more hope in pursuing an alternative solution for a stable supply of quality food, an urban garden project called community gardens (contributing to building communities) was created with a goal of promoting the social and cultural nature of AFNs [16]. While the literature available on AFNs is primarily stressing on quality, locality, and ecology, Sage [37] analyzed the case of good food network in South West Ireland, using an approach of more qualitative insight into understanding the relationship between the participating members of a short food supply chain. They suggest that this food network is an economy of regard reflected through personal relations and social connection. Deeply embedded in local communities, social capital and embeddedness are commonly used in the literature discussing the benefits of the collaboration between members of an AFN being the knowledge exchange and social relations and learning [21]. Collaborative CSAs, also known as cCSAs, as another example, are different from single farmer CSAs, and other than financial incentives, several capitals, notably social capital, are the core values of cCSAs apart from democracy in food provisioning [38]. Human capital is another benefit of CSAs through the accumulation of hands-on knowledge and experiences [38].

Wittman et al. [15] discuss farmers' market as an example of social economy in North America. As part of the social economy, the farmers' markets in western Canada are not after a greater amount of profit, but to fulfill the social objectives and reinvest the profit generated into further investments in the markets' infrastructure [15]. The economic influence of the farmers' markets is not despised by their social mission. These markets had enjoyed a significant growth in operating revenue, created by direct sales, and resulted in multiplier effects three times greater [15].

1.5 Food quality and safety concerns

AFNs are a representative of economy of quality, with food quality and safety as a prerequisite [8]. With support from the national level, short food supply chains have been rediscovered for their diverse forms of businesses. In an industry-wide crisis in agri-food networks, Taylor [39] used the value chain analysis technique in the study of two pork supply chains in the UK. The pressing need to become more integrated urges the red meat chains to make unprecedented changes to their traditional business model [39] and build trust between different echelons in their supply chains and establish cooperation between the supply chain members to attract consumers to choose the domestically produced pork products than imported.

Local food production is more accessible to consumers, and this increases the health and security of the food [22]. The popularity of short food supply chain in Belgium is another sign of its influence in Europe. Similarly, short food supply chains are also potentially prone to contamination in the environment where the food was produced or harvested. Microbiological safety and quality aspects of the chains are specifically concerned with the microbiological safety aspects of food and food production [17]. Even short food supply chains with a lower level of complexity have links which can potentially pose some threats to the food safety. While practices and policies such as Good Agricultural Practices (GAP), European Hygiene Legislation, Hazardous Analysis and Critical Control Points (HACCP), and Good Manufacturing Practices are enforced, but still, sources of contamination are likely to undermine the safety of at least of the links of the supply chains [17].

Currently, the farmers in SFSCs have individual experiences with selecting the quality of raw materials, and they also keep an eye on the production that takes place. Therefore, the microbiological risks can be contained in most cases. Shorter chain and smaller scale help farms reduce the cycle of turnover, and consumers can purchase the food items in a short time after production. Smaller farms have fewer kinds of food being produced, therefore lower the possibility of cross-contamination. Additionally, smaller farms have a tradition of maintaining food safety and quality [17]. On the other hand, some farms have both fresh plant and animal productions, which are more likely to lead to cross-contamination through irrigation water or crops [17]. Also, the lack of professional food safety knowledge is potentially undermining the farmers' ability to avoid microbiological risks.

From the consumer's perspective, local food reduces the risk of food safety and security to them, because local foods are usually less processed and fresher and there is a high level of transparency. Consumers can either reach their sites of production or talk to growers to obtain information about local grown foods [40].

2. Demographics, motivations, and impacts of consumers participating in AFNs

People started to realize that local food system can substantially benefit a variety of local actors. For instance, it promotes the development of the local economy, reduces the distance of transportation, and facilitates the sustainable development [41]. Abate [1] believes that the local food economy can connect and involve the different local actors in the local community, revitalize businesses that may be hardly surviving and provide them with business opportunities, and also promote the sense of identity which is similar to the viewpoint of Oberholtzer and Grow [42]. It can also create job opportunities, increase the local income, and diversify the local economy. Feenstra et al. [43] argued that the farmers' market provided the best opportunity for farmers to develop their businesses, improved their skills on dealing with customer relations and other marketing practices, and also encouraged them to add values to their products or services which also benefit their customers. Hughes and Boys [44] also discussed about the economic impacts of local food. They believe that the local food system will bring multiple benefits to different actors in the system, such as vendors, local labors, and farmers' markets. Furthermore, the localization can also contribute to the higher quality of life, because he believed that the regional growth has been a new economic development pattern in the modern world. Meanwhile, the localization of food as a marketing strategy can promote the local businesses and entrepreneurism.

The motivators of consumers to attend the farmers' market is also of interest for researchers to study. From theoretical perspective, Conner et al. [45] believed that local food basically creates three values for consumers. First is the shortened distance of travel from the production place to the market. Second is the higher quality of food which mainly indicates the freshness. Third is the social interactions that refer to face to face communication. Also, the perceptions of quality are different in the various regions, ranging from environmental sustainability and animal welfare to rural tradition, local knowledge, and culture [13]. A number of studies have shown the consumers' interest in foods with local attributes. Generally, researches constantly showed the strong willingness of consumers to buy in farmers' markets, and those shoppers have similar demographic characteristics, but their motivating factors can be slightly different between different regions. Based on the research in different regions of the US, Baker et al. [5] suggested that the most important reasons of consumers using farmers' markets are accessing fresh local foods and supporting local agriculture. But consumers in Manitoba, Canada, have demonstrated different motivations, based on the research by Food Matters Manitoba [46]; the major driver of attending farmers' markets of consumers in Manitoba was to support the local economy, and the second most important is to help local farmers. The interest in the food quality was only placed at third. However, based on the research for consumers across the whole country, they perceive the top motivator for buying local food products was freshness of food, and supporting local agriculture business was reported as the second most important factor [46]. Therefore, Manitoban consumers have more concern on the community development than other places in Canada. Byker et al. [24] argues that based on the studies in different areas, some motivating factors to participate in AFNs are consistent, including freshness and high quality of food, food safety and security, pesticide-free, support local agriculture, and social interactions. Some other factors can vary from different regions or different consumers; for instance, convenience is an important factor for some consumers but not for all. The price of a product was also concerned by a part of consumers. Conner et al. [45] found similar phenomenon; some consumers perceived local foods as high quality and willing to pay more for the benefit. However, some consumers think local foods should have lower price. Specific lifestyle also motivates some consumers to shop at farmers' markets, such as cooking, baking, or interest in some specific types of food.

Besides those factors, Hunt [47] linked the demographic factors with the motivating factors to shop in farmers' market. He believed that the social interaction is a significant motivator for consumers based on their demographic characteristics. His research was based on more than 200 consumers in the farmers' market. About 98% of the respondents agreed that they had fun while interacting with other people in farmers' market, 94% of them talk with vendors, and 81% of people meet people they know in the market. Therefore, farmers' market can be a platform for those seniors to interact with people and increase their satisfaction of shopping experience. He also claimed that 45% of them know the farmers' market by word of mouth. This can also be an effective marketing strategy of those farmers' markets. In another study, Wittman et al. [15] suggests that personal interactions between vendors and consumers is partially why farmers' markets are one of the most important market channels for local food. Some respondents prefer more choices of marketing channels, but the authentic relationships built through direct marketing can hardly be paralleled by other channels [15]. Such direct communication allows vendors to better share the stories behind the fresh produces, enabling the patrons to be more knowledgeable about the total cost of the produces, and therefore the reason for these patrons to pay a premium price becomes more justified [15].

A number of researchers had put their efforts on characterizing farmers' market consumers. From the demographic perspective, National farmers' market impact study report from 2009 [48] suggested that 72% of consumers are female while 28% are male in Canada, and 70% of them are 40 years and older, and about 30% of them are between 50 and 64. Byker et al. [24] found that the percentage of female consumers in different states of the US was ranging from 64 to 77%, and their average age was over 40 years old. Researchers also found most of consumers are well-educated regardless of their age and gender. The study by Conner et al. [45] on consumers showed that the average education level was college. Hunt [47] also had similar outcomes on his research. Brown [22] described most consumers as middle aged, middle income or above, well educated women. In another study, it is reported that younger people from age 21 to 29 are not interested in supporting local farmers, and people with lower incomes even have no perceptions of farmers' markets [49].

Johnson and Endres [9] reported consumer's desire for fresh, high quality, and pesticide-free foods as the top reasons for purchasing locally produced food products. Second is to minimize the environmental footprint in the production process and reduce transportation emissions. Third is to reduce the cost of farmers in transportation, processing, and packaging foods and help them to achieve a higher profitability.

From managerial point of view, based on the survey conducted by Oberholtzer and Grow [42], most managers of farmers' market believed that the impacts brought by farmers' market to the community can be in many aspects. They described that a farmers' market provides a platform for all kinds of social and economic practices. This benefits to the form of community and increases the individuals' sense of belonging to the community. Meanwhile, its basic function also enables it to provide consumers with fresh and inexpensive food.

3. Vendors' perspective on selling products in farmers' market

Although researches have reported growing consumer interest in local food, most statistics has shown that the sales volume of farmers' market continuously represent only a small portion of total food sales. Onozaka et al. [50] found that 83% of consumers believed their primary source of buying foods is supermarket. The conventional retailers are still primary channel for consumers to buy foods [18]. This fact shows inconsistency with the benefits associated with farmers' markets in the community and high demand and willingness of consumers on shopping in farmers' market. In addition, based on the Canadian survey on vendors in 2008, 42% of vendors reported more than a half of their incomes are generated in the farmers' market. In other words, income from the farmers' markets is not major a source of income for most farm owners in Canada.

Some researchers described that the primary motivator of local food movement is to reduce the carbon emission and promote community development and improve reciprocity [40, 51]. As an essential constituent of the economy, the agricultural sector in Greece, for example, was primarily located in isolated rural areas with scarce resources contributing significantly to the sustainable rural development, but still face economic difficulties [52]. Researchers also found that the vendors' interests on selling their products in local market are not aligned with the consumers' demands to buy local. Schneider and Francis [41] conducted a survey that examined the consumers' interests on local food and vendors' interest on selling local. Their result showed that more than a half (50.7%) of consumers are very interested in farmers' market and 12.9% of the total consumers are extremely interested. However, in terms of farmers' interest, 65.2% of them are not interested in selling products through farmers' market, and only 2.7% of them are extremely interested. The results indicated that the farm owners prefer to sell their crops in other regions rather than in local market.

Previous studies indicated that there are both benefits and challenges for farmers to sell their products in farmers' markets. In terms of benefits, from vendor's perspective, first, to sell in local market can shorten the distance they travel, which means it reduces the transportation cost [47]. Second, selling in local market enables them to have more interactions with consumers, and improve their operations by understanding consumers' demands. Based on the Canadian vendor survey, 25% of vendors agreed the farmers' markets provide them with a friendly and attractive environment to market their products [48]. Moreover, they believe that farmers' markets provide them with more business opportunities which help them to thrive. Hunt [47] suggested that farmers who sell in the farmers' markets prioritized the social interactions over the profits. Based on his study of vendors in farmers' markets, the most important motivator for them to sell in farmers' market is to build relationship between consumers, then is to generate profit. Actually, he found that 94% of shoppers in farmers' markets would talk to vendors, two-thirds of them would make friendly gossip, more than a half (55%) of them discuss the production methods with vendors, and 44% of them had dialogues about sampling products. Social interactions also incorporate communications between vendors. Beckie et al. [18] claimed farmers' market plays a role of clustering vendors in western provinces of Canada. They suggested that in western Canada, vendors collaborate with each other to achieve their common goals including improved profitability, diversified customer bases, or increased creativity in marketing practices. This kind of clustering enables them to exchange their knowledges, which provides the knowledge mobilization. Knowledge mobilization is particularly important in food sector because it comes with health and safety concerns as well as environmental concerns. It also broadens the labor base and involves more actors in the supply chain. Those interactions have a lot of implications to vendors. First, it can help vendors build solid connections with consumers and improve their loyalty. Second, interactions with consumers and farmers in local market can promote social learning, lead to the innovation on farming practices, and help vendors improve the quality of products [53].

Nevertheless, some papers also identified some challenges and barriers for farmers to sell in the local market. Robinson and Farmer [25] claimed that for vendors, the biggest challenge is consumers' perception. As mentioned previously, some consumers are not willing to pay premium for the benefits of local food. But in fact, their cost can be higher than those nonlocal producers as farmers' markets charge them membership fee. They are also struggling on marketing their foods which are also costly, and if they cannot maintain their sales volume at a certain level, they may not be able to cover their costs. Attending farmers' markets will cost vendors in different forms [4]. They argued that major costs include time, gas, equipment and supplies, and different kinds of fees like insurance, permit fees, etc. This is one of the reasons that the consumers in farmers' markets are always more educated and with higher income. Because they are more likely to pay premium for the local foods. Therefore, based on the Robinson and Farmer's [25] description, the first challenge for vendors is to convince more potential consumers to pay for their products with price premium. The second barrier is the gap between consumer demand and supply in terms of season. Research suggests that most of the consumers in farmers' markets are loyal consumers and they visit farmers' market frequently, spend a lot in there, and have solid personal relationships with vendors [45]. Canadian researches on consumers also indicated that consumers are not satisfied with the limited season and hours for the farmers' markets because their demands cannot be fulfilled. But due to the limitation of technique and knowledge for most small-scale farm owners, they are not capable for the season extension practices. This limitation increases the potential of losing loyal customers. The third is the limited ability to promote the awareness to a larger base of population. Farmers' Markets Canada [48] also found that the top two reasons for people to not shop at farmers' markets are the inconvenience of location and the lack of awareness, and according to Colasanti et al. [49], the major reasons for less awareness of farmers' markets are largely due to the ineffective promotion strategy and the inconvenience of location. According to Wittman et al. [15], many customers would choose to support locally sourced food when there is a greater variety of products and easier accessibility for shopping.

The primary challenges facing farmers' markets in Canada are exerted by the conventional food networks, competitive prices and scaling-up of production, and an unpredictable environment [15]. As a labor-intensive industry with its ethical standards, vendors at farmers' markets must pay a living wage to their employees, and at the same time, the investment in environmentally sustainable practices further raises the cost of production [15]. Specific constraints are exerting pressure to the further development of SFSC, primarily in the shortage of necessary knowledge and skills, lack of entrepreneurial culture in farmers, or administrative burden which may baffle small farmers [13].

4. Development of short food supply chain in the Canadian context

Conner et al. [45] conducted a research based on 70 farmers' markets, 3174 shoppers, and 487 vendors across the country. They found that in Canada, Farmers' markets are making significant contributions to the economy and communities ranging from $1.55 million to $3.09 million annually. They are the key players in promoting Canadian agricultural products, facilitating vendors' capabilities and developing labor improvement and accessibility. However, they found that there are still growth opportunities for them since a majority of consumers are using grocery stores rather than farmer's market.

4.1 Alberta and British Colombia provinces

Wittman et al. [15] surveyed some farmers' markets in Alberta and British Colombia. These markets are identified as strategic venues linking producers and consumers of local food. Different from conventional food networks in the organization of production, distribution, and consumption, the farmers' markets in North America are also examples of the social economy. With the support from private sector and the government, many Canadian farmers' markets are a combination of the local food system and social economy [15]. The constraints on the term "local" are not strictly defined, as situations in these two provinces are varied when compared with their counterparts in Europe. The qualified vendors in a Canadian farmers' market must meet the requirements of producing food in a local region. Some say their scale limits the further development of farmers' markets. However, scholars disagree with this common misconception. Scaling up farmers' markets could benefit them from economies of scale, but such action could potentially bring negative power disparities and environmental impacts resulting from conventionalization [15]. Like the situation in the United States, the amount of organic food distributed through direct sales outlets was under 10% in 2006 [15]. One major obstacle facing the development of farmers' markets in Canada was the disparity between policy and practice [15]. The restriction on the local and authenticity is a perfect example of this disparity.

Wittman et al. [15] analyzed the potential barriers and challenges undermining the potential growth of farmers' markets in Alberta and British Colombia provinces. The participants were interviewed with two overarching questions, "How can farmers' markets in western Canada scale up their role in supporting the advancement of local food systems within a social economy framework?" and "What are the barriers impeding farmers' markets from acting as transformational agents for a more sustainable food system?". The feedback from the participants reflected a diverse way of understanding the local food systems. For instance, the way how participants perceive the role of farmers' markets was largely determined by the dynamics of supply and demand relationship and the definition of boundaries of authenticity in the farmers' markets [15].

4.2 Province of Nova Scotia

4.2.1 Case study 1: Select Nova Scotia

Select Nova Scotia is a local food program initiated by the government to promote the food grown in this province [26]. To discover the actual effectiveness of this program, Knight conducted an online survey to assess the awareness of Select Nova Scotia and particularly the respondents' perception of the local benefits and barriers as well as purchase motivation and behavior [26]. As one out of at least four Canadian provinces supporting local food initiatives, Select Nova Scotia has been financially supported by the provincial government since 2007. Through sponsoring various campaigns and events, it not only aims to promote and educate consumers about local food, but also exploits development opportunities of the regional food program [26]. Knight evaluated the effectiveness of this program, particularly the awareness levels and impacts it has on consumer preferences and consumption [26]. Based on several economic assessments, this case study works as a benchmark for the future studies of Select Nova Scotia. Primary or shared grocery shoppers residing in Nova Scotia were chosen, and randomization was applied to eliminate possible biases [26]. The respondents were divided into six focus groups based on three types of profiles: food skills and interest, social, and social-demographics.

A third of the respondents were already aware of the Select Nova Scotia and knew about this initiative through primarily retail venues such as farmers' markets followed by advertisements on mass media. Through factor analyses, benefits were categorized into four groups: societal, attribute, social, and price [26]. Respondents with a different demographic background and food skills and interests would position every single one of the four types of benefits rather poles apart. Also, gender is another influencing factor. Availability, price, location, and appearance are the four aspects of barriers which would affect consumers' decisions to buy locally sourced food. Similarly, respondents with different characteristics would perceive the significance of each of these barriers differently. Based on the frequency of buying, different respondents can be divided into three buy-local groups, high, medium and low. Respondents initially aware of the program are more likely to be in the high group, while other social-demographic aspects also affect the categorization of these respondents. Using multinomial logistic regression method, it was found out that the respondents' awareness of this initiative contributed positively to their motivation to buy locally [26]. The respondents' awareness of this initiative was as good as stated goals, regardless of its short history. Some critics doubted that the respondents that primarily covered the buy-local groups were already aware of the initiative. However, the cross tabulations suggest that only less than half respondents in the high group were already aware of it, while the other two groups had a lower percentage. It is recommended in this study that initiatives such as Select Nova Scotia should put greater emphasis on targeting medium and low buying groups through better managing the benefits and barriers.

4.2.2 Case study 2: Nova Scotia Farmers' Market Economic Impact Study 2013

With a rich history of farming, fishing, and artisan production, Nova Scotians have a habit of visiting and shopping at farmers' markets [54]. The farmers' markets in Nova Scotia had enjoyed significant growth since 2004.

An economic impact study on Nova Scotia Farmers' Markets was carried out in 2013 to figure out the reasons behind the three-fold growth in less than a decade. Twenty-one out of 43 Nova Scotian Farmers' Markets participated in this study. A number of patrons visited, and the average amount of money they spent was collected on the day. The dominant motivations behind these patrons' shopping behavior were supporting local businesses and demand for locally sourced, sustainable food [54].

Farmers' markets in Nova Scotia take various forms as they are in both rural and urban areas, opening in specific seasons or all year round [54]. Participating markets are grouped with their respective sizes; small ones have less than 20 vendors, and large ones can have over 40 vendors altogether [54]. The average attendance at large markets are correspondingly higher than those of medium- and small-sized ones. Among all the large markets, attendance at the Farmers' Market in Sydney, Cape Breton, ranked the third lowest [54]. The number of shopping groups counted is not limited to a group of patrons or households, but individuals as well. On average, a market can generate $22,654 worth of revenue for the region [54]. By market, Halifax Seaport enjoys highest average spending, equivalent to $50 per shopping group, and Cape Breton Farmers' Market in Sydney ranks the 11th in all the participating markets.

Vendor mix also has a huge impact on the average spending per shopping group [54]. It stands for the selection of vendors and products, which therefore lead to a higher spending of the patrons. If the farmers' markets could provide more complementary products, they could potentially generate more sales since many of the patrons have limited time for grocery shopping. Compared with the total shopping group, or household, spending on food, the farmers' market could have better growth.

The economic impacts farmers' markets can generate are both direct, direct sales made in the markets, and indirect, money spent at surrounding businesses. The total economic impact is estimated to be $1,881,573, calculated by multiplying the average dollar amount by average market attendance [54]. Small markets enjoyed greater indirect economic impact as they are more common in smaller regions, in which patrons often shop at one time, therefore help to boost the sales of nearby stores.

Forty-five percent of respondents suggest that supporting local food movement was their primary reason for shopping, followed by 30% found that shopping was a fun and social experience [54]. Also, the quality products farmers' markets provide, and the unique selection of products than elsewhere is also important factors for patrons to shop at the markets [54]. More importantly, the patrons at the market were not very price sensitive. They expect a higher value rather than lower price. Crawford [54] suggested that these farmers' markets should get their stakeholders increasingly involved through meetings, presentations, to explore opportunities for their future development.

5. Discussion

Recent years have shown increasing interest in short food supply chains or alternative food networks (AFNs) [4, 5, 7, 8, 24]. As direct channels of food distribution, AFNs drew the participants closer and enabled the redistribution of power across the food chain [29].

AFNs can be categorized into three kinds: direct sales by individuals, e.g., U-picks and farm gate sales; collective direct sales, e.g., farmers market; and partnerships, e.g., Community Supported Agriculture (CSAs). Contrary to global food supply chains, AFNs have facilitated the mobilization of locally produced fresh products [10] by connecting suppliers, e.g., farmers, with consumers [11]. The transition from passive end-consumers to proactive citizen-consumers clearly represents a new type of relationship between the producers and consumers of food [19]. Furthermore, consumers are becoming more conscious about environmentally friendly production and distribution systems [32], and interested in finding out about the origin of their food [16, 33]. In terms of environmental benefits, local food economy shortens the travel distance of food [45] and reduces the carbon emissions incurred in the transportation process. Furthermore, Abate [1] believes that the local food economy can connect and involve different local actors in the local community, revitalize businesses that may be hardly surviving and provide them with business opportunities, and also promote the sense of identity which is similar to the viewpoint of Oberholtzer and Grow [42]. It can also create job opportunities, increase the local income, and diversify the local economy.

Based on the research in different regions of the US on consumers' motivations to buy local, Baker et al. [5] suggested that the most important reasons of consumers using farmers' markets are access to fresh local foods and support local agriculture. But consumers in Manitoba, Canada, demonstrated different motivations based on the findings from the Food Matters Manitoba research study [46]; the major driver for attending farmers' market in Manitoba was to support the local economy, and the second most important factor was to help local farmers. The interest in the food quality was only placed at third. However, based on the research conducted on different consumers from across the country, i.e., Canada, it was discovered that consumers perceive freshness of food as the top motivator for buying local food products, followed by supporting local agriculture businesses as the second most important factor [48]. Besides these, Hunt [47] linked the demographic factors with the motivating factors to shop in farmers' market. He believes that social interaction is a significant motivator for consumers based on their demographic characteristics. His research was based on more than 200 consumers in the farmers' market. About 98% of the respondents agreed that they had fun while interacting with other people in farmers' market; 94% of them talk with vendors and 81% of people meet people they know in the market. Johnson and Endres [9] reported consumer's desire for fresh, high quality, and pesticide-free foods as the top reasons for purchasing locally produced food products. Second is to minimize the environmental footprint in the production process and reduce transportation emissions. Third is to help farmers achieve higher profitability by supporting their food production and distribution system which requires less investment in processing and packaging of food products. From a food safety perspective, the local food system reduces the risk of food safety and security because locally produced food items are usually fresher and less processed, and there is a high level of transparency. Consumers can either reach their sites of production or talk to growers to obtain information about local grown foods [40]. From managerial point of view, based on the survey conducted by Oberholtzer and Grow [42], most managers of farmers' market believe that the impacts brought by farmers' market to the community can be in many aspects. They described that a farmers' market provides a platform for all kinds of social and economic practices. This contributes to the formation of linked community and increases the individuals' sense of belonging in the community. Meanwhile, its basic function also enables it to provide consumers with fresh and inexpensive food.

There are both benefits and challenges for farmers to sell their products in farmers' markets. In terms of benefits, from vendors' perspective, first, selling food products in local markets shortens the distance they travel, which implies reduced transportation costs [47]. Second, selling in local markets enables them to have more interactions with consumers and improves their operations by understanding consumers' demands. Based on the Robinson and Farmer's [25] description, the first challenge for vendors is to convince more potential consumers to pay a premium price for their products. The second barrier is the gap between consumer demand and supply in terms of season. Farmers' Markets Canada [48] also found that the top two reasons for people to not shop at farmers' markets are the inconvenience of location and the lack of awareness. According to Colasanti et al. [49], the major reasons for less awareness of farmers' markets are largely due to the ineffective promotion strategy and the inconvenience of location. According to Wittman et al. [15], many customers would choose to support locally sourced food when there is a greater variety of products and easier accessibility for shopping.

6. Findings

In the past decade, there has been a growing interest in sustainable food production and distribution systems which promote food citizenship, food democracy, social capital and embeddedness, and sustainability. By deploying a systematic review approach, this chapter provides a holistic exploration of Alternate Food Networks (AFNs) which contributes to further mobilization of locally produced products as well as informing the channels of distribution in short food supply chains.

Our research demonstrates that there is a common set of motivating factors for consumers to engage in AFNs; however, there is disparity in the significance level of these factors to consumers in different regions. The implication of this finding mainly concerns the selection of a channel by consumers to participate in buy-local activities. For instance, if the social interaction aspect of buying local is not of high importance to some consumers in a specific region, they may consider shopping from an online farmers' market with a delivery or pick-up option instead of farmers' market. This phenomenon can lead to transformation or addition of business practices in currently existing platforms which connect suppliers, e.g., farmers, with consumers. In this regard, farmers' market in some regions may consider adding different channels of distribution including online ordering with delivery option to their platform.

Besides investigating the economic and societal benefits of AFNs and consumers' motivations to participate in buy-local activities, we also explore the farmers' perspectives on engaging in sell-local activities. Farmers suggest that reduced transportation costs, more interactions with consumers, and improving their marketing strategies as well as operations by understanding consumers' demands are among the benefits of selling at farmers' market [18, 47]. Furthermore, farmers' markets provide farmers with a friendly and attractive environment to market their products [48] and facilitate their collaboration with other vendors to achieve their common goals including improved profitability [18]. Besides, this clustering enables them to exchange their knowledge which provides the knowledge mobilization. The implications of those interactions are helping vendors build solid connections with consumers, promote social learning, lead the innovation on farming practices, and improve the quality of products [18]. In terms of challenges of selling local, Robinson and Farmer's [25] suggest that vendors need to convince more potential consumers to pay a premium price for their products. The second barrier is the gap between consumer demand and supply in terms of season. Farmers' Markets Canada [48] also found that the top two reasons for people to not shop at farmers' markets are the inconvenience of location and the lack of awareness. Overcoming these challenges can enhance consumers' access to fresh and locally produced products.

7. Recommendation for future research

Farmers' markets are established venues for mobilization of locally produced food products with societal benefits such as supporting community economic development and sustainable food production and distribution systems in addition to promoting food democracy, food citizenship, social and human capital, and facilitating knowledge mobilization between farmers and vendors.

The social interaction aspect of participating in alternative food networks (AFNs) has been cited in the literature as one of the motivating factors for farmers to engage in sell-local activities in the farmers' markets. Similarly, some studies suggest that the social interaction aspect of buying local is a motivating factor for consumers to participate in buy-local activities in farmers' market. COVID-19 pandemic and the need for social-distancing affected brick and mortar businesses in general, and in our context farmers' market. There was a shift in selection of a channel to access locally produced food products for safety reasons, and placing online orders with delivery option at the door was offered to consumers. Future research can explore consumers' preferences on channel selection to participate in buy-local activities post COVID-19 pandemic. To capture one aspect of the potential impacts of COVID-19 pandemic on AFNs, we propose an investigation on whether the social interaction is still an influential factor in consumers decision to buy local, meaning whether consumers will resume their shopping at the farmers' market post-pandemic. This can have business implications in alternative food networks (AFNs) as it is possible that the importance of the social interaction aspect of buying local will be replaced by the convenience of receiving fresh locally produced food products at consumers' doorstep via online ordering process.

Note

This chapter includes excerpts from Applied Research Projects submitted to Cape Breton University by Yang Xing and Cheng Zheng in partial fulfillment of the requirements of the degree of Master of Business Administration.

Author details

Sahand Ashtab*, Yang Xing and Cheng Zheng
Shannon School of Business, Cape Breton University, Sydney, Canada

*Address all correspondence to: sahand_ashtab@cbu.ca

References

[1] Abate G. Local food economies: Driving forces, challenges, and future prospects. Journal of Hunger & Environmental Nutrition. 2008;**3**(4):384-399

[2] Blythman J. Shopped: The Stocking Power of British Supermarkets. London: Fourth estate; 2004

[3] Cone CA, Myhre A. Community supported agriculture: A sustainable alternative to industrial agriculture? Human Organisation. 2000;**59**(2):187-197

[4] Greenbelt Farmers Market Network. Tracking Market Farmer and Vendor Performance 2009-2015 [Internet]. 2015. Available from: https://www.greenbeltmarkets.ca/wp-content/uploads/resources/2016-Tracking-Market-Farmer-Report-web.pdf

[5] Baker D, Hamshaw K, Kolodinsky J. Who shops at the market? Using consumer surveys to grow farmers' markets: Findings from a regional market in northwestern Vermont. Journal of Extension. 2009;**47**(6):1-9

[6] Miller A. Making markets mainstream [Internet]. 2014. Available from: https://sustainablefoodtrust.org/articles/farmers-market-moving-it-mainstream/

[7] Brunori G, Rossi A, Malandrin V. Co-producing transition: Innovation processes in farms adhering to solidarity-based purchase groups (GAS) in Tuscany, Italy. International Journal of Sociology of Agriculture and Food. 2010;**18**(1):28-53

[8] Chiffoleau Y, Millet-Amrani S, Canard A. From short food supply chains to sustainable agriculture in urban food systems: Food democracy as a vector of transition. Agriculture. 2016;**6**(57):1-18

[9] Johnson NR, Endres AB. Small producer, big hurdles: Barriers facing producers of local foods. Hamline Journal of Public Law & Policy. 2011;**33**(1):49-122

[10] Pearson D, Henryks J, Trott A, Jones P, Parker G, Dumaresq D, et al. Local food: Understanding consumer motivations in innovative retail formats. British Food Journal. 2011;**113**(7):886-899

[11] Hinrichs C. Embeddedness and local food systems: Notes on two types of direct agricultural market. Journal of Rural Studies. 2000;**16**(3):295-303

[12] Jarosz L. The city in the country: Growing alternative food networks in metropolitan areas. Journal of Rural Studies. 2008;**24**(3):231-244

[13] Marie-Laure AG. Short Food Supply Chains and Local Food Systems in the EU. European Parliamentary Research Service; 2016. pp. 1-10

[14] Hance A, Schumacher G. Resources for Farmers' Markets [Internet]. 2004. Available from: http://www.nenw.org/farmersmarkets/

[15] Wittman HMB, Beckie M, Hergesheimer C. Linking local food systems and the social economy? Future roles for farmers' Markets in Alberta and British Columbia. Rural Sociology. 2012;**77**(1):36-61

[16] Zagata L. We want farmers' markets!' Case study of emerging civic food networks in the Czech Republic. International Journal of Sociology of Agriculture and Food. 2012;**19**(3):347-364

[17] Verraes C, Uyttendaele M, Sindic M, Berkvens D, Herman L, Daube AC. Microbiological safety and quality aspects of the short supply chain. British Food Journal. 2015;**117**(9):2250-2261

[18] Beckie MA, Kennedy EH, Wittman H. Scaling up alternative food networks: Farmers' markets and the role of clustering in western Canada. Agriculture and Human Values. 2012;**29**(3):333-345

[19] Renting H, Marsden TK, Banks J. Understanding alternative food networks: Exploring the role of short food supply chains in rural development. Environment and Planning. 2012;**35**:393-411

[20] Hamprecht J, Corsten D, Noll M, Meier E. Controlling the sustainability of food supply chains. Supply Chain Management. 2005;**10**(1):7-10

[21] Tregear A, Cooper S. Embeddedness, social capital and learning in rural areas: The case of producer cooperatives. Journal of Rural Studies. 2016;**44**:101-110

[22] Brown A. Farmers' market research 1940-2000: An inventory and review. American Journal of Alternative Agriculture. 2002;**17**(4):167-176

[23] Smithers J, Lamarche J, Joseph AE. Unpacking the terms of engagement with local food at the farmers' market: Insights from Ontario. Journal of Rural Studies. 2008;**24**:337-350

[24] Byker C, Shanks J, Misyak S, Serrano E. Characterizing farmers' market shoppers: A literature review. Journal of Hunger & Environmental Nutrition. 2012;**7**(1):38-52

[25] Robinson J, Farmer J. Selling Local: Why Local Food Movements Matter. Bloomington, Indiana: Indiana University Press; 2017

[26] Knight AJ. Evaluating local food programs: The case of select Nova Scotia. Evaluation and Program Planning. 2013;**36**:29-39

[27] Martinez SW, Hand M, Da Pra M, Pollack S, Ralston K, Smith T, et al. Local Food Systems; Concepts, Impacts, and Issues. U.S. Department of Agriculture, Economic Research Service; 2010:1-80

[28] Ilbery B, Maye D. Retailing local food in the Scottish–English borders: A supply chain perspective. Geoforum. 2006;**37**(3):352-367

[29] Claire Lamine MD. The civic and social dimensions of food production and distribution in alternative food networks in France and southern Brazil. International Journal of Sociology of Agriculture and Food. 2012;**19**(3):383-401

[30] Alternative food networks & why food & farming matters to us? [Internet]. 2011. Available from Transition Rutland: https://transitionrutland.wordpress.com/what-are-alternative-food-networks-and-why-does-food-and-farming-matter-to-us/

[31] Inga-Lena Darkow BF. Sustainability in food service supply chains: Future expectations from European industry experts. Supply Chain Management: An International Journal. 2015;**20**(2):163-178

[32] Monteiro CA, Levy RB, Claro RM, Castro IRD, Cannon G. A new classification of foods based on the extent and purpose of their processing. Cadernos de Saúde Pública. 2010;**26**(11):2039-2049

[33] Seyfang G. Avoiding Asda? Exploring consumer motivations in local organic food networks. Local Environment. 2008;**13**(3):187-201

[34] McMichael P. The power of food. Agriculture and Human Values. 2000;**17**:21-33

[35] Weatherell C, Tregear A, Allinson J. In search of concerned consumer: UK public perceptions of food, farming and

buying local. Journal of Rural Studies. 2003;**19**(2):233-244

[36] D'amico M, Vita GD, Chinnici G, Pappalardo G, Pecorino AB. Short food supply chain and locally produced wines: Factors affecting consumer behaviour. Italian Journal of Food Science. 2014;**26**:329-334

[37] Sage C. Social embeddedness and relations of regard alternative 'good foo' networks in south-west Ireland. Journal of Rural Studies. 2003;**19**:47-60

[38] Flora C, Bregendahi C. Collaborative community-supported agriculture: Balancing community capitals for producers and consumers. International Journal of Sociology of Agriculture and Food. 2012;**19**(3):329-346

[39] Taylor DH. Strategic considerations in the development of lean Agri-food supply chains: A case study of the UK pork sector. Supply Chain Management: An International Journal. 2006;**11**(3):271-280

[40] Brain R. The local food movement: Definitions, benefits, and resources. USU Extension Publication. 2012:1-4

[41] Schneider ML, Francis CA. Marketing locally produced foods: Consumer and farmer opinions in Washington County. Nebraska. Renewable Agriculture and Food Systems. 2005;**20**(4):252-260

[42] Oberholtzer L, Grow S. Producer Only Farmers' Markets in the mid-Atlantic Region: A Survey of Market Managers. Arlington, VA: Henry A. Wallace Centre for Agricultural and Environmental Policy at Winrock International; 2003

[43] Feenstra GW, Lewis C, Hinrichs CC, Gillespie GW, Hilchey D. Entrepreneurial outcomes and Enterprise size in US retail farmers' markets. American Journal of Alternative Agriculture. 2003;**18**(1):46-55

[44] Hughes DW, Boys KA. What we know and don't know about the economic development benefits of local food systems [Internet]. Choices. 2015. Available from: https://www.choicesmagazine.org/choices-magazine/theme-articles/community-economics-of-local-foods/what-we-know-and-dont-know-about-the-economic-development-benefits-of-local-food-systems

[45] Conner DS, Montri AD, Montri DN, Hamm MW. Consumer demand for local produce at extended season farmers' markets: Guiding farmer marketing strategies. Renewable Agriculture and Food Systems. 2009;**24**(4):251-259

[46] Food Matters Manitoba. Growing for the Future-Building a Local, Sustainable Food Economy in Manitoba [Internet]. 2012. Available from: https://foodsecurecanada.org/sites/foodsecurecanada.org/files/mbfoodassessment_sm.pdf

[47] Hunt AR. Consumer interactions and influences on farmers' market vendors. Renewable Agriculture and Food Systems. 2007;**22**(1):54-66

[48] Experience Renewal Solutions Inc. National farmers' market impact study 2009 report [Internet]. 2009. Available from: https://www.unbc.ca/sites/default/files/sections/david-connell/farmers-markets/nationalfarmers marketimpactstudy2009.pdf

[49] Colasanti KJ, Conner DS, Smalley SB. Understanding barriers to farmers' market patronage in Michigan: Perspectives from marginalized populations. Journal of Hunger & Environmental Nutrition. 2010;**5**(3):316-338

[50] Onozaka Y, Nurse G, McFadden DT. Local food consumers: How motivations

and perceptions translate to buying behaviour. Choices. 2010;**25**(1):1-6

[51] Feenstra GW. Local food systems and sustainable communities. American Journal of Alternative Agriculture. 1997;**12**(1):28-36

[52] Sergaki P, Nastis S. Collective entrepreneurship in agriculture and its contribution to sustainable rural development in Greece. Journal of the Geographical Institute "Jovan Cvijić" SASA. 2011;**61**(2):109-128

[53] Hinrichs C, Gillespie GW, Feenstra GW. Social learning and innovation at retail farmers' markets. Rural Sociology. 2004;**69**(1):31-58

[54] Crawford K. Nova Scotia farmers' markets economic impact study [Internet]. 2013. Available from: https://farmersmarketsns.files.wordpress.com/2014/12/cultivating_full_report_screen.pdf